SYSTEMS OF ORDINARY DIFFERENTIAL EQUATIONS

Harper's Series in Modern Mathematics
I. N. Herstein and Gian-Carlo Rota, Editors

Differential Geometry Louis Auslander
Abstract Methods in Partial Differential Equations Robert W. Carroll
The Calculus: An Introduction Casper Goffman
Calculus of Several Variables Casper Goffman
Introduction to Real Analysis Casper Goffman
Systems of Ordinary Differential Equations: An Introduction Jack L. Goldberg and Arthur J. Schwartz
Constructive Real Analysis A. A. Goldstein
Finite Groups Daniel Gorenstein
Introduction to Calculus Donald Greenspan
Elementary Probability Theory Melvin Hausner
Introduction to Linear Algebra Peter Kahn
A Course in Numerical Analysis D. Melvin Lieberstein
Symbolic Logic and the Real Number System A. H. Lightstone
Foundations of Algebra and Analysis: An Elementary Approach Anthony R. Lovaglia and Gerald C. Preston
Cohomology Operations and Application in Homotopy Theory Robert E. Mosher and Martin C. Tangora
Ordered Topological Vector Spaces Anthony L. Peressini

SYSTEMS OF ORDINARY DIFFERENTIAL EQUATIONS: An Introduction

JACK L. GOLDBERG
ARTHUR J. SCHWARTZ
The University of Michigan

HARPER & ROW, PUBLISHERS
NEW YORK, EVANSTON, SAN FRANCISCO, LONDON

Standard Book Number: 06-042384-6

Library of Congress Catalog Card Number: 70-174527

to Sandy
Danny
Betty
Anne
Michael
Ruth

CONTENTS

PREFACE

The study of differential equations provides one of the main bridges between abstract mathematics (particularly infinitesimal calculus and linear algebra) and many of its applications to science and engineering. It is also important in the further study of certain branches of mathematics. It is not only a useful subject, but one which possesses elegance and structure, the result of a long history of remarkable effort on the part of many mathematicians—an effort which still continues.

The basic problem studied in differential equations may be stated as follows: Given information about the derivatives of a function, what can we deduce about the function itself? For example, if we are told that $f(t)$ is a real-valued function of the real variable t such that $f'(t) = 0$ for all t, we may deduce (via the Mean Value Theorem) that $f(t)$ is constant *and* that any constant function is consistent with the given information. A second example is somewhat more interesting: If we are told that $f(t)$ is such that $f'(t) = 2f(t)$ for all t, then $f(t)$ must be of the form $f(t) = ke^{2t}$ where $K = f(0)$. It is not very hard to see that $f(t) = Ke^{2t}$ is consistent with $f'(t) = 2f(t)$; however, some effort *is* required to show that $f(t)$ *must be* of this form (see Theorem 2.1.1).

The problems above deal with derivatives of functions of a single real variable, that is, *ordinary* differential equations, as *opposed to* problems involving *partial* derivatives of functions of several variables whose study is called *partial* differential equations. We will also be concerned with ordinary derivatives of higher order: $f''(t)$, $f'''(t)$, etc., and vector-valued functions.

This book is intended for a *first course* in ordinary differential equations.

In addition to basic integral and differential calculus, an acquaintance with elementary linear algebra is very useful but not absolutely necessary, since the first chapter reviews this topic. We have written this book for a wide audience and it is our hope that *anyone* with the requisite knowledge of calculus and the desire to learn this subject will profit from using this text.

As mentioned above, Chapter 1 is a review of linear algebra. We urge the reader or instructor to move through this chapter rapidly. The concepts in the chapter will be reinforced by application in succeeding chapters. We remark, also, that the material on the inner products and norms, Section 1.9, is not heavily used and may be omitted.

Chapters 2 through 5 form the core of the book. It has been the authors' experience, with a wide variety of students, that this material offers little difficulty. The last three chapters are independent of each other. A one-semester course will probably provide time to cover one or possibly two of these chapters, a choice which may be made by the instructor or the class. The diagram below describes this situation.

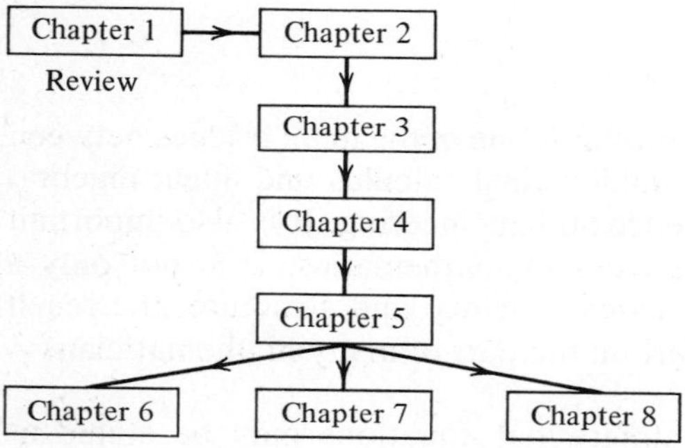

Starred sections are rather abstract and involve deeper arguments. They are not quite in the mainstream of the book and, thus, may be omitted. On the other hand, they do contain material of interest.

The problem sets at the end of each section are generally divided into two parts. The first part consists of routine problems which the student should be able to work out if he has comprehended the material of the foregoing section; the second part consists of more challenging problems which often extend the material of the section.

Much of the pedagogical and some of the mathematical development of this book is original, but the literature of differential equations is extensive, and many of these ideas have found their way into this book. We specifically mention our debt to Hurewicz's *Lectures on Ordinary Differential Equations*, Pontryagin's *Ordinary Differential Equations*, and Hochstadt's *Differential Equations: A Modern Approach*.

It is our great pleasure to acknowledge Professor Wilfred Kaplan, who read the manuscript, used it in his course, and offered many valuable comments. We also wish to thank Mrs. Dorothy Lentz and Miss Kathy Kamm for their assistance in the typing of the manuscript.

Jack L. Goldberg
Arthur J. Schwartz

SYSTEMS OF ORDINARY DIFFERENTIAL EQUATIONS

Chapter 1
VECTORS AND MATRICES

1.1. THE SOLUTION OF SIMULTANEOUS EQUATIONS

Ancient Chinese mathematicians recorded the use of matrices in the solution of simultaneous equations as early as 250 B.C. Our modern theory, based on the concept of a linear transformation, is generally dated from the appearance of a memoir by the British mathematician Arthur Cayley in 1858.

Times were appropriate for the rapid expansion of Cayley's ideas and by the middle of this century linear algebra, as the subject is now known, had become a cornerstone for much of modern mathematics. Our interest in this extensive branch of mathematics is connected with its oldest application—the solution of systems of linear equations.

Consider, for example, the following equations:

$$\begin{aligned} x \quad\; + \; z &= 1 \\ 2x + y + \; z &= 0 \\ x + y + 2z &= 1 \end{aligned} \tag{1.1.1}$$

We can solve (1.1.1) by elimination. We proceed systematically, eliminating the first unknown, x, from the second and third equations using the first equation. This results in the system

$$\begin{aligned} x \quad + z &= \;\;\, 1 \\ y - z &= -2 \\ y + z &= \;\;\, 0 \end{aligned} \tag{1.1.2}$$

Next we eliminate y from the third equation and get

$$\begin{aligned} x \quad + \; z &= \;\;\, 1 \\ y - \; z &= -2 \\ 2z &= \;\;\, 2 \end{aligned} \tag{1.1.3}$$

We can now read off $z = 1$ from the third equation and from this deduce $y = -1$ from the second equation and $x = 0$ from the first equation.

It should be clear that the elimination process depends on the coefficients of the equations and not on the unknowns. We could have collected all the coefficients in (1.1.1) in a rectangular array

$$\begin{bmatrix} 1 & 0 & 1 & 1 \\ 2 & 1 & 1 & 0 \\ 1 & 1 & 2 & 1 \end{bmatrix} \tag{1.1.4}$$

and eliminated x and y using the rows of the array as though they were equations. For instance, (-2) times each entry in the first row added, entry by entry, to the second row and (-1) times each entry in the first row added to the third row yields the array

$$\begin{bmatrix} 1 & 0 & 1 & 1 \\ 0 & 1 & -1 & -2 \\ 0 & 1 & 1 & 0 \end{bmatrix} \tag{1.1.5}$$

which exhibits the coefficients of (1.1.2). The zeros in the first column of (1.1.5) refer to the fact that x no longer appears in any equation but the first. The elimination of y from the third equation requires the replacement of the 1 in the third row, second column, by 0. We do this by subtracting the second row from the third and thus obtain the coefficients of (1.1.3) displayed in the array

$$\begin{bmatrix} 1 & 0 & 1 & 1 \\ 0 & 1 & -1 & -2 \\ 0 & 0 & 2 & 2 \end{bmatrix} \tag{1.1.6}$$

Once the equations have been manipulated this far it is not essential to perform any further simplification. For the sake of completeness we observe that dividing the third row by 2 (which amounts to dividing the equation $2z = 2$ by 2), then adding the new third row to the second and subtracting it from the first, leads to the array

$$\begin{bmatrix} 1 & 0 & 0 & 0 \\ 0 & 1 & 0 & -1 \\ 0 & 0 & 1 & 1 \end{bmatrix} \tag{1.1.7}$$

This corresponds to the equations

$$\begin{aligned} x &= 0 \\ y &= -1 \\ z &= 1 \end{aligned}$$

All this prompts a definition.

1.1.1. DEFINITION *We say a system of equations is row-reduced when any one of the following three steps is performed*

(a) *the interchange of any two rows*
(b) *the addition of a multiple of one row to another*
(c) *the multiplication of a row by a nonzero constant*
These steps are known as row operations.

Note that a row operation replaces a given set of equations with a new set with precisely the same solutions. In this connection, the word "nonzero" in (c) is critical. The rectangular array of coefficients of the unknowns is called the *coefficient matrix*, while the array given in (1.1.4) is known as the *augmented matrix*. The coefficient matrix of (1.1.1) is

$$\begin{bmatrix} 1 & 0 & 1 \\ 2 & 1 & 1 \\ 1 & 1 & 2 \end{bmatrix}$$

1.1.1. Example

Use row operations to solve the set of equations

$$\begin{aligned} t+x+y+z&=1 \\ t-x-y+z&=0 \\ 2t+x+y-z&=2 \end{aligned} \tag{1.1.8}$$

Solution

The matrix of the coefficients augmented by the right-hand side is

$$\begin{bmatrix} 1 & 1 & 1 & 1 & 1 \\ 1 & -1 & -1 & 1 & 0 \\ 2 & 1 & 1 & -1 & 2 \end{bmatrix} \tag{1.1.9}$$

We eliminate x from the second and third equation by subtracting the first row from the second and twice the first row from the third. Thus (1.1.9) becomes

$$\begin{bmatrix} 1 & 1 & 1 & 1 & 1 \\ 0 & -2 & -2 & 0 & -1 \\ 0 & -1 & -1 & -3 & 0 \end{bmatrix} \tag{1.1.10}$$

If we now subtract from the third row of (1.1.10), one-half the second row, we obtain

$$\begin{bmatrix} 1 & 1 & 1 & 1 & 1 \\ 0 & -2 & -2 & 0 & -1 \\ 0 & 0 & 0 & -3 & \frac{1}{2} \end{bmatrix} \tag{1.1.11}$$

which corresponds to the equations

$$\begin{aligned} t+\ x+\ y+\ z&=\ 1 \\ -2x-2y\qquad&=-1 \\ -3z&=\ \tfrac{1}{2} \end{aligned}$$

Thus $z=-\frac{1}{6}$ and if we set $y=k$ then $x=\frac{1}{2}-k$ and $t=\frac{2}{3}$. Note that the choice of y is completely arbitrary. One might have decided to let x be arbitrarily given, say $x=c$. Then $y=\frac{1}{2}-c$. The solution of (1.1.8) may be written in either form.

1.1.2. Example

Solve the equations

$$\begin{aligned} x \quad\quad + z &= 1 \\ 2x \quad\quad + z &= 0 \\ x + y + z &= 1 \end{aligned}$$

Solution

The augmented matrix is

$$\begin{bmatrix} 1 & 0 & 1 & 1 \\ 2 & 0 & 1 & 0 \\ 1 & 1 & 1 & 1 \end{bmatrix}$$

Then

$$\begin{bmatrix} 1 & 0 & 1 & 1 \\ 2 & 0 & 1 & 0 \\ 1 & 1 & 1 & 1 \end{bmatrix} \rightarrow \begin{bmatrix} 1 & 0 & 1 & 1 \\ 0 & 0 & -1 & -2 \\ 0 & 1 & 0 & 0 \end{bmatrix} \rightarrow \begin{bmatrix} 1 & 0 & 0 & -1 \\ 0 & 0 & -1 & -2 \\ 0 & 1 & 0 & 0 \end{bmatrix} \quad (1.1.12)$$

where the arrows denote the application of various row operations. The array (1.1.12) yields $x=-1$, $z=2$, and $y=0$. This is the solution.

Exercise 1.1

Part 1

Solve the following systems of equations by row-reducing the augmented matrices

1. $\begin{aligned} x \quad\quad - 2z - t &= 0 \\ x + y + 2z - 3t &= 0 \end{aligned}$

2. $\begin{aligned} x \quad\quad + 2z - t &= 0 \\ y \quad\quad\quad\quad &= 0 \\ x + y + 3z - t &= 0 \end{aligned}$

3. $\begin{aligned} x \quad\quad + 3z + t &= 0 \\ -2t &= 0 \\ x + y + 3z \quad\quad &= 0 \end{aligned}$

4. $\begin{aligned} x + 2y - z &= 0 \\ 2x + y + z &= 0 \\ 3x + 3y \quad\quad &= 0 \end{aligned}$

5. $\begin{aligned} x + 2y - z &= 2 \\ 2x + y + z &= -1 \\ 3x + 3y \quad\quad &= 1 \end{aligned}$

6. $\begin{aligned} -x \quad\quad + z &= -1 \\ x + y \quad\quad &= 0 \\ z &= 0 \end{aligned}$

1.2. THE ARITHMETIC OF MATRICES

We have seen the convenience afforded by the simple expedient of operations performed on the array of coefficients of a system of equations rather than on the equations themselves. Further work along these lines supports this view more substantially. Ultimately mathematicians thought of giving these arrays an existence of their own apart from their connection with simultaneous equations. It is this aspect we explore now. We agree to call any rectangular array a *matrix*. The arrays

$$[1,0,-1,0] \qquad \begin{bmatrix} 1 \\ 2 \\ \pi \\ -7 \end{bmatrix} \qquad \begin{bmatrix} 0 & 1 \\ 1 & 1 \\ 1 & 0 \end{bmatrix} \qquad \begin{bmatrix} 0 & 0 & 0 \\ 0 & 0 & 0 \\ 0 & 0 & 0 \end{bmatrix} \tag{1.2.1}$$

are examples of matrices. The number of elements in a given matrix is specified by giving the number of rows and columns in that order. These numbers are called the *dimensions* of the matrix. Thus the first matrix in (1.2.1) would be called 1×4 (read "one by four"), the second 4×1, the third 3×2, and the last 3×3. In this book Roman capitals always refer to matrices. Let the $m \times n$ matrices A and B be given by

$$\mathrm{A} = \begin{bmatrix} a_{11} & a_{12} & \cdots & a_{1n} \\ a_{21} & a_{22} & \cdots & a_{2n} \\ \cdot & & & \cdot \\ \cdot & & & \cdot \\ \cdot & & & \cdot \\ a_{m1} & a_{m2} & \cdots & a_{mn} \end{bmatrix} \qquad \mathrm{B} = \begin{bmatrix} b_{11} & b_{12} & \cdots & b_{1n} \\ b_{21} & b_{22} & & b_{2n} \\ \cdot & & & \cdot \\ \cdot & & & \cdot \\ \cdot & & & \cdot \\ b_{m1} & b_{m2} & & b_{mn} \end{bmatrix} \tag{1.2.2}$$

1.2.1. DEFINITION *The matrices* A *and* B *are said to be equal and written* A = B *if*

$$a_{ij} = b_{ij} \tag{1.2.3}$$

for each $i = 1, 2, \ldots, m$ *and each* $j = 1, 2, \ldots, n$.

Implicit in the definition of equality is the assumption that the dimensions of A and B are the same. The equality of two matrices implies the equality of mn numbers, the entries of the matrices. In addition to matrix equality, we define addition and multiplication of matrices by a constant.

1.2.2. DEFINITION *For the matrices* A *and* B *given by* (1.2.2) *and for any constant* k,

$$\mathrm{A}+\mathrm{B} = \begin{bmatrix} a_{11}+b_{11} & a_{12}+b_{12} & \cdots & a_{1n}+b_{1n} \\ a_{21}+b_{21} & a_{22}+b_{22} & \cdots & a_{2n}+b_{2n} \\ \cdot & & & \cdot \\ \cdot & & & \cdot \\ \cdot & & & \cdot \\ a_{m1}+b_{m1} & a_{m2}+b_{m2} & \cdots & a_{mn}+b_{mn} \end{bmatrix} \tag{1.2.4}$$

$$k\mathrm{A} = \begin{bmatrix} ka_{11} & ka_{12} & \cdots & ka_{1n} \\ ka_{21} & ka_{22} & & ka_{2n} \\ \cdot & \cdot & & \cdot \\ \cdot & \cdot & & \cdot \\ \cdot & \cdot & & \cdot \\ ka_{m1} & ka_{m2} & \cdots & ka_{mn} \end{bmatrix} \tag{1.2.5}$$

We denote by O any matrix whose entries are all zeros. The dimensions of O will generally be clear from the context. With O, A, B, and kA defined, it is now easy to verify the algebraic rules listed below.

1.2.3. THEOREM *Assume that* A, B, C, *and* O *have the same dimensions and that h and k are numbers. Then*

(a) $A+B=B+A$
(b) $A+(B+C)=(A+B)+C$
(c) $A+O=A$
(d) $A+(-1)A=O$
(e) $0A=O$
(f) $k(hA)=(kh)A$
(g) $k(A+B)=kA+kB$
(h) $(k+h)A=kA+hA$

If we understand by $B-A$, a matrix such that $(B-A)+A=B$, then (d) enables us to find such a matrix and provides a definition of subtraction. For,

$$\begin{aligned}[B+(-1)A]+A &= B+[(-1)A+A]\\ &= B+[A+(-1)A]\\ &= B+O\\ &= B\end{aligned}$$

1.2.4. DEFINITION *The transpose of the matrix* A, *written* A^T, *is a matrix whose columns are the rows of* A *and whose rows are the columns of* A, *in the same order.*

1.2.1. Example

The following illustrate the definition of the transpose of a matrix.

(a) $A=[1,-1,1]$ $\qquad A^T=\begin{bmatrix}1\\-1\\1\end{bmatrix}$

(b) $A=\begin{bmatrix}0 & 1\\-1 & 3\end{bmatrix}$ $\qquad A^T=\begin{bmatrix}0 & -1\\1 & 3\end{bmatrix}$

1.2.5. DEFINITION *A matrix is symmetric if it is equal to its own transpose, that is,* $A=A^T$. *A matrix is antisymmetric if* $A=-A^T$.

Note that a symmetric or antisymmetric matrix must be a square matrix. Two more special matrices occur with enough frequency to warrant special names. Denoting by the *main diagonal* or, shortly, the *diagonal* of a square matrix, the entries $a_{11}, a_{22}, \ldots, a_{nn}$, we have the following definitions.

1.2.6. DEFINITION *A diagonal matrix is a square matrix whose entries off the diagonal are all zeros. The diagonal matrix with only* 1's *on the diagonal is called the identity matrix, written* I.

1.2.7. DEFINITION *A square matrix all of whose entries below the diagonal are zeros is an upper triangular matrix.*

1.2.2. Example

(a) A diagonal matrix is an upper triangular matrix.

(b) O and I are diagonal matrices and therefore upper triangular also.

(c) The transpose of a diagonal matrix is a diagonal matrix and is always symmetric.

(d) $\begin{bmatrix} 0 & 1 \\ 0 & 0 \end{bmatrix}$ and $\begin{bmatrix} 1 & 1 & 1 \\ 0 & 1 & 1 \\ 0 & 0 & 1 \end{bmatrix}$ are upper triangular matrices.

1.2.8. DEFINITION *A square matrix all of whose entries above the main diagonal are zeros is called a lower triangular matrix.*

The transpose of an upper triangular matrix is a lower triangular matrix and conversely. Also note that diagonal matrices are the only matrices which are simultaneously upper and lower triangular.

1.2.9. THEOREM *Every square matrix can be row-reduced to an upper triangular matrix by repeated use of the row operation* (b) *alone.*

PROOF Let A be an arbitrary $n \times n$ matrix. Consider the entry a_{11}. If it is not zero the first row may be used to clear all the nonzero entries in the first column. If $a_{11} = 0$ and there is a nonzero entry in the first column, say a_{i1}, then add the ith row to the first row obtaining a matrix in which the entry in the first row, first column is nonzero. Proceed to clear the remaining elements in the first column by use of (b). If all the entries in the first column happen to be zeros we need do nothing to the first column. We now ignore the first row and first column in what follows. The remaining $(n-1) \times (n-1)$ matrix is operated on similarly to the original matrix. This process can be continued until A is reduced to an upper triangular matrix. This completes the proof.

1.2.10. COROLLARY *If the matrix* A *can be row-reduced to an upper triangular matrix with nonzero entries on the main diagonal, then* A *can be row-reduced by applications of row-operation* (b) *to a diagonal matrix with nonzero diagonal entries.*

1.2.11. COROLLARY *If the matrix* A *can be reduced to an upper triangular matrix with at least one zero entry on the diagonal then* A *can be reduced to an upper triangular matrix with at least one row of zeros by application of row operation* (b) *alone.*

PROOF We set the proof as an exercise: Exercise 1.2, Part 2, Problem 4.

Exercise 1.2

Part 1

1. Prove Theorem 1.2.3 for arbitrary 3×3 matrices A, B, C, and O.
2. What matrices are simultaneously symmetric and antisymmetric?
3. Explain why a system of equations whose matrix of coefficients is upper triangular can be solved without further simplification, provided there is a solution. Write an example illustrating the case with no solutions.
4. Reduce the following matrices to upper triangular form by repeated use of row operation (b):

(a) $\begin{bmatrix} 1 & 0 & 0 \\ -2 & 2 & 0 \\ 1 & 3 & -1 \end{bmatrix}$ (b) $\begin{bmatrix} 0 & 1 & 0 \\ 1 & 0 & 0 \\ 0 & 0 & 1 \end{bmatrix}$

(c) $\begin{bmatrix} a & b \\ c & d \end{bmatrix}$ $a \neq 0$ (d) $\begin{bmatrix} a & b \\ c & d \end{bmatrix}$ $a = 0$

(e) $\begin{bmatrix} 1 & 2 & 1 \\ 2 & 4 & -2 \\ 0 & 0 & 1 \end{bmatrix}$

5. For arbitrary A, prove $(kA)^T = kA^T$.
6. Prove a symmetric (or antisymmetric) matrix must be a square matrix.

Part 2

1. Prove Theorem 1.2.3 for arbitrary $n \times n$ matrices.
2. Prove $(A^T)^T = A$.
3. Prove $(A + B)^T = A^T + B^T$.
4. Suppose A is an upper triangular matrix with a zero entry on the diagonal. Prove that A can be row-reduced by means of (b) alone to a matrix with a row of zeros.
5. Suppose A is an upper triangular matrix with no zero entries on the diagonal. Prove that A can be row-reduced to a diagonal matrix by means of (b) alone.
6. By means of Problems 4 and 5, prove that any square matrix can be row-reduced to either a diagonal matrix or an upper triangular matrix with a row of zeros by use of row operation (b) alone. Do you think it possible to row-reduce a square matrix to both forms by different sequences of (b)?

In the next five problems use only Theorem 1.2.3 and the definitions, $A - B = A + (-1)B$, $-B = O - B$. Avoid reference to the entries of A and B. Prove:

7. $C + (A - B) = (C + A) - B$
8. $-B = (-1)B$
9. $A - B = -(B - A)$
10. $-(-B) = B$
11. $-(kA) = k(-A) = (-k)A$

1.3. MATRIX MULTIPLICATION

Multiplication of matrices is more complicated than addition and subtraction and far less "natural." To define AB we require the number of columns of

A to be equal to the number of rows of B. This is called the compatibility requirement.

1.3.1. DEFINITION *Suppose*

$$A = \begin{bmatrix} a_{11} & a_{12} & \cdots & a_{1k} \\ a_{21} & a_{22} & & a_{2k} \\ \cdot & \cdot & & \cdot \\ \cdot & \cdot & & \cdot \\ \cdot & \cdot & & \cdot \\ a_{m1} & a_{m2} & \cdots & a_{mk} \end{bmatrix} \quad \text{and} \quad B = \begin{bmatrix} b_{11} & b_{12} & \cdots & b_{1n} \\ b_{21} & b_{22} & & b_{2n} \\ \cdot & \cdot & & \cdot \\ \cdot & \cdot & & \cdot \\ \cdot & \cdot & & \cdot \\ b_{k1} & b_{k2} & \cdots & b_{kn} \end{bmatrix}$$

Then AB = C *means that the entry in the ith row and the jth column of* C, *namely,* c_{ij} *is given by*

$$c_{ij} = a_{i1}b_{1j} + a_{i2}b_{2j} + \cdots + a_{ik}b_{kj} \tag{1.3.1}$$

Here $i = 1,2,\ldots,m$ and $j = 1,2,\ldots,n$.

Thus C is $m \times n$ when A is $m \times k$ and B is $k \times n$. This definition looks more complicated than it really is. A few practice problems illustrate the definition quite well. Study the following examples.

1.3.1. Example

Compute the product C = AB, where

$$A = \begin{bmatrix} 1 & 2 \\ 3 & 4 \end{bmatrix} \quad \text{and} \quad B = \begin{bmatrix} 1 & -1 \\ 0 & 2 \end{bmatrix}$$

Solution

To obtain the first row of C (1.3.1) tells us to use the first row of A with each of the columns of B. Then,

$$C = \begin{bmatrix} 1\cdot1+2\cdot0 & 1\cdot(-1)+2\cdot2 \\ \underline{\qquad} & \underline{\qquad} \end{bmatrix} = \begin{bmatrix} 1 & 3 \\ - & - \end{bmatrix}$$

The second row of C is obtained by using the second row of A. Thus,

$$C = \begin{bmatrix} 1 & 3 \\ 3\cdot1+4\cdot0 & 3(-1)+4\cdot2 \end{bmatrix} = \begin{bmatrix} 1 & 3 \\ 3 & 5 \end{bmatrix}$$

1.3.2. Example

Compute D = BA using the matrices of the preceding example. Show BA ≠ AB.

Solution

$$\begin{bmatrix} 1 & -1 \\ 0 & 2 \end{bmatrix} \begin{bmatrix} 1 & 2 \\ 3 & 4 \end{bmatrix} = \begin{bmatrix} 1\cdot1+(-1)\cdot3 & 1\cdot2+(-1)\cdot4 \\ 0\cdot1+\quad 2\cdot3 & 0\cdot2+\quad 2\cdot4 \end{bmatrix}$$
$$= \begin{bmatrix} -2 & -2 \\ 6 & 8 \end{bmatrix}$$

1.3.3. Example

Compute AB where

$$A=\begin{bmatrix}1 & 0 & 1\\ 0 & 0 & 1\end{bmatrix} \quad \text{and} \quad B=\begin{bmatrix}1 & 0 & 0 & \alpha\\ 0 & 0 & 0 & \beta\\ 0 & 0 & 0 & \gamma\end{bmatrix}$$

Solution

$$AB=\begin{bmatrix}1 & 0 & 0 & \alpha+\gamma\\ 0 & 0 & 0 & \gamma\end{bmatrix}$$

Note that BA is not defined. Certain other difficulties plague this definition of matrix multiplication. For instance, $AB = O$ does not imply $A = O$ or $B = O$. Witness

$$\begin{bmatrix}1 & 0\\ 2 & 0\end{bmatrix}\begin{bmatrix}0 & 0\\ 1 & 1\end{bmatrix}=\begin{bmatrix}0 & 0\\ 0 & 0\end{bmatrix}$$

Directly related to this phenomenon is the violation of the "law of cancellation": $AB = AC$ does not imply $B = C$. For $AB = AC$ is equivalent to $A(B-C) = O$ and we have seen that neither A nor $B-C$ need be the zero matrix. For example:

$$\begin{bmatrix}1 & 0\\ 2 & 0\end{bmatrix}\begin{bmatrix}1 & 1\\ 1 & 1\end{bmatrix}=\begin{bmatrix}1 & 0\\ 2 & 0\end{bmatrix}\begin{bmatrix}1 & 1\\ 0 & 0\end{bmatrix}$$

In view of all this, the student might wonder why we choose to call the operation defined by (1.3.1) multiplication? The answer hinges on three points: first, in spite of some difficulties many properties of "multiplication" still hold; second, no other multiplication has been shown to be of any use; and finally, this product of matrices is connected with the product of linear transformations, a subject of some importance which we do not discuss in this book. We conclude this section with some further examples illustrating matrix multiplication.

1.3.4. Example

Find C given

$$C=\begin{bmatrix}1 & 3\\ -1 & 0\end{bmatrix}\begin{bmatrix}x\\ y\end{bmatrix}$$

Solution

$$C=\begin{bmatrix}x+3y\\ -x\end{bmatrix}$$

This example illustrates a connection between systems of equations and the multiplication of matrices. For instance,

$$\begin{bmatrix} 1 & 3 & 0 \\ -1 & 2 & 1 \\ 0 & 1 & 1 \end{bmatrix} \begin{bmatrix} x \\ y \\ z \end{bmatrix} = \begin{bmatrix} 1 \\ 1 \\ 0 \end{bmatrix}$$

means (multiplying the left-hand side)

$$\begin{bmatrix} x+3y \\ -x+2y+z \\ y+z \end{bmatrix} = \begin{bmatrix} 1 \\ 1 \\ 0 \end{bmatrix}$$

and by the definition of equality of two matrices,

$$\begin{aligned} x+3y &= 1 \\ -x+2y+\ z &= 1 \\ y+\ z &= 0 \end{aligned}$$

1.3.5. Example

Write the system

$$\begin{aligned} x+y-\ z &= 0 \\ x-y+\ z &= 1 \\ y-2z &= 3 \end{aligned}$$

in matrix form.

Solution

The product of the coefficient matrix and the column matrix whose entries are the unknowns is the key to this problem. One may verify that

$$\begin{bmatrix} 1 & 1 & -1 \\ 1 & -1 & 1 \\ 0 & 1 & -2 \end{bmatrix} \begin{bmatrix} x \\ y \\ z \end{bmatrix} = \begin{bmatrix} 0 \\ 1 \\ 3 \end{bmatrix}$$

is the required representation.

The rules listed in Theorem 1.2.3 can now be supplemented by a few more.

1.3.2. THEOREM *Suppose* A, B, C, O, *and* I *are matrices for which each of the products listed below exist. Then,*

(a) A(BC) = (AB)C
(b) A(B + C) = AB + AC
(c) AI = IA = A (1.3.2)
(d) OA = AO = O
(e) k(AB) = (kA)B = A(kB)

PROOFS The reader is asked to supply the proofs in Exercise 1.3, Part 2, Problem 1.

Exercise 1.3

Part 1

1. Find the products:

(a) $\begin{bmatrix} 1 & 3 \\ 3 & 1 \end{bmatrix}\begin{bmatrix} 1 & 4 \\ 4 & 1 \end{bmatrix}$

(b) $\begin{bmatrix} 1 & 1 & 0 \\ 0 & 1 & 2 \\ 0 & 0 & 1 \end{bmatrix}\begin{bmatrix} 1 & -2 & 1 \\ 0 & 2 & 2 \\ 0 & 0 & 1 \end{bmatrix}$

(c) $\begin{bmatrix} 2 & 0 & 0 \\ 0 & 1 & 0 \\ 0 & 0 & -1 \end{bmatrix}\begin{bmatrix} x \\ y \\ z \end{bmatrix}$

(d) $[a, \ b, \ c]\begin{bmatrix} a \\ b \\ c \end{bmatrix}$

(e) $\begin{bmatrix} -6 & 7 \\ 7 & -8 \end{bmatrix}\begin{bmatrix} 8 & 7 \\ 7 & 6 \end{bmatrix}$

(f) $\begin{bmatrix} 2 & 5 \\ 1 & 3 \end{bmatrix}\begin{bmatrix} 11 & 30 \\ -4 & -11 \end{bmatrix}\begin{bmatrix} 3 & -5 \\ -1 & 2 \end{bmatrix}$

(g) $[0][1, \ 7, \ -2]$

2. Verify $A(B+C) = AB + AC$, when

$$A = \begin{bmatrix} 1 & 2 \\ 3 & -1 \end{bmatrix} \qquad B = \begin{bmatrix} 0 & 1 \\ 2 & 3 \end{bmatrix} \qquad C = \begin{bmatrix} 1 & 1 \\ 0 & 1 \end{bmatrix}$$

3. Find A^2, A^3, A^4 where

$$A = \begin{bmatrix} 0 & 1 & 1 \\ 0 & 0 & 1 \\ 0 & 0 & 0 \end{bmatrix}$$

4. Find a formula for A^n where

$$A = \begin{bmatrix} 1 & 1 \\ 0 & 1 \end{bmatrix}$$

5. Let

$$A = \begin{bmatrix} 1 & 2 \\ 0 & 1 \end{bmatrix}$$

Compute

(a) $3A^2 - 9A + 6I$
(b) $3(A - I)(A - 2I)$
(c) $3(A - 2I)(A - I)$

6. Verify $(A+I)^3 = A^3+3A^2+3A+I$ for
   (a) $A = I$
   (b) $A = O$

   (c) $A = J_3 \equiv \begin{bmatrix} 1 & 1 & 1 \\ 1 & 1 & 1 \\ 1 & 1 & 1 \end{bmatrix}$

   (d) $A = \begin{bmatrix} 1 & 0 & -1 \\ 1 & 2 & 2 \\ -1 & 1 & 0 \end{bmatrix}$

7. Expand

   (a) $[x_1, x_2, x_3] \begin{bmatrix} 1 & -1 & 0 \\ 0 & 1 & 1 \\ 0 & 1 & -1 \end{bmatrix} \begin{bmatrix} x_1 \\ x_2 \\ x_3 \end{bmatrix}$

   (b) $[x_1, x_2, x_3] \begin{bmatrix} 2 & 0 & 0 \\ 0 & 1 & 0 \\ 1 & 1 & 1 \end{bmatrix} \begin{bmatrix} x_1 \\ x_2 \\ x_3 \end{bmatrix}$

8. Following Example 1.3.5, write the following systems in matrix form.

   (a) $\begin{aligned} x+y \quad &= 0 \\ y+z &= 0 \\ x+y+z &= 0 \end{aligned}$ (b) $\begin{aligned} x &= 0 \\ y &= 0 \\ z &= 0 \end{aligned}$

   (c) $\begin{aligned} x+\ y+z+t &= -1 \\ -2y \qquad -t &= \ \ 0 \\ z+t &= \ \ 5 \end{aligned}$

Part 2

1. Verify Theorem 1.3.2 for general $n \times n$ matrices A, B, O, and I.
2. Prove that AB is upper triangular if A and B are.
3. Prove $(AB)^T = B^T A^T$.
4. Find examples of 3×3 matrices A and B such that $AB = O$ but $A \neq O$ and $B \neq O$.
5. Use your answers to Problem 4 to construct examples of matrices A, B, and C such that $AB = AC$ but $B \neq C$.
6. If $AB = BA$ prove that A and B are square matrices with the same dimensions.

1.4. THE INVERSE OF A MATRIX

The quotient of two matrices is not a concept defined for matrices. In its place and to serve similar purposes we introduce the notion of the *inverse*.

1.4.1. DEFINITION *If for a square matrix* A, *there exists a matrix* A^{-1} *for which*

$$AA^{-1} = A^{-1}A = I \tag{1.4.1}$$

then we call A^{-1} *the inverse of* A *and say* A *is invertible.*

The identity matrix, I, is its own inverse. The "zero" matrix, O, is an example of a matrix which is not invertible because $OB = O$ for any matrix, B. Matrices which do not have inverses are called *singular* or *noninvertible*.

The importance of the inverse can be seen by considering the following problem:

Given two matrices A and B, find a third matrix C, such that

$$AC = B \tag{1.4.2}$$

Assuming A is a square matrix and that $AC = B$ is dimensionally consistent, the multiplication of both sides of (1.4.2) by A^{-1} leads to

$$A^{-1}B = A^{-1}(AC) = (A^{-1}A)C = IC = C$$

On the other hand the computation $A(A^{-1}B) = B$ proves that $C = A^{-1}B$ is a matrix such that $AC = B$. Note especially the important case where B (and therefore C) are column matrices. Then $AC = B$ may be interpreted as the matrix representation of a set of simultaneous equations whose coefficient matrix is A. Thus if the coefficient matrix is invertible, the product $A^{-1}B$ represents a column matrix whose entries are the solutions of the simultaneous equations. Turned about, this latter observation can be used to compute A^{-1}. The next two examples explain this idea.

1.4.1. Example

Find A^{-1} when

$$A = \begin{bmatrix} 1 & 2 \\ -1 & 1 \end{bmatrix}$$

Solution

Here we have

$$\begin{bmatrix} 1 & 2 \\ -1 & 1 \end{bmatrix} A^{-1} = \begin{bmatrix} 1 & 0 \\ 0 & 1 \end{bmatrix} \qquad \text{where } A^{-1} = \begin{bmatrix} a & c \\ b & d \end{bmatrix}$$

which yields two systems of two equations in two unknowns; namely,

$$\begin{bmatrix} 1 & 2 \\ -1 & 1 \end{bmatrix}\begin{bmatrix} a \\ b \end{bmatrix} = \begin{bmatrix} 1 \\ 0 \end{bmatrix} \qquad \text{and} \qquad \begin{bmatrix} 1 & 2 \\ -1 & 1 \end{bmatrix}\begin{bmatrix} c \\ d \end{bmatrix} = \begin{bmatrix} 0 \\ 1 \end{bmatrix}$$

We now augment A with both right-hand sides and proceed to solve both systems at once by row reduction of the matrix,

$$\begin{bmatrix} 1 & 2 & 1 & 0 \\ -1 & 1 & 0 & 1 \end{bmatrix}$$

We select our operations so as to reduce A to an identity matrix. Thus adding the first row to the second and then $-\frac{2}{3}$ of the second to the first yields

$$\begin{bmatrix} 1 & 0 & \frac{1}{3} & -\frac{2}{3} \\ 0 & 3 & 1 & 1 \end{bmatrix}$$

Dividing the second row by 3 gives

$$\begin{bmatrix} 1 & 0 & \frac{1}{3} & -\frac{2}{3} \\ 0 & 1 & \frac{1}{3} & \frac{1}{3} \end{bmatrix}$$

We interpret this to mean

$$\begin{bmatrix} a \\ b \end{bmatrix} = \begin{bmatrix} \frac{1}{3} \\ \frac{1}{3} \end{bmatrix} \quad \text{and} \quad \begin{bmatrix} c \\ d \end{bmatrix} = \begin{bmatrix} -\frac{2}{3} \\ \frac{1}{3} \end{bmatrix}$$

and hence

$$A^{-1} = \begin{bmatrix} a & c \\ b & d \end{bmatrix} = \begin{bmatrix} \frac{1}{3} & -\frac{2}{3} \\ \frac{1}{3} & \frac{1}{3} \end{bmatrix}$$

We have found a matrix X, such that $AX = I$. One may verify that $XA = I$ also. It is true in general that for $n \times n$ matrices A and B, $AB = I$ if and only if $BA = I$.

1.4.2. Example

Invert the matrix

$$A = \begin{bmatrix} 1 & 0 & 1 \\ 2 & 1 & 1 \\ 1 & 1 & 2 \end{bmatrix}$$

Solution

We row-reduce A augmented by I, namely,

$$\begin{bmatrix} 1 & 0 & 1 & 1 & 0 & 0 \\ 2 & 1 & 1 & 0 & 1 & 0 \\ 1 & 1 & 2 & 0 & 0 & 1 \end{bmatrix}$$

$$\rightarrow \begin{bmatrix} 1 & 0 & 1 & 1 & 0 & 0 \\ 0 & 1 & -1 & -2 & 1 & 0 \\ 0 & 0 & 2 & 1 & -1 & 1 \end{bmatrix}$$

$$\rightarrow \begin{bmatrix} 1 & 0 & 0 & \frac{1}{2} & \frac{1}{2} & -\frac{1}{2} \\ 0 & 1 & 0 & -\frac{3}{2} & \frac{1}{2} & \frac{1}{2} \\ 0 & 0 & 1 & \frac{1}{2} & -\frac{1}{2} & \frac{1}{2} \end{bmatrix}$$

Hence

$$A^{-1} = \begin{bmatrix} \frac{1}{2} & \frac{1}{2} & -\frac{1}{2} \\ -\frac{3}{2} & \frac{1}{2} & \frac{1}{2} \\ \frac{1}{2} & -\frac{1}{2} & \frac{1}{2} \end{bmatrix}$$

The matrix A in this example is the matrix of coefficients in (1.1.1). If we set

$$B = \begin{bmatrix} 1 \\ 0 \\ 1 \end{bmatrix} \quad \text{and} \quad C = \begin{bmatrix} x \\ y \\ z \end{bmatrix}$$

we have represented (1.1.1) as AC = B and from the remarks below Equation (1.4.2),

$$A^{-1}B = \begin{bmatrix} \frac{1}{2} & \frac{1}{2} & -\frac{1}{2} \\ -\frac{3}{2} & \frac{1}{2} & \frac{1}{2} \\ \frac{1}{2} & -\frac{1}{2} & \frac{1}{2} \end{bmatrix} \begin{bmatrix} 1 \\ 0 \\ 1 \end{bmatrix}$$

$$= \begin{bmatrix} 0 \\ -1 \\ 1 \end{bmatrix} = \begin{bmatrix} x \\ y \\ z \end{bmatrix}$$

confirming $x = 0$, $y = -1$, and $z = 1$.

The method illustrated in the previous example leads to a general criterion for the existence of A^{-1}. The proof of this criterion depends upon the following result.

1.4.2. LEMMA *A homogeneous system of equations with more unknowns than equations always has a solution with at least one unknown not zero.*

The system of equations described in the lemma is known as an *underdetermined* system and the proof of this remark is surprisingly long. It is avoided here. We set a special case of the proof in Exercise 1.4, Part 2, Problem 1.

1.4.3. THEOREM *Suppose the square matrix* A *is row-reduced to the upper triangular matrix* A′. *If* A′ *has no zeros on the diagonal then* A *is invertible. If* A′ *has at least one zero on the diagonal then* A *is singular.*

PROOF If A′ has no zero entries on its diagonal, then by Corollary 1.2.10 A can be reduced to a diagonal matrix with no zeros on the diagonal and from there reduced to the identity. Suppose the sequence of row operations that reduces A to I reduces I to B. Then the n sets of n equations in n unknowns, represented by AX = I, has the same solutions as the equivalent set represented by IX = B. Therefore X = B and AB = I. From the remark preceding Example 1.4.2, B is the inverse of A.

On the other hand, if A′ has a zero entry on the diagonal, then by Corollary 1.2.11 A can be reduced to A″, a matrix with a row of zeros. The homogeneous system represented by the matrix equation AX = O, where X and O are column matrices, has the same solutions as the equivalent system represented by A″X = O. This latter system contains more unknowns than equations because the matrix A″ has a row of zeros. Thus, from the lemma, X can be found such that AX = O and X ≠ O. If A^{-1} were to exist then it would follow from $A^{-1}(AX) = A^{-1}O = O$ that X = O. Therefore A is singular, as was to be proved.

We now restate this theorem in contrapositive form.

1.4.4. COROLLARY *If* A *is invertible, every upper triangular matrix obtained from* A *by row operations has no zeros on its diagonal. If* A *is singular, then every upper triangular matrix obtained from* A *by row operations has at least one zero entry on its diagonal.*

1.4.3. Example

Show that A is singular, where

$$A = \begin{bmatrix} 2 & 2 & 1 \\ 3 & 3 & -2 \\ 1 & 1 & -3 \end{bmatrix}$$

Solution

The reader can verify that

$$\begin{bmatrix} 2 & 2 & 1 & 1 & 0 & 0 \\ 1 & 1 & -3 & -1 & 1 & 0 \\ 0 & 0 & 0 & 1 & -1 & 1 \end{bmatrix}$$

is obtainable via row reductions from A augmented with I. That A^{-1} fails to exist is deduced from the fact that the existence of A^{-1} would imply the contradictory relationship

$$\begin{bmatrix} 2 & 2 & 1 \\ 1 & 1 & -3 \\ 0 & 0 & 0 \end{bmatrix} A^{-1} = \begin{bmatrix} 1 & 0 & 0 \\ -1 & 1 & 0 \\ 1 & -1 & 1 \end{bmatrix} \tag{1.4.3}$$

(See Exercise 1.4, Part 1, Problems 7 and 8.)

1.4.5. THEOREM *For any square invertible matrices* A *and* B

(a) *there is but one inverse of* A
(b) $(AB)^{-1} = B^{-1}A^{-1}$
(c) $(A^{-1})^{-1} = A$

PROOF If $AB = I$ then $A^{-1}(AB) = A^{-1}I$, which means $B = A^{-1}$ and (a) is proved. For the proof of (b) study the equations

$$(AB)(B^{-1}A^{-1}) = A(BB^{-1})A^{-1} = I$$
$$(B^{-1}A^{-1})(AB) = B^{-1}(A^{-1}A)B = I$$

Part (c) is trivial.

1.4.6. THEOREM *If* A^{-1} *exists, the equation* $AX = B$ *has the unique solution,* $X = A^{-1}B$.

PROOF This is just a restatement of the conclusion drawn previously from Equation (1.4.2). We record it here as a theorem for easy reference. We often use this result when X and B are column matrices.

If $B = 0$ is a column matrix, Theorem 1.4.6 asserts that $X = 0$ is the sole solution of $AX = 0$. Much more than this is true.

1.4.7. THEOREM *Suppose* $AX = O$, *with* X *and* O *column matrices. Then* $AX = 0$ *has the unique solution* $X = O$ *if* A *is invertible and infinitely many solutions if* A *is singular.*

PROOF The first conclusion is a consequence of Theorem 1.4.6 and the fact that $AO = O$. If we assume that A is singular then the argument given in the second part of the proof of Theorem 1.4.3 can be applied to show that $AX = O$ has a nonzero solution, say $X_1 \neq O$. But kX_1 is a solution for every choice of the constant k. Thus there are infinitely many solutions when A is singular.

Exercise 1.4

Part 1

1. If A^{-1} exists and A is $n \times n$, what are the dimensions of A^{-1}? Must A^{-1} be square?
2. If A is $n \times n$ and $AC = B$ and A^{-1} exists, then A^{-1} and B have dimensions which allow $A^{-1}B$. Why? Is the same true for BA^{-1}?
3. If $AC = B$, A is $n \times n$ and C is a column matrix, what are the dimensions of B?
4. Use row operations to decide whether the following matrices are singular. When the inverse does exist find it.

(a) $\begin{bmatrix} -1 & 0 & 1 \\ 1 & 1 & 0 \\ 0 & 0 & 1 \end{bmatrix}$ (b) $\begin{bmatrix} 2 & 0 & 1 \\ 0 & 3 & 4 \\ 0 & 0 & 7 \end{bmatrix}$

(c) $\begin{bmatrix} 1 & 2 \\ -2 & 1 \end{bmatrix}$ (d) $\begin{bmatrix} 0 & 0 & 1 \\ 0 & 1 & 0 \\ 1 & 0 & 0 \end{bmatrix}$

(e) $\begin{bmatrix} \cos\theta & \sin\theta \\ -\sin\theta & \cos\theta \end{bmatrix}$ (f) $\begin{bmatrix} 2 & 0 & 0 \\ 4 & -1 & 0 \\ 0 & 1 & -1 \end{bmatrix}$

(g) $\begin{bmatrix} 1 & 0 & 2 \\ 0 & 1 & 0 \\ 0 & 5 & 0 \end{bmatrix}$

5. Verify that

$$\begin{bmatrix} 1 & 2 \\ -1 & 1 \end{bmatrix} \quad \text{and} \quad \begin{bmatrix} \frac{1}{3} & -\frac{2}{3} \\ \frac{1}{3} & \frac{1}{3} \end{bmatrix}$$

are inverses of each other.

6. Same as Problem 5 for the matrices

$$\begin{bmatrix} 1 & 0 & 1 \\ 2 & 1 & 1 \\ 1 & 2 & 2 \end{bmatrix} \quad \text{and} \quad \begin{bmatrix} 0 & \frac{2}{3} & -\frac{1}{3} \\ -1 & \frac{1}{3} & \frac{1}{3} \\ 1 & -\frac{2}{3} & \frac{1}{3} \end{bmatrix}$$

7. In Equation (1.4.3) show that

$$\begin{bmatrix} 2 & 2 & 1 \\ 1 & 1 & -3 \\ 0 & 0 & 0 \end{bmatrix} A^{-1} = \begin{bmatrix} - & - & - \\ - & - & - \\ 0 & 0 & 0 \end{bmatrix}$$

regardless of the entries of A^{-1}.

Part 2

1. Prove Lemma 1.4.2 for the case of 1 equation in 3 unknowns; for 2 equations in 3 unknowns.
2. Redo Example 1.4.3 mirroring the proof of Theorem 1.4.3. That is, find a nonzero solution to $AX = O$.
3. Show that the product of two matrices is nonsingular if both are nonsingular. State the contrapositive.
4. State the converse of the assertion given in Problem 3. Prove it. State the contrapositive.
5. Find the inverse of

$$\left[\begin{array}{cc|cc} 1 & 2 & 0 & 0 \\ 2 & 1 & 0 & 0 \\ \hline 0 & 0 & 3 & 1 \\ 0 & 0 & 2 & 3 \end{array}\right]$$

6. Generalize Problem 5 as follows:
   If A and B are invertible, what is the inverse of

$$C = \left[\begin{array}{c|c} A & O \\ \hline O & B \end{array}\right]$$

   where A is $n \times n$, B is $m \times m$, and the remaining O matrices have dimensions that make C, $(n+m) \times (n+m)$.

1.5. DETERMINANTS OF $n \times n$ MATRICES

The reader has probably encountered determinants of 2×2 and 3×3 matrices and may recall the formulas:

$$\det \begin{bmatrix} a_{11} & a_{12} \\ a_{21} & a_{22} \end{bmatrix} = a_{11}a_{22} - a_{12}a_{21} \tag{1.5.1}$$

$$\det \begin{bmatrix} a_{11} & a_{12} & a_{13} \\ a_{21} & a_{22} & a_{23} \\ a_{31} & a_{32} & a_{33} \end{bmatrix} = \begin{matrix} a_{11}a_{22}a_{33} + a_{12}a_{23}a_{31} \\ +\, a_{13}a_{21}a_{32} - a_{13}a_{22}a_{31} \\ -\, a_{12}a_{21}a_{33} - a_{11}a_{23}a_{32} \end{matrix} \tag{1.5.2}$$

perhaps under a different notation.†

†Determinants are often written as square arrays bordered by vertical lines such as

$$\begin{vmatrix} a_{11} & a_{12} \\ a_{21} & a_{22} \end{vmatrix}$$

For many reasons we select the notation used above.

The determinant has many uses. We cite a few:

(1) The system

$$\begin{aligned} a_{11}x + a_{12}y + a_{13}z &= 0 \\ a_{21}x + a_{22}y + a_{23}z &= 0 \\ a_{31}x + a_{32}y + a_{33}z &= 0 \end{aligned}$$

has a nontrivial solution (x, y, and z *not* all zero) if and only if the determinant of the coefficient matrix vanishes, that is, if and only if

$$\det \begin{bmatrix} a_{11} & a_{12} & a_{13} \\ a_{21} & a_{22} & a_{23} \\ a_{31} & a_{32} & a_{33} \end{bmatrix} = 0$$

(2) The construction of the solutions to simultaneous equations using Cramer's Rule involves the quotients of various determinants.

(3) If V_1 and V_2 are column matrices with two entries defining the sides of a parallelogram, as drawn in Figure 1.5.1, and if $[V_1, V_2]$ denotes the matrix with these as columns, then

$$|\det [V_1, V_2]| = \text{area of the parallelogram}$$

Figure 1.5.1

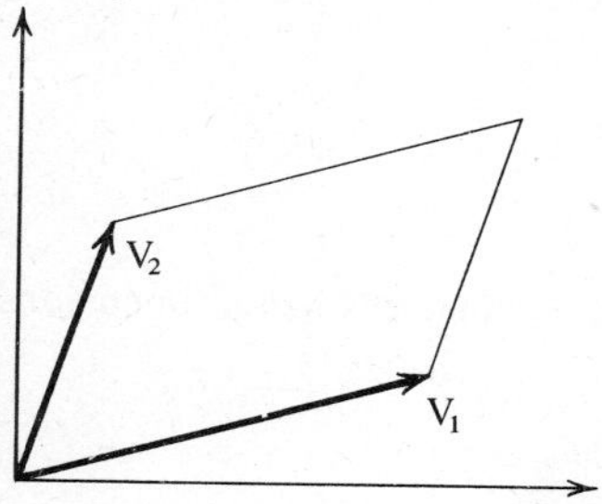

(4) If V_1, V_2, V_3 are 3×1 column matrices defining the sides of a parallelopiped (see Figure 1.5.2) and if $[V_1, V_2, V_3]$ denotes the matrix with these as columns, then

$$|\det [V_1, V_2, V_3]| = \text{volume of the parallelopiped}$$

The determinant can be defined for any square matrix in such a way that these applications, among many others, are preserved in higher dimension.

In order to state the definition of the determinant for the $n \times n$ matrix A,

$$A = \begin{bmatrix} a_{11} & \cdots & a_{1n} \\ a_{21} & \cdots & a_{2n} \\ \cdot & & \cdot \\ \cdot & & \cdot \\ \cdot & & \cdot \\ a_{n1} & \cdots & a_{nn} \end{bmatrix} \tag{1.5.3}$$

Figure 1.5.2

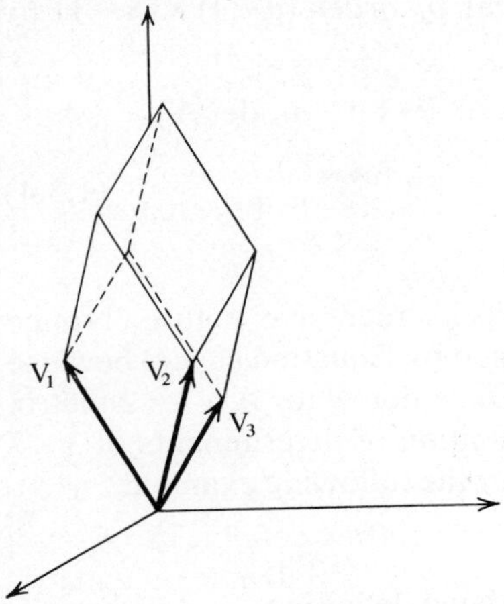

we construct n matrices each with dimensions $(n-1)\times(n-1)$ by omitting the first row of A and any one of the n columns. Set

$$A_1 = \begin{bmatrix} a_{22} & a_{23} & \cdots & a_{2n} \\ a_{32} & a_{33} & \cdots & a_{3n} \\ \cdot & \cdot & & \cdot \\ \cdot & \cdot & & \cdot \\ \cdot & \cdot & & \cdot \\ a_{n2} & a_{n3} & \cdots & a_{nn} \end{bmatrix}$$

$$A_2 = \begin{bmatrix} a_{21} & a_{23} & \cdots & a_{2n} \\ a_{31} & a_{33} & \cdots & a_{3n} \\ \cdot & \cdot & & \cdot \\ \cdot & \cdot & & \cdot \\ \cdot & \cdot & & \cdot \\ a_{n1} & a_{n3} & \cdots & a_{nn} \end{bmatrix}$$

and so on, until at the end,

$$A_n = \begin{bmatrix} a_{21} & a_{22} & \cdots & a_{2(n-1)} \\ a_{31} & a_{32} & \cdots & a_{3(n-1)} \\ \cdot & \cdot & & \cdot \\ \cdot & \cdot & & \cdot \\ \cdot & \cdot & & \cdot \\ a_{n1} & a_{n2} & \cdots & a_{n(n-1)} \end{bmatrix}$$

In general, then, A_i is an $(n-1)\times(n-1)$ matrix obtained from A by omitting its first row and ith column.

1.5.1. DEFINITION *The determinant of the* $n \times n (n > 1)$ *matrix* A *is defined in terms of determinants of matrices of order* $(n-1) \times (n-1)$ *by the relation*

$$\det A = a_{11} \det A_1 - a_{12} \det A_2 + \cdots + (-1)^{n+1} a_{1n} \det A_n$$
$$= \sum_{k=1}^{n} (-1)^{k+1} a_{1k} \det A_k \tag{1.5.4}$$

and if A *is the* 1×1 *matrix* $A = [a_{11}]$, $\det A = a_{11}$.

The essential feature of this definition is its recursive nature. Notice that the determinant of a 2×2 matrix is defined by Equation (1.5.4) because each matrix A_k $(k = 1,2)$ is a 1×1 matrix. Once det A for A a 2×2 matrix is known, Equation (1.5.4) provides the definition of determinants of 3×3 matrices, and so on. As an illustration consider the following example.

1.5.1. Example

Show that Equation (1.5.1) follows from the above definition.

Solution

We have

$$\det \begin{bmatrix} a_{11} & a_{12} \\ a_{21} & a_{22} \end{bmatrix} = a_{11} \det [a_{22}] - a_{12} \det [a_{21}]$$
$$= a_{11}a_{22} - a_{12}a_{21}$$

The definition of determinant given above is not suitable for computational purpose except under special circumstances. To find a more convenient method we present a brief excursion into the theory of determinants.†

1.5.2. THEOREM *If* A *is any square matrix and* A^T *is its transpose then*

$$\det A^T = \det A$$

1.5.3. THEOREM *If* A *is lower or upper triangular then* det A *is the product of the diagonal entries of* A.

PROOF If A is lower triangular, then

$$\det A = a_{11} \det A_1 + 0 \det A_2 + \cdots + 0 \det A_n$$
$$= a_{11} \det A_1$$

Since A_1 is also lower triangular this evaluation continues to leave only one term until finally

$$\det A = a_{11} \cdots a_{nn}$$

†For a fuller treatment of determinants including proofs of Theorems 1.5.2 and 1.5.4, the reader is referred to F. M. Stein, *Introduction to Matrices and Determinants*, Wadsworth, Belmont, Calif., 1967.

If A is upper triangular, A^T is lower triangular and by Theorem 1.5.2 $\det A = \det A^T$. From what has been proved above, $\det A^T$ is the product of the diagonal entries of A^T which are, of course, the diagonal entries of A. This completes the proof.

1.5.4. THEOREM *If A′ is obtained from A by addition of a multiple of one row to another then*

$$\det A = \det A'$$

This last result offers the optimal means for computing det A. Contrast the labor in the next two examples.

1.5.2. Example

Calculate det A where

$$A = \begin{bmatrix} 1 & 2 & 1 & 3 \\ -1 & 1 & 3 & 2 \\ 1 & 0 & 2 & 3 \\ -1 & 1 & 1 & 4 \end{bmatrix}$$

Solution

We have

$$\det A = \det \begin{bmatrix} 1 & 2 & 1 & 3 \\ 0 & 3 & 4 & 5 \\ 0 & -2 & 1 & 0 \\ 0 & 3 & 2 & 7 \end{bmatrix}$$

by various applications of Theorem 1.5.4 using multiples of row 1. Similarly,

$$\det A = \det \begin{bmatrix} 1 & 2 & 1 & 3 \\ 0 & 3 & 4 & 5 \\ 0 & -2 & 1 & 0 \\ 0 & 3 & 2 & 7 \end{bmatrix} = \det \begin{bmatrix} 1 & 2 & 1 & 3 \\ 0 & 3 & 4 & 5 \\ 0 & 0 & \frac{11}{3} & \frac{10}{3} \\ 0 & 0 & -2 & 2 \end{bmatrix}$$

$$= \det \begin{bmatrix} 1 & 2 & 1 & 3 \\ 0 & 3 & 4 & 5 \\ 0 & 0 & \frac{11}{3} & \frac{10}{3} \\ 0 & 0 & 0 & \frac{42}{11} \end{bmatrix} = (1)(3)\left(\tfrac{11}{3}\right)\left(\tfrac{42}{11}\right)$$

$$= 42$$

1.5.3. Example

Calculate det A for A of Example 1.5.2 by use of Definition 1.5.1.

Solution

We have

$$\det A = (1)\begin{bmatrix} 1 & 3 & 2 \\ 0 & 2 & 3 \\ 1 & 1 & 4 \end{bmatrix} - (2)\begin{bmatrix} -1 & 3 & 2 \\ 1 & 2 & 3 \\ -1 & 1 & 4 \end{bmatrix}$$

$$+ (1)\begin{bmatrix} -1 & 1 & 2 \\ 1 & 0 & 3 \\ -1 & 1 & 4 \end{bmatrix} - (3)\begin{bmatrix} -1 & 1 & 3 \\ 1 & 0 & 2 \\ -1 & 1 & 1 \end{bmatrix}$$

$$= (1)\det A_1 - (2)\det A_2 + (1)\det A_3 - (3)\det A_4$$

We find

$$\det A_1 = 1(8-3) - 3(0-3) + 2(0-2) = 10$$
$$\det A_2 = (-1)(8-3) - 3(4+3) + 2(1+2) = -20$$
$$\det A_3 = (-1)(0-3) - (1)(4+3) + 2(1-0) = -2$$
$$\det A_4 = (-1)(0-2) - (1)(1+2) + 3(1-0) = 2$$

Hence,

$$\det A = 1 \cdot 10 - 2 \cdot (-20) + 1(-2) - (3)(2)$$
$$= 42$$

A complication may occur if a diagonal entry vanishes at some point. This difficulty is easily remedied as the next example shows.

1.5.4. Example

Calculate det A where

$$A = \begin{bmatrix} 0 & 1 & 1 \\ 1 & 0 & 0 \\ 0 & 0 & 1 \end{bmatrix}$$

Solution

We begin by adding the 2nd row to the first and then proceed along the lines indicated in Example 1.5.2. That is,

$$\det A = \det\begin{bmatrix} 1 & 1 & 1 \\ 1 & 0 & 0 \\ 0 & 0 & 1 \end{bmatrix} = \det\begin{bmatrix} 1 & 1 & 1 \\ 0 & -1 & -1 \\ 0 & 0 & 1 \end{bmatrix}$$

$$= (1)(-1)(1) = -1$$

1.5.5. Example

Show: $\det A = 0$ if any row or column of A is a row or column of zeros.

Solution

If the row of zeros is the first row of A the conclusion is a trivial consequence of Equation (1.5.4). If the row of zeros is the ith row we simply add the first row to the ith row making the first and ith rows equal. Now subtracting the

ith row from the first yields a matrix whose first row is a row of zeros and whose determinant is equal to the determinant of A. The fact det A^T = det A completes the proof.

Besides its computational expediency, Theorem 1.5.4 has important theoretical implications. The next two theorems illustrate this.

1.5.5. THEOREM *A is singular if and only if* det A = 0.

PROOF If A is singular it can be row-reduced to triangular form with at least one zero entry on the diagonal. This follows from Theorem 1.4.3. In fact, the row reduction may be effected by use of the row operation prescribed in Theorem 1.5.4. By Theorem 1.5.3 and Theorem 1.5.4 det A = 0. Similarly if A is invertible, A can be row-reduced to upper triangular form with no zero entries on the diagonal. Hence A^{-1} exists implies det A $\neq 0$ by these same two theorems.

A direct corollary follows.

1.5.6. COROLLARY *The homogeneous system* AX = O *where* X *and* O *are column matrices and* A *is* $n \times n$ *has a solution* X $\neq$ O *if and only if* det A = 0.

PROOF The system AX = O has a unique solution if and only if A is invertible by Theorem 1.4.6. Hence, by Theorem 1.5.5 it has a unique solution if and only if det A $\neq 0$. Since X = O is always a solution, we conclude AX = O has a nonzero solution if and only if det A = 0.

We conclude this brief introduction to determinants with one more useful theorem.

1.5.7. THEOREM *If* A *and* B *are* $n \times n$ *matrices then*

$$\det(AB) = (\det A)(\det B)$$

The reader is again referred to F. M. Stein's book for a proof of this result (see footnote on p. 22).

Exercise 1.5

Part 1

1. Compute the determinants for the following matrices:

(a) $\begin{bmatrix} 1 & 1 & 0 \\ 2 & 0 & 2 \\ 0 & 5 & 5 \end{bmatrix}$ (b) $\begin{bmatrix} 0 & \cdots & 0 & a_{1n} \\ 0 & \cdots & a_{2(n-1)} & a_{2n} \\ \vdots & & \vdots & \vdots \\ \vdots & \iddots & \vdots & \vdots \\ \vdots & \iddots & \vdots & \vdots \\ a_{n1} & \cdots & a_{n(n-1)} & a_{nn} \end{bmatrix}$

(c) $\begin{bmatrix} -1 & 2 & 1 & 1 \\ 0 & 1 & 0 & 0 \\ 0 & 1 & 1 & -1 \\ 2 & 0 & 0 & 3 \end{bmatrix}$ (d) $\begin{bmatrix} 2 & 1 & 3 \\ 0 & 1 & 1 \\ 1 & 0 & -1 \end{bmatrix}$

2. Find three values of λ for which

$$\det \begin{bmatrix} 2-\lambda & 2 & 0 \\ 1 & 2-\lambda & 1 \\ 1 & 2 & 1-\lambda \end{bmatrix} = 0$$

3. Verify $\det(AB) = \det A \det B$ when

$$A = \begin{bmatrix} 3 & 1 & 4 \\ -1 & 0 & 1 \\ 1 & 2 & 1 \end{bmatrix} \qquad B = \begin{bmatrix} 0 & 1 & 1 \\ -3 & 2 & 0 \\ 1 & 0 & 1 \end{bmatrix}$$

4. Find

$$\det \begin{bmatrix} 1 & x_1 \\ 1 & x_2 \end{bmatrix} \qquad \det \begin{bmatrix} 1 & x_1 & x_1^2 \\ 1 & x_2 & x_2^2 \\ 1 & x_3 & x_3^2 \end{bmatrix}$$

5. Find

$$\det \begin{bmatrix} -\lambda & 1 \\ -a_0 & -a_1-\lambda \end{bmatrix}$$

6. Consider the matrix

$$A = \begin{bmatrix} 1 & 2 & 1 \\ -1 & 2 & 3 \\ 0 & 1 & 2 \end{bmatrix}$$

(a) Compute det A by means of Equation (1.5.2)
(b) Compute det A by row reduction.
(c) Compute det A by means of repeated application of Definition 1.5.1.

Part 2

1. Prove: if A is nonsingular then $\det(A^{-1}) = (\det A)^{-1}$.
2. Prove: $\det(B^{-1}AB) = \det A$.
3. Prove: $\det(A^n) = (\det A)^n$.
4. Prove: $\det(kA) = k^n \det A$, where A is $n \times n$.
5. Prove: If two rows or two columns of A are proportional then $\det A = 0$.
6. Use determinants to prove:
AB is nonsingular if and only if A and B are nonsingular. Does the truth of this proposition imply any of the following?
(i) AB is singular if either A or B is singular.
(ii) A is singular implies AB is singular.
7. Derive Equation (1.5.2) by application of Definition 1.5.1.

1.6. VECTOR SPACES OF *n*-TUPLES

The reader is familiar with the notion of a force or a velocity as examples of the usage of the term "vector" in physics. The graphical portrayal of vectors as arrows is a reflection of their historical use in representing quantities having magnitude and direction. We desire a definition which is not restricted to three or less dimensions, which does not depend upon pictures for proofs and yet still retains enough geometric content to stimulate our intuition. Toward this aim we call a column matrix or row matrix a *vector* and identify its n components as the n coordinates of a point in n-dimensional Euclidean space. Thus if $n \leq 3$ the vector may be identified with the directed line segment extending from the origin to the point in question. Under this interpretation the addition of vectors corresponds to the "parallelogram law" of addition of forces or velocities. (To emphasize the interpretation of a column matrix and a row matrix as a vector we alter our matrix notation and write **a**, **b**, **c**, **x**, **y**, for the column and row matrices previously denoted by A, B, C, X, Y, respectively.) Consider Figure 1.6.1.

Figure 1.61

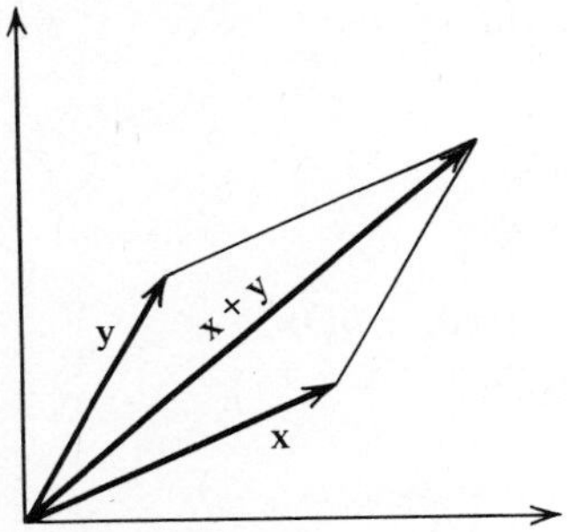

In this and the following sections we consider vectors from this geometric standpoint. In particular, we introduce and study special sets of vectors known as vector spaces.

1.6.1. DEFINITION *A nonempty set of vectors $\mathscr{V}$ is said to be a vector space if*

$$\begin{aligned} &\text{(a) } \mathbf{x} \in \mathscr{V} \textit{ and } \mathbf{y} \in \mathscr{V} \textit{ implies } \mathbf{x}+\mathbf{y} \in \mathscr{V}. \\ &\text{(b) } \mathbf{x} \in \mathscr{V} \textit{ implies } k\mathbf{x} \in \mathscr{V} \textit{ for every scalar } k. \end{aligned} \tag{1.6.1}$$

If the scalars used in (b) and the entries of the vectors are restricted to be real constants the vector space is called a real vector space; if the scalars are selected from the field of complex numbers, we refer to $\mathscr{V}$ as a complex vector space. The abbreviation $\mathbf{x} \in \mathscr{V}$ is to be read "$\mathbf{x}$ belongs to $\mathscr{V}$" or "$\mathbf{x}$ is a member of the set of elements denoted by $\mathscr{V}$." Now if $\mathscr{V}$ is a vector space (a) and (b) imply

(c)† $\mathbf{0} \in \mathscr{V}$

(d) $a_1\mathbf{x}_1 + x_2\mathbf{x}_2 + \cdots + a_k\mathbf{x}_k \in \mathscr{V}$, for any choice of scalars, $a_1, a_2, \ldots, a_k$, if $\mathbf{x}_1 \in \mathscr{V}, \mathbf{x}_2 \in \mathscr{V}, \ldots, \mathbf{x}_k \in \mathscr{V}$

†The $n \times 1$ column matrix with zero entries is written O as are all zero matrices. However, when we wish to interpret this matrix as a vector, we write **0**.

Remarks

Since $\mathbf{x}+\mathbf{y}$ is defined, $\mathbf{x}$ and $\mathbf{y}$ must have the same number of entries. Hence if $\mathbf{x}$ is an n-tuple and $\mathbf{x} \in \mathscr{V}$, then $\mathscr{V}$ consists of n-tuples and thus $\mathbf{0} = [0, 0, \ldots, 0]$ is an n-tuple. (For typographical reasons we shall often write a column matrix as a row matrix as we have done with $\mathbf{0}$ above.)

A vector $\mathbf{x}_0$, defined by the relationship

$$\mathbf{x}_0 = a_1\mathbf{x}_1 + a_2\mathbf{x}_2 + \cdots + a_k\mathbf{x}_k \tag{1.6.2}$$

is said to be a *linear combination* of $\mathbf{x}_1, \mathbf{x}_2, \ldots, \mathbf{x}_k$.

1.6.2. THEOREM *The set of all linear combinations of $\mathbf{x}_1, \mathbf{x}_2, \ldots, \mathbf{x}_k$, denoted by $\langle \mathbf{x}_1, \mathbf{x}_2, \ldots, \mathbf{x}_k \rangle$, is a vector space known as the span of $\mathbf{x}_1, \mathbf{x}_2, \ldots, \mathbf{x}_k$.*

PROOF The set of all linear combinations certainly contains $\mathbf{x}+\mathbf{y}$ and $k\mathbf{x}$ if it contains $\mathbf{x}$ and $\mathbf{y}$. For if

$$\mathbf{x} = a_1\mathbf{x}_1 + a_2\mathbf{x}_2 + \cdots + a_k\mathbf{x}_k$$
$$\mathbf{y} = b_1\mathbf{x}_1 + b_2\mathbf{x}_2 + \cdots + b_k\mathbf{x}_k$$

then

$$\mathbf{x}+\mathbf{y} = (a_1+b_1)\mathbf{x}_1 + (a_2+b_2)\mathbf{x}_2 + \cdots + (a_k+b_k)\mathbf{x}_k$$
$$k\mathbf{x} = (ka_1)\mathbf{x}_1 + (ka_2)\mathbf{x}_2 + \cdots + (ka_k)\mathbf{x}_k$$

This shows that $\mathbf{x}+\mathbf{y} \in \langle \mathbf{x}_1, \mathbf{x}_2, \ldots, \mathbf{x}_k \rangle$ and $k\mathbf{x} \in \langle \mathbf{x}_1, \mathbf{x}_2, \ldots, \mathbf{x}_k \rangle$.

1.6.1. Example

Show that the set of all n-tuples is a vector space.

Solution

Certainly, if $\mathbf{x}$ and $\mathbf{y}$ are n-tuples, so are $\mathbf{x}+\mathbf{y}$ and $k\mathbf{x}$. This is all that is needed. However, an alternative argument is instructive. Consider the vector space $\langle \mathbf{e}_1, \mathbf{e}_2, \ldots, \mathbf{e}_n \rangle$ where

$$\mathbf{e}_1 = \begin{bmatrix} 1 \\ 0 \\ 0 \\ \cdot \\ \cdot \\ \cdot \\ 0 \end{bmatrix} \quad \mathbf{e}_2 = \begin{bmatrix} 0 \\ 1 \\ 0 \\ \cdot \\ \cdot \\ \cdot \\ 0 \end{bmatrix} \cdots \mathbf{e}_n = \begin{bmatrix} 0 \\ 0 \\ \cdot \\ \cdot \\ \cdot \\ 0 \\ 1 \end{bmatrix} \tag{1.6.3}$$

are n-tuples written as vectors. Then if $\mathbf{x} = [x_1, x_2, \ldots, x_n]$ is an arbitrary n-tuple

$$\mathbf{x} = x_1\mathbf{e}_1 + x_2\mathbf{e}_2 + \cdots + x_n\mathbf{e}_n \tag{1.6.4}$$

Thus

$$\mathbf{x} \in \langle \mathbf{e}_1, \mathbf{e}_2, \ldots, \mathbf{e}_n \rangle$$

We denote the space of all n-tuples of real numbers, regarded as vectors, by $\mathscr{R}^n$ and of all n-tuples of complex numbers by $\mathscr{C}^n$. Then we may have

$$\mathscr{R}^n = \langle \mathbf{e}_1, \mathbf{e}_2, \ldots, \mathbf{e}_n \rangle \qquad (1.6.5)$$

or

$$\mathscr{C}^n = \langle \mathbf{e}_1, \mathbf{e}_2, \ldots, \mathbf{e}_n \rangle$$

depending on the field of scalars. By this we mean

$$\mathscr{R}^n = \{a_1\mathbf{e}_1 + a_2\mathbf{e}_2 + \cdots + a_n\mathbf{e}_n | a_1, \ldots, a_n \quad \text{real}\}$$

and

$$\mathscr{C}^n = \{a_1\mathbf{e}_1 + a_2\mathbf{e}_2 + \cdots + a_n\mathbf{e}_n | a_1, \ldots, a_n \quad \text{complex}\}$$

The context will make clear from which set we select our scalars.

An important notion is the concept of a subspace.

1.6.3. DEFINITION *A nonempty subset $\mathscr{W}$ of the vectors in a vector space $\mathscr{V}$ is a subspace of $\mathscr{V}$ if $\mathscr{W}$ is a vector space with the same scalars as in $\mathscr{V}$.*

Remarks

Not all subsets of $\mathscr{V}$ are vector spaces nor are all vector spaces which are subsets of $\mathscr{V}$ also subspaces of $\mathscr{V}$. We require in this definition that $\mathscr{W}$, considered in its own right, is a vector space over the same scalars as in $\mathscr{V}$. Note that $\mathscr{R}^n$ is not a subspace of $\mathscr{C}^n$, since if $\mathbf{x}_0 \in \mathscr{R}^n$, $i\mathbf{x}_0 \notin \mathscr{R}^n$.

Exercise 1.6

Part 1

1. Which of the vectors $\mathbf{a} = [2, -1, 1]$, $\mathbf{b} = [0, 0, 1]$, $\mathbf{c} = [-1, 0, 1]$ is in the span $\langle \mathbf{x}, \mathbf{y} \rangle$ of $\mathbf{x} = [1, 1, 1]$ and $\mathbf{y} = [-1, 0, 1]$?
2. Which of the vectors $[2, 0, 1]$, $[-1, 1, 2]$, $[0, 0, 0]$ are in the span $\langle \mathbf{x}, \mathbf{y} \rangle$ of $\mathbf{x} = [1, 0, -1]$ and $\mathbf{y} = [-1, 1, 0]$?
3. Show that the set of vectors $\mathscr{S} = \{[0, -b, b]\}$ where b is any real number, is a vector space.
4. Show that the set of vectors $\mathscr{S} = \{[0, 0, k, 1-c]\}$ is a vector space, where k and c are real numbers.
5. Which of the following sets are vector spaces?
   (a) $\mathscr{S} = \{[k, 1], k$ is any real number$\}$
   (b) $\mathscr{S} = \{[k, 0, l], k$ and l are any real numbers$\}$

Part 2

1. Prove that the set of vectors consisting of the zero vector alone is a vector space.

2. Prove that the set of vectors common to any two subspaces of $\mathscr{R}^n$ is a vector space.
3. Prove that the set of all points lying on a plane through the origin is a subspace of $\mathscr{R}^3$.
4. Prove that a line in $\mathscr{R}^2$ is a subspace of $\mathscr{R}^2$ if and only if it passes through the origin.
5. What is the span of a single, nonzero vector $\mathbf{x}$? If this vector is a 2-tuple describe $\langle \mathbf{x} \rangle$ geometrically. If $\mathbf{x}$ is a 3-tuple describe $\langle \mathbf{x} \rangle$.
6. If $\mathbf{x}$ and $\mathbf{y}$ are 3-tuples describe geometrically the set $\langle \mathbf{x}, \mathbf{y} \rangle$ by identifying the coordinates of a point with the entries of the vector. Be sure to consider the cases: (1) One or both of the given vectors are zero. (2) One vector is a scalar multiple of the other.

1.7. LINEAR INDEPENDENCE

Suppose the vector $\mathbf{y}$ is defined by the sum

$$\mathbf{y} = a_1\mathbf{x}_1 + a_2\mathbf{x}_2 + \cdots + a_k\mathbf{x}_k \qquad k \geqq 1 \tag{1.7.1}$$

Then, as before, $\mathbf{y}$ is a *linear combination* of $\mathbf{x}_1, \mathbf{x}_2, \ldots, \mathbf{x}_k$. The scalars $a_1, a_2, \ldots, a_k$ may be real or complex numbers. If all the scalars are zero then (1.7.1) is called *trivial* and, of course, $\mathbf{y} = \mathbf{0}$. On the other hand, $\mathbf{y}$ may be the zero vector without all the scalars zero as seen in the sum, $\mathbf{0} = 2[1, 1] - [2, 2]$. When a linear combination is the zero vector without all the scalars being zero we call the combination *nontrivial.*

1.7.1. DEFINITION *A sequence of vectors (each an n-tuple) is said to be linearly dependent if* $\mathbf{0}$ *is a nontrivial linear combination of these vectors, that is,*

$$\mathbf{0} = a_1\mathbf{x}_1 + a_2\mathbf{x}_2 + \cdots + a_k\mathbf{x}_k \qquad k \geqq 1 \tag{1.7.2}$$

where at least one scalar is not zero. If the given vectors are not linearly dependent they are called linearly independent.

A useful consequence of this definition is the following theorem.

1.7.2. THEOREM *If* $\mathbf{x}_1, \mathbf{x}_2, \ldots, \mathbf{x}_k$ *are linearly independent and*

$$\mathbf{0} = a_1\mathbf{x}_1 + a_2\mathbf{x}_2 + \cdots + a_k\mathbf{x}_k \tag{1.7.3}$$

then

$$a_1 = a_2 = \cdots = a_k = 0$$

PROOF If even one scalar in Equation (1.7.3) were not zero the sum would be nontrivial by definition. But then the given vectors would be linearly dependent, a contradiction.

1.7.1. Example

Demonstrate the linear dependence of the following vectors

(a) $[1, 1, 0], [-1, 1, 0], [0, 1, 0]$

(b) $\mathbf{0}, \mathbf{x}_1, \mathbf{x}_2, \mathbf{x}_3$
(c) $\mathbf{x}_1, \mathbf{x}_2 - \mathbf{x}_1, 2\mathbf{x}_1 + \mathbf{x}_2$

Solution

First of all

$$\begin{bmatrix}1\\1\\0\end{bmatrix}+\begin{bmatrix}-1\\1\\0\end{bmatrix}-2\begin{bmatrix}0\\1\\0\end{bmatrix}=\begin{bmatrix}0\\0\\0\end{bmatrix}$$

Next,

$$1\cdot\mathbf{0}+0\cdot\mathbf{x}_1+0\cdot\mathbf{x}_2+0\cdot\mathbf{x}_3=0 \qquad \text{and} \qquad 3\mathbf{x}_1+1(\mathbf{x}_2-\mathbf{x}_1)-(2\mathbf{x}_1+\mathbf{x}_2)=\mathbf{0}$$

shows the dependence of the vectors given in (b) and (c), respectively.

1.7.2. Example

Show that [1, 1, 0, 1], [1, 0, 0, 1], [1, −1, 0, 1], and [0, 0, 1, 0] is a linearly dependent set by finding the scalars so that (1.7.2) holds.

Solution

Label the given vectors, $\mathbf{x}_1$, $\mathbf{x}_2$, $\mathbf{x}_3$, and $\mathbf{x}_4$ successively. We now reduce the matrix

$$\begin{bmatrix}1 & 1 & 0 & 1 & \mathbf{x}_1\\ 1 & 0 & 0 & 1 & \mathbf{x}_2\\ 1 & -1 & 0 & 1 & \mathbf{x}_3\\ 0 & 0 & 1 & 0 & \mathbf{x}_4\end{bmatrix} \tag{1.7.4}$$

The rows of this matrix are the given vectors while the last column is simply used to keep track of the various row operations used in the reduction. Note, for instance, the last column of the matrix given as (1.7.5):

$$\begin{bmatrix}1 & 1 & 0 & 1 & \mathbf{x}_1\\ 0 & -1 & 0 & 0 & \mathbf{x}_2-\mathbf{x}_1\\ 0 & -2 & 0 & 0 & \mathbf{x}_3-\mathbf{x}_1\\ 0 & 0 & 1 & 0 & \mathbf{x}_4\end{bmatrix} \tag{1.7.5}$$

exhibits the row operations used to go from (1.7.4) to (1.7.5). Finally, from (1.7.5) we obtain

$$\begin{bmatrix}1 & 1 & 0 & 1 & \mathbf{x}_1\\ 0 & -1 & 0 & 0 & \mathbf{x}_2-\mathbf{x}_1\\ 0 & 0 & 0 & 0 & \mathbf{x}_3-\mathbf{x}_1-2(\mathbf{x}_2-\mathbf{x}_1)\\ 0 & 0 & 1 & 0 & \mathbf{x}_4\end{bmatrix} \tag{1.7.6}$$

The third row of (1.7.6) shows that

$$[0,0,0,0]=\mathbf{x}_3-\mathbf{x}_1-2(\mathbf{x}_2-\mathbf{x}_1)$$

or more neatly,

$$\mathbf{x}_1 - 2\mathbf{x}_2 + \mathbf{x}_3 = \mathbf{0}$$

which is the required nontrivial sum. The point illustrated by this last example can be put in the form of a theorem.

1.7.3. THEOREM *If the matrix whose rows are the vectors,* $\mathbf{x}_1, \mathbf{x}_2, \ldots, \mathbf{x}_k$ *can be row-reduced to a matrix with a row of zeros, then* $\mathbf{x}_1, \mathbf{x}_2, \ldots, \mathbf{x}_k$ *form a linearly dependent sequence.*

PROOF Some critical steps of the proof are outlined in Exercise 1.7, Part 2, Problems 4 and 5.

Note the following three simple consequences of this theorem.

1.7.4. COROLLARY *The vectors* $\mathbf{x}_1, \mathbf{x}_2, \ldots, \mathbf{x}_k$ *are linearly dependent if any are the zero vector.*

PROOF The reader should supply an argument.

1.7.5. COROLLARY *If* $\mathbf{x}_1, \ldots, \mathbf{x}_k$ *are n-tuples and* $k > n$, *the set is linearly dependent.*

PROOF The matrix with rows $\mathbf{x}_1, \mathbf{x}_2, \ldots, \mathbf{x}_k$ has more rows than columns by hypothesis. Row reductions will yield either a row of zeros among the first n rows or a diagonal matrix with no zeros on the diagonal matrix of dimension $n \times n$. In the latter case the $n+1$ row can be cleared to zero and hence in either case the conclusion follows from Theorem 1.7.3.

1.7.6. COROLLARY *If* $\mathbf{x}_1, \mathbf{x}_2, \ldots, \mathbf{x}_n$ *are n-tuples then they are linearly independent if and only if the determinant of the matrix formed with these vectors as rows is not zero.*

PROOF Let X be the matrix whose rows are the vectors $\mathbf{x}_1, \mathbf{x}_2, \ldots, \mathbf{x}_n$. Let Y be any matrix constructed from X by row operations of the type allowed by Theorem 1.5.4. Then

$$\det X = \det Y \tag{1.7.7}$$

by this theorem. Now a matrix Y can be found with a row of zeros if and only if the given vectors are linearly dependent. However, if a row of Y is all zeros then $\det Y = 0$ and from (1.7.7), $\det X = 0$. If no Y has a row of zeros X may be reduced to a diagonal matrix with no zeros on the diagonal and from (1.7.1), $\det X \neq 0$.

1.7.3. Example

Find those numbers λ for which $[1-\lambda, 0, 0]$, $[1, 1-\lambda, 0]$, and $[1, 1, 1-\lambda]$ are linearly dependent.

Solution

Consider the matrix

$$A(\lambda) = \begin{bmatrix} 1-\lambda & 0 & 0 \\ 1 & 1-\lambda & 0 \\ 1 & 1 & 1-\lambda \end{bmatrix}$$

Then $\det A(\lambda) = (1-\lambda)^3 = 0$ only when $\lambda = 1$.

1.7.4. Example

The n-tuples

$$\mathbf{e}_1 = \begin{bmatrix} 1 \\ 0 \\ \cdot \\ \cdot \\ \cdot \\ 0 \end{bmatrix} \quad \mathbf{e}_2 = \begin{bmatrix} 0 \\ 1 \\ 0 \\ \cdot \\ \cdot \\ 0 \end{bmatrix} \cdots \mathbf{e}_n = \begin{bmatrix} 0 \\ \cdot \\ \cdot \\ \cdot \\ 0 \\ 1 \end{bmatrix}$$

are linearly independent.

Solution

The relevant matrix is the $n \times n$ identity.

Exercise 1.7

Part 1

1. Which of the following sequences of vectors are linearly dependent?
   (a) $[1, 0, 1], [1, 1, -1], [-1, 1, -3]$
   (b) $[1, 1, 0, 1], [-1, 2, 0, 0], [0, 0, 1, 0], [1, -1, -1, 1]$
   (c) $[-1, 0, 0, 1], [2, -1, 1, 1], [0, -1, 1, 3]$
2. Find k so that the following vectors are dependent:

$$[k, 0, 1], [1, k, -1], [-1, 1, k]$$

3. If $\mathbf{x}_1, \mathbf{x}_2, \mathbf{x}_3, \mathbf{x}_4$ is a linearly independent sequence, is $\mathbf{x}_1, \mathbf{x}_2, \mathbf{x}_3$? Why? Generalize.
4. If $\mathbf{x}_1, \mathbf{x}_2, \mathbf{x}_3, \mathbf{x}_4$ is a linearly dependent sequence, is $\mathbf{x}_1, \mathbf{x}_2, \mathbf{x}_3$? Explain.
5. For each of the following sequences of linearly dependent vectors, express $\mathbf{0}$ as a nontrivial linear combination.
   (a) Problem 1(a), this set
   (b) Problem 1(c), this set
   (c) $[-1, 2, 0, 0], [1, 2, -1, 0], [1, 1, 0, 1], [1, 5, -1, 1]$
   (d) $[1, 1, 0, 1], [1, 0, 0, 1], [0, 0, 1, 0], [1, -1, 0, 1]$

Part 2

1. A sequence consisting of a single vector, $\mathbf{x} \neq \mathbf{0}$, is linearly independent. Why?
2. Given a linearly dependent sequence of at least two vectors none of which is the zero vector such that

$$\mathbf{0} = a_1\mathbf{x}_1 + a_2\mathbf{x}_2 + \cdots + a_k\mathbf{x}_k$$

is nontrivial. Prove at least two of the scalars are not zero.

3. If the nonzero vectors $\mathbf{x}_1, \mathbf{x}_2, \ldots, \mathbf{x}_k$ are linearly dependent then there is a vector in this set which is a nontrivial linear combination of the other vectors; prove this fact.

4.* Let $\mathscr{S} = \{\mathbf{x}_1, \mathbf{x}_2, \ldots, \mathbf{x}_k\}$, $k > 1$, be a sequence of vectors. Let $\mathscr{T}$ be a second sequence constructed from $\mathscr{S}$ so that

$$\mathscr{T} = \{\mathbf{x}_1, \ldots, \mathbf{x}_{i-1}, \mathbf{x}_i - \alpha\mathbf{x}_j, \mathbf{x}_{i+1}, \ldots, \mathbf{x}_k\} \qquad i > j$$

That is, the vector $\mathbf{x}_i$ in $\mathscr{S}$ is replaced by $\mathbf{x}_i - \alpha\mathbf{x}_j$ to form $\mathscr{T}$.
(a) Prove: If $\mathscr{S}$ is linearly dependent so is $\mathscr{T}$.
Hint: Assume $a_1\mathbf{x}_1 + a_2\mathbf{x}_2 + \cdots + a_k\mathbf{x}_k = \mathbf{0}$ is nontrivial. Then

$$\begin{aligned}\mathbf{0} &= a_1\mathbf{x}_1 + \cdots + a_k\mathbf{x}_k = a_1\mathbf{x}_1 + \cdots + a_i(\mathbf{x}_j - \alpha\mathbf{x}_j) + a_i\alpha\mathbf{x}_j + \cdots + a_k\mathbf{x}_k \\ &= a_1\mathbf{x}_1 + \cdots + (a_j + a_i\alpha)\mathbf{x}_j + \cdots + a_i(\mathbf{x}_i - \alpha\mathbf{x}_j) + \cdots + a_k\mathbf{x}_k\end{aligned}$$

Now deduce that $\mathscr{T}$ is linearly dependent.
(b) Prove: If $\mathscr{S}$ is linearly independent so is $\mathscr{T}$.
Hint: The statement (b) is equivalent to: If $\mathscr{T}$ is linearly dependent so is $\mathscr{S}$. (Why?) Now modify the argument given in (a).

5.* (a) Interpret the conclusions drawn from (a) and (b) in Problem 4 in terms of the rows of a matrix after repeated row operations of the form: Add a multiple of one row to a second row.
(b) Prove Theorem 1.7.3.

1.8. BASES AND DIMENSION

The concept of linear independence and the span of a set of vectors are combined in the notion of a basis for a vector space.

1.8.1. DEFINITION *We call the sequence of linearly independent vectors,* $\mathbf{x}_1, \mathbf{x}_2, \ldots, \mathbf{x}_k$ *a basis for the vector space* $\mathscr{V}$ *if*

$$\langle \mathbf{x}_1, \mathbf{x}_2, \ldots, \mathbf{x}_k \rangle = \mathscr{V}$$

The linearly independent vectors $\mathbf{x}_1, \mathbf{x}_2, \ldots, \mathbf{x}_n$ of Example 1.7.4 provide us with our first example of a basis, for, as was seen in Example 1.6.1, this set spans $\mathscr{R}^n$ and $\mathscr{C}^n$ and is thus a basis for both spaces. We now state without proof a fundamental result.

1.8.2. THEOREM *Let* $\mathscr{V}$ *be a subspace of* $\mathscr{R}^n$. *Then there is always a basis for* $\mathscr{V}$ *and every basis for* $\mathscr{V}$ *has the same number of vectors.*

Remarks

(1) This theorem guarantees the existence of at least one basis for $\mathscr{V}$. Notice that, although $\mathbf{e}_1, \mathbf{e}_2, \ldots, \mathbf{e}_n$ are linearly independent and surely span $\mathscr{R}^n$ which contains $\mathscr{V}$, they are not necessarily a basis for $\mathscr{V}$ since they are not in general elements of $\mathscr{V}$. For instance, let $\mathscr{V} = \langle [1, 1] \rangle$. Then $\mathscr{V}$ is one-dimensional but neither $\mathbf{e}_1 = [1, 0]$ nor $\mathbf{e}_2 = [0, 1]$ belong to $\mathscr{V}$.

(2) The common number of basis vectors in each basis is known as the

dimension of $\mathscr{V}$ and is written $\dim \mathscr{V}$. Clearly, $\dim \mathscr{R}^n = n$. It can be shown that $\dim \mathscr{V} \leqq n$. It is not immediately clear how to determine a basis or the dimension of a given subspace. A helpful result in this connection is the following theorem which asserts, essentially, that the span of a given set of vectors is unaltered if the set is diminished by excluding a linearly dependent vector.

1.8.3. THEOREM *Let* $\mathbf{x}_k$ *be a linear combination of* $\mathbf{x}_1, \mathbf{x}_2, \ldots, \mathbf{x}_{k-1}, k \geqq 2$. *Then* $\langle \mathbf{x}_1, \mathbf{x}_2, \ldots, \mathbf{x}_k \rangle = \langle \mathbf{x}_1, \mathbf{x}_2, \ldots, \mathbf{x}_{k-1} \rangle$.

PROOF The proof is easy. The hypothesis requires

$$\mathbf{x}_k = a_1\mathbf{x}_1 + a_2\mathbf{x}_2 + \cdots + a_{k-1}\mathbf{x}_{k-1}$$

If $\mathbf{x}_0 \in \langle \mathbf{x}_1, \mathbf{x}_2, \ldots, \mathbf{x}_k \rangle$ then

$$\begin{aligned} \mathbf{x}_0 &= b_1\mathbf{x}_1 + b_2\mathbf{x}_2 + \cdots + b_k\mathbf{x}_k \\ &= b_1\mathbf{x}_1 + b_2\mathbf{x}_2 + \cdots + b_k(a_1\mathbf{x}_1 + a_2\mathbf{x}_2 + \cdots + a_{k-1}\mathbf{x}_{k-1}) \\ &= (b_1 + b_k a_1)\mathbf{x}_1 + (b_2 + b_k a_2)\mathbf{x}_2 + \cdots + (b_{k-1} + b_k a_{k-1})\mathbf{x}_{k-1} \end{aligned}$$

This proves that $\mathbf{x}_0 \in \langle \mathbf{x}_1, \mathbf{x}_2, \ldots, \mathbf{x}_{k-1} \rangle$. On the other hand, if $\mathbf{x}_0 \in \langle \mathbf{x}_1, \mathbf{x}_2, \ldots, \mathbf{x}_{k-1} \rangle$ then surely $\mathbf{x}_0 \in \langle \mathbf{x}_1, \mathbf{x}_2, \ldots, \mathbf{x}_k \rangle$.

The usefulness of this theorem is demonstrated in the next three examples.

1.8.1. Example

Find the dimension and basis for the vector space spanned by [1, 1, 0, 1], [1, 0, 0, 1], [1, −1, 0, 1], and [0, 0, 1, 0].

Solution

These vectors were examined earlier in Example 1.7.2 where we discovered that various relationships between them could be expressed by the matrix (1.7.6), namely,

$$\begin{bmatrix} 1 & 1 & 0 & 1 & \mathbf{x}_1 \\ 0 & -1 & 0 & 0 & \mathbf{x}_2 - \mathbf{x}_1 \\ 0 & 0 & 0 & 0 & \mathbf{x}_3 - \mathbf{x}_1 - 2(\mathbf{x}_2 - \mathbf{x}_1) \\ 0 & 0 & 1 & 0 & \mathbf{x}_4 \end{bmatrix}$$

From this matrix we read that $\mathbf{x}_3$ is a combination of $\mathbf{x}_1$ and $\mathbf{x}_2$ and that $\mathbf{x}_4$ is not a linear combination of $\mathbf{x}_1$ and $\mathbf{x}_2 - \mathbf{x}_4$. From this it follows that $\mathbf{x}_4$ is not a linear combination of $\mathbf{x}_1$ and $\mathbf{x}_2$ (why?). Hence $\mathbf{x}_1 = [1, 1, 0, 1]$, $\mathbf{x}_2 = [1, 0, 0, 1]$, and $\mathbf{x}_4 = [0, 0, 1, 0]$ form a basis for the span of the four given vectors. The dimension is 3.

1.8.2. Example

Find the dimension and a basis for the vector space $\mathscr{K}$ spanned by [1, 1, 0, 1], [1, 0, 0, 1], [1, −1, 0, 1], [1, 2, 0, 1].

Solution

Let $\mathbf{x}_1 = [1, 1, 0, 1]$, $\mathbf{x}_2 = [1, 0, 0, 1]$, $\mathbf{x}_3 = [1, -1, 0, 1]$, and $\mathbf{x}_4 = [1, 2, 0, 1]$. Then

$$\begin{bmatrix} 1 & 1 & 0 & 1 & \mathbf{x}_1 \\ 1 & 0 & 0 & 1 & \mathbf{x}_2 \\ 1 & -1 & 0 & 1 & \mathbf{x}_3 \\ 1 & 2 & 0 & 1 & \mathbf{x}_4 \end{bmatrix} \rightarrow \begin{bmatrix} 1 & 1 & 0 & 1 & \mathbf{x}_1 \\ 0 & -1 & 0 & 0 & \mathbf{x}_2 - \mathbf{x}_1 \\ 0 & -2 & 0 & 0 & \mathbf{x}_3 - \mathbf{x}_1 \\ 0 & 1 & 0 & 0 & \mathbf{x}_4 - \mathbf{x}_1 \end{bmatrix}$$

$$\rightarrow \begin{bmatrix} 1 & 0 & 0 & 1 & \mathbf{x}_1 + (\mathbf{x}_2 - \mathbf{x}_1) = \mathbf{x}_2 \\ 0 & -1 & 0 & 0 & \mathbf{x}_2 - \mathbf{x}_1 \\ 0 & 0 & 0 & 0 & \mathbf{x}_3 - \mathbf{x}_1 - 2(\mathbf{x}_2 - \mathbf{x}_1) = \mathbf{x}_3 - 2\mathbf{x}_2 + \mathbf{x}_1 \\ 0 & 0 & 0 & 0 & \mathbf{x}_4 - \mathbf{x}_1 + (\mathbf{x}_2 - \mathbf{x}_1) = \mathbf{x}_4 + \mathbf{x}_2 - 2\mathbf{x}_1 \end{bmatrix}$$

Thus $\dim \mathscr{K} = 2$ and $\mathbf{x}_1$ and $\mathbf{x}_2$ form a basis. We can also conclude that $\mathbf{x}_3 - 2\mathbf{x}_2 + \mathbf{x}_1 = \mathbf{0}$ and $\mathbf{x}_4 + \mathbf{x}_2 - 2\mathbf{x}_1 = \mathbf{0}$.

The set of all solutions to $\mathrm{A}\mathbf{x} = \mathbf{0}$ supplies an important example of a vector space.

1.8.4. THEOREM *The set of all solutions of the homogeneous system* $\mathrm{A}\mathbf{x} = \mathbf{0}$ *is a vector space.*

PROOF Since $\mathbf{0}$ is a solution the set of solutions is nonempty. If $\mathbf{x}_1$ and $\mathbf{x}_2$ are any two solutions of $\mathrm{A}\mathbf{x} = \mathbf{0}$ then $k_1\mathbf{x}_1 + k_2\mathbf{x}_2$ is a solution for any scalars k_1 and k_2 since

$$\mathrm{A}(k_1\mathbf{x}_1 + k_2\mathbf{x}_2) = k_1\mathrm{A}\mathbf{x}_1 + k_2\mathrm{A}\mathbf{x}_2 = \mathbf{0}$$

We call this vector space the *solution space* of $\mathrm{A}\mathbf{x} = \mathbf{0}$. Let us denote the solution space by $\mathscr{H}$. Since $\mathscr{H}$ is a set of n-tuples (assuming A is $m \times n$) its dimension is n or less, because $n+1$ or more n-tuples are always linearly dependent. If $\mathbf{x}_1, \mathbf{x}_2, \ldots, \mathbf{x}_k$, $(k \leqq n)$ is a basis for $\mathscr{H}$ then every vector in $\mathscr{H}$ – that is, every solution of $\mathrm{A}\mathbf{x} = \mathbf{0}$ – is a linear combination of these vectors. For this reason it is customary to call

$$c_1\mathbf{x}_1 + c_2\mathbf{x}_2 + \cdots + c_k\mathbf{x}_k \tag{1.8.1}$$

the *general solution* of $\mathrm{A}\mathbf{x} = \mathbf{0}$.

1.8.3. Example

Describe geometrically the vector space of solutions of $\mathrm{A}\mathbf{x} = \mathbf{0}$. Find a basis of the solution space and write the general solution, where

$$\mathrm{A} = \begin{bmatrix} 1 & -2 & 1 \\ 2 & -4 & -2 \\ -1 & 2 & -1 \end{bmatrix} \quad \text{and} \quad \mathbf{x} = \begin{bmatrix} x \\ y \\ z \end{bmatrix}$$

Solution

We use row reduction to find the equivalent system of equations

$$\begin{bmatrix} 1 & -2 & 1 \\ 0 & 0 & 0 \\ 0 & 0 & 0 \end{bmatrix} \begin{bmatrix} x \\ y \\ z \end{bmatrix} = \begin{bmatrix} 0 \\ 0 \\ 0 \end{bmatrix}$$

and hence $x-2y+z=0$. Geometrically, the set of points satisfying this equation lie on a plane through the origin. We may select z and y arbitrarily. Pick $z=k_1$, $y=k_2$ and then $x=2k_2-k_1$ and

$$\begin{bmatrix} x \\ y \\ z \end{bmatrix} = \begin{bmatrix} -k_1+2k_2 \\ 0+\ k_2 \\ k_1+0 \end{bmatrix} = k_1 \begin{bmatrix} 1 \\ 0 \\ 1 \end{bmatrix} + k_2 \begin{bmatrix} 2 \\ 1 \\ 0 \end{bmatrix} \tag{1.8.2}$$

Therefore $[1,0,1]$ and $[2,1,0]$ form a basis for $\mathscr{H}$ and the dimension of $\mathscr{H}$ is 2. Equation (1.8.1) expresses the general solution.

1.8.4. Example

Solve the system

$$\mathbf{Ax} = \begin{bmatrix} 1 & 0 & 0 & 0 & 0 \\ 0 & 1 & 0 & 0 & 0 \\ 0 & 0 & 1 & 0 & 0 \end{bmatrix} \mathbf{x} = \mathbf{0}$$

and discuss the solution space.

Solution

Since $\mathbf{x}$ is a 5-tuple, $\mathscr{H}$ is a subspace of $\mathscr{R}^5$. Set $\mathbf{x}=[x_1,x_2,x_3,x_4,x_5]$. Then $\mathbf{Ax}=[x_1,x_2,x_3]=[0,0,0]$. Therefore a necessary and sufficient condition that $\mathbf{x}$ be a solution is

$$\mathbf{x} = \begin{bmatrix} 0 \\ 0 \\ 0 \\ x_4 \\ x_5 \end{bmatrix} = x_4 \begin{bmatrix} 0 \\ 0 \\ 0 \\ 1 \\ 0 \end{bmatrix} + x_5 \begin{bmatrix} 0 \\ 0 \\ 0 \\ 0 \\ 1 \end{bmatrix}$$

From the latter equation $\mathscr{H} = \langle [0,0,0,1,0],\ [0,0,0,0,1] \rangle$ and the dimension of $\mathscr{H}$ is 2.

In contrast to $\mathbf{Ax}=0$, the set of all solutions of the inhomogeneous system $\mathbf{Ax}=\mathbf{b}\neq\mathbf{0}$ is not a vector space because, among other things, $\mathbf{0}$ is not a solution. However, the set of all solutions of $\mathbf{Ax}=\mathbf{b}$ is intimately connected with the solutions of $\mathbf{Ax}=\mathbf{0}$.

1.8.5. THEOREM *If* $\mathbf{x}_p$ *is any solution of* $\mathbf{Ax} = \mathbf{b}$ *and* $\mathbf{x}_1, \mathbf{x}_2, \ldots, \mathbf{x}_k$ *is a basis for the solution space of* $\mathbf{Ax} = \mathbf{0}$ *then*

$$\mathbf{x} = \mathbf{x}_p + c_1\mathbf{x}_1 + c_2\mathbf{x}_2 + \cdots + c_k\mathbf{x}_k \tag{1.8.3}$$

is a solution of $\mathbf{Ax} = \mathbf{b}$ *and every solution is of this form.*

PROOF First of all, $\mathbf{Ax} = \mathbf{A}(\mathbf{x}_p + c_1\mathbf{x}_1 + \cdots + c_k\mathbf{x}_k) = \mathbf{Ax}_p + \mathbf{A}(c_1\mathbf{x}_1 + \cdots + c_k\mathbf{x}_k) = \mathbf{b}$ shows that (1.8.3) is always a solution of $\mathbf{Ax} = \mathbf{b}$. If $\mathbf{z}$ is any solution of $\mathbf{Ax} = \mathbf{b}$ then $\mathbf{z} - \mathbf{x}_p$ is clearly a solution of $\mathbf{Ax} = \mathbf{0}$, since $\mathbf{A}(\mathbf{z} - \mathbf{x}_p) = \mathbf{Az} - \mathbf{Ax}_p = \mathbf{b} - \mathbf{b} = \mathbf{0}$. But then $\mathbf{z} - \mathbf{x}_p$ is a linear combination of $\mathbf{x}_1, \mathbf{x}_2, \ldots, \mathbf{x}_k$ which means that for some $c_1, c_2, \ldots, c_k$, $\mathbf{z} - \mathbf{x}_p = c_1\mathbf{x}_1 + c_2\mathbf{x}_2 + \cdots + c_k\mathbf{x}_k$ and hence $\mathbf{z}$ is of the form (1.8.3), concluding the proof.

In view of this theorem, (1.8.3) is known as the *general solution* of $\mathbf{Ax} = \mathbf{b}$.

1.8.5. Example

Find the general solution of

$$\mathbf{Ax} = \begin{bmatrix} 1 & 0 & 0 & 0 & 0 \\ 0 & 1 & 0 & 0 & 0 \\ 0 & 0 & 0 & 0 & 0 \end{bmatrix} \mathbf{x} = \mathbf{b} = \begin{bmatrix} 1 \\ 1 \\ 0 \end{bmatrix}$$

Solution

We first find $\mathscr{H}$ for $\mathbf{Ax} = \mathbf{0}$. In this case

$$\mathbf{x} = \begin{bmatrix} 0 \\ 0 \\ x_3 \\ x_4 \\ x_5 \end{bmatrix} = x_3 \begin{bmatrix} 0 \\ 0 \\ 1 \\ 0 \\ 0 \end{bmatrix} + x_4 \begin{bmatrix} 0 \\ 0 \\ 0 \\ 1 \\ 0 \end{bmatrix} + x_5 \begin{bmatrix} 0 \\ 0 \\ 0 \\ 0 \\ 1 \end{bmatrix}$$

and $\mathscr{H} = \langle [0, 0, 1, 0, 0], [0, 0, 0, 1, 0], [0, 0, 0, 0, 1] \rangle$.

By inspection, $\mathbf{x}_p = [1, 1, 0, 0, 0]$ is a solution of $\mathbf{Ax} = \mathbf{b}$. From Theorem 1.8.5, every solution of $\mathbf{Ax} = \mathbf{b}$ is therefore in the form

$$\mathbf{x} = \begin{bmatrix} 1 \\ 1 \\ 0 \\ 0 \\ 0 \end{bmatrix} + c_1 \begin{bmatrix} 0 \\ 0 \\ 1 \\ 0 \\ 0 \end{bmatrix} + c_2 \begin{bmatrix} 0 \\ 0 \\ 0 \\ 1 \\ 0 \end{bmatrix} + c_3 \begin{bmatrix} 0 \\ 0 \\ 0 \\ 0 \\ 1 \end{bmatrix}$$

$$= \begin{bmatrix} 1 \\ 1 \\ c_1 \\ c_2 \\ c_3 \end{bmatrix}$$

Remarks

Theorem 1.8.5 does not assert that $\mathbf{Ax} = \mathbf{b}$ has any solutions at all. Note that $\mathbf{x}_p$ is assumed to exist as part of the hypothesis. If A is a square matrix three cases can occur: (1) A is nonsingular, in which case $\mathbf{x} = A^{-1}\mathbf{b}$ is the unique solution of $\mathbf{Ax} = \mathbf{b}$; (2) A is singular and $\mathbf{x}_p$ exists. This is the most interesting case; (3) A is singular and $\mathbf{Ax} = \mathbf{b}$ has no solutions. This case occurs, for instance, in

$$\begin{bmatrix} 1 & 0 & 0 \\ 0 & 1 & 0 \\ 0 & 0 & 0 \end{bmatrix} \mathbf{x} = \begin{bmatrix} 1 \\ 1 \\ 1 \end{bmatrix}.$$

Exercise 1.8

Part 1

1. Find a basis and the dimension of the following vector spaces:

   (a) $\langle [1, 1, 0, 1], [1, 0, 0, 1], [0, 0, 1, 0], [1, -1, 0, 1] \rangle$
   (b) $\langle [-1, 2, 0, 0], [1, 2, -1, 0], [1, 1, 0, 1], [1, 5, -1, 1] \rangle$
   (c) $\langle [1, 0, 1], [1, 1, 1], [-1, -1, 3] \rangle$

2. Find a basis and the dimension of the solution space in the following problems. Describe the solution spaces geometrically

   (a) $\begin{bmatrix} 1 & -2 & -1 \\ 2 & -4 & -2 \\ 0 & 1 & 1 \end{bmatrix} \begin{bmatrix} x \\ y \\ z \end{bmatrix} = \begin{bmatrix} 0 \\ 0 \\ 0 \end{bmatrix}$

   (b) $\begin{bmatrix} -1 & 1 & 0 \\ 0 & 0 & 1 \\ 1 & 0 & 0 \end{bmatrix} \begin{bmatrix} x \\ y \\ z \end{bmatrix} = \begin{bmatrix} 0 \\ 0 \\ 0 \end{bmatrix}$

3. Use Theorem 1.8.5 to find the general solution if any to Problem 2(a) above when the right-hand side is

   (a) $[1, 1, 1]$ (b) $[1, 2, -1]$

4. Find a basis and the dimension of the solution spaces in the following problems.

   (a) $\begin{bmatrix} 1 & 0 & 2 & -1 \\ 0 & 1 & 0 & 0 \\ 1 & 1 & 3 & -1 \end{bmatrix} \mathbf{x} = \mathbf{0}$

   (b) $\begin{bmatrix} 1 & 0 & 3 & 1 \\ 0 & 1 & 0 & -2 \\ 1 & 1 & 3 & 0 \end{bmatrix} \mathbf{x} = \mathbf{0}$

   (c) $\begin{bmatrix} 1 & 0 & -2 & -1 \\ 1 & 1 & 2 & -3 \end{bmatrix} \mathbf{x} = \mathbf{0}$

5. Find the general solutions.

   (a) $\begin{bmatrix} 1 & 2 & -1 \\ 2 & 1 & 1 \\ 3 & 3 & 0 \end{bmatrix} \mathbf{x} = \mathbf{0}$

(b) $\begin{bmatrix} 1 & 2 & -1 \\ 2 & 1 & 1 \\ 3 & 3 & 0 \end{bmatrix} \mathbf{x} = \begin{bmatrix} 2 \\ -1 \\ 1 \end{bmatrix}$

(c) $\begin{bmatrix} -1 & 0 & 1 \\ 1 & 1 & 0 \\ 0 & 0 & 1 \end{bmatrix} \mathbf{x} = \begin{bmatrix} -1 \\ 1 \\ 0 \end{bmatrix}$

6. Find a basis for the solution space of the systems given below. State the dimension and find the general solution.

(a) $\begin{aligned} x + y &= 0 \\ 2x + 2y &= 0 \end{aligned}$ (b) $\begin{bmatrix} 1 & 0 & 0 \\ 0 & 1 & 0 \\ 0 & 0 & 0 \end{bmatrix} \mathbf{x} = \mathbf{0}$

(c) $\begin{bmatrix} 0 & 1 & 0 \\ 0 & 0 & 1 \\ 0 & 0 & 0 \end{bmatrix} \mathbf{x} = \mathbf{0}$ (d) $\begin{bmatrix} 0 & 1 & 0 \\ 0 & 0 & 0 \\ 0 & 0 & 0 \end{bmatrix} \mathbf{x} = \mathbf{0}$

7. Find a vector $\mathbf{x}_3$ so that $\mathbf{x}_1 = [0, 1, 1]$, $\mathbf{x}_2 = [1, 0, 2]$, $\mathbf{x}_3$ forms a basis for $\mathscr{R}^3$.

Part 2

1. Interpret the statements:
   (a) A basis is a minimal spanning set.
   (b) A basis is a maximal linearly independent set.
2. Suppose A is $n \times n$. Suppose the n-tuples $\mathbf{x}_1, \mathbf{x}_2, \ldots, \mathbf{x}_k$ form a linearly dependent set. Prove that $A\mathbf{x}_1, A\mathbf{x}_2, \ldots, A\mathbf{x}_k$ is linearly dependent. State the converse. Is the converse true?
3.* If $\mathbf{x}_1, \mathbf{x}_2, \ldots, \mathbf{x}_k$ span $\mathscr{V}$ and every vector $\mathbf{v} \in \mathscr{V}$ has the *unique* representation

$$\mathbf{v} = a_1\mathbf{x}_1 + a_2\mathbf{x}_2 + \cdots + a_k\mathbf{x}_k$$

then $\mathbf{x}_1, \mathbf{x}_2, \ldots, \mathbf{x}_k$ is a basis for $\mathscr{V}$. Prove this theorem.

1.9. VECTOR NORMS

In the previous sections we ignored questions which dealt with the notion of distance. However, if the correspondence between n-tuples on the one hand and vector geometry on the other hand is to be made complete we must introduce metric concepts. There are three interconnected ideas: the inner product, the norm, and the distance function.

Let **x** and **y** be vectors in $\mathscr{C}$ with $\mathbf{x} = [x_1, x_2, \ldots, x_n]$ and $\mathbf{y} = [y_1, y_2, \ldots, y_n]$. The complex number†

$$\mathbf{x} \cdot \mathbf{y} = x_1\bar{y}_1 + x_2\bar{y}_2 + \cdots + x_n\bar{y}_n \tag{1.9.1}$$

is called the *inner product* of **x** and **y**. If **x** and **y** have real components then

†If $a = \alpha + i\beta$ is a complex number, α, β real, then $\bar{a} = \alpha - i\beta$ is called the complex conjugate of a; $a = \bar{a}$ if and only if a is real, that is, $\beta = 0$.

(1.9.1) reduces to $x_1y_1 + x_2y_2 + \cdots + x_ny_n$ since $y_i = \bar{y}_i$ for real y_i. Theorem 1.9.1 follows from Equation (1.9.1).

1.9.1. THEOREM

(a) $\mathbf{x} \cdot \mathbf{y} = \overline{\mathbf{y} \cdot \mathbf{x}}$
(b) $\mathbf{x} \cdot (a\mathbf{y} + b\mathbf{z}) = \bar{a}(\mathbf{x} \cdot \mathbf{y}) + \bar{b}(\mathbf{x} \cdot \mathbf{z})$
(c) $\mathbf{x} \cdot \mathbf{x} \geqslant 0$
(d) *If* $\mathbf{x} \cdot \mathbf{x} = 0$ *then* $\mathbf{x} = \mathbf{0}$

We write

$$\|\mathbf{x}\| = \sqrt{(\mathbf{x} \cdot \mathbf{x})} \tag{1.9.2}$$

and call $\|\mathbf{x}\|$ the *norm* of $\mathbf{x}$.

1.9.2. THEOREM *For* $\mathbf{x}$ and $\mathbf{y} \in \mathscr{C}^n$ and *k a scalar*,

(a) $\|\mathbf{x}\| \geqslant 0$; $\|\mathbf{x}\| = 0$ *implies* $\mathbf{x} = \mathbf{0}$
(b) $\|k\mathbf{x}\| = |k|\,\|\mathbf{x}\|$
(c) $|\mathbf{x} \cdot \mathbf{y}| \leqslant \|\mathbf{x}\|\,\|\mathbf{y}\|$

PROOF Parts (a) and (b) follow from Theorem 1.9.1 with no effort. Part (c), sometimes called the *Schwarz Inequality*, is more subtle. The proof is simple but not at all "natural." Let λ be any complex number and consider

$$\begin{aligned} 0 &\leqslant (\mathbf{x} + \lambda\mathbf{y}) \cdot (\mathbf{x} + \lambda\mathbf{y}) \\ &= (\mathbf{x} \cdot \mathbf{x}) + (\mathbf{x} \cdot \lambda\mathbf{y}) + (\lambda\mathbf{y} \cdot \mathbf{x}) + (\lambda\mathbf{y} \cdot \lambda\mathbf{y}) \\ &= \|\mathbf{x}\|^2 + \bar{\lambda}(\mathbf{x} \cdot \mathbf{y}) + \lambda(\overline{\mathbf{x} \cdot \mathbf{y}}) + |\lambda|^2\,\|\mathbf{y}\|^2 \end{aligned} \tag{1.9.3}$$

Since this inequality is valid for all complex λ, it is true for the special (clever) choice

$$\lambda = -\frac{(\mathbf{x} \cdot \mathbf{y})}{\|\mathbf{y}\|^2} \tag{1.9.4}$$

(Note: if $\|\mathbf{y}\|^2 = 0$, (c) holds trivially).

We substitute (1.9.4) into (1.9.3) and obtain

$$\begin{aligned} 0 \leqslant \|\mathbf{x}\|^2 - \frac{(\overline{\mathbf{x} \cdot \mathbf{y}})}{\|\mathbf{y}\|^2}(\mathbf{x} \cdot \mathbf{y}) - \frac{(\mathbf{x} \cdot \mathbf{y})}{\|\mathbf{y}\|^2}(\overline{\mathbf{x} \cdot \mathbf{y}}) \\ + \frac{|\mathbf{x} \cdot \mathbf{y}|^2}{\|\mathbf{y}\|^4}\|\mathbf{y}\|^2 = \|\mathbf{x}\|^2 - \frac{|\mathbf{x} \cdot \mathbf{y}|^2}{\|\mathbf{y}\|^2} \end{aligned}$$

Hence the theorem follows upon taking square roots.

As a corollary,

$$\|\mathbf{x} + \mathbf{y}\| \leqslant \|\mathbf{x}\| + \|\mathbf{y}\| \tag{1.9.5}$$

because

$$\begin{aligned}\|\mathbf{x}+\mathbf{y}\|^2 &= (\mathbf{x}+\mathbf{y}) \cdot (\mathbf{x}+\mathbf{y}) = \|\mathbf{x}\|^2 + (\mathbf{x}\cdot\mathbf{y}) + (\mathbf{y}\cdot\mathbf{x}) + \|\mathbf{y}\|^2 \\ &= \|\mathbf{x}\|^2 + 2 \text{ real part of } (\mathbf{x}\cdot\mathbf{y}) + \|\mathbf{y}\|^2 \\ &\leqslant \|\mathbf{x}\|^2 + 2\,|\mathbf{x}\cdot\mathbf{y}| + \|\mathbf{y}\|^2 \leqslant \|\mathbf{x}\|^2 + 2\|\mathbf{x}\|\,\|\mathbf{y}\| + \|\mathbf{y}\|^2 \\ &= (\|\mathbf{x}\| + \|\mathbf{y}\|)^2\end{aligned}$$

For the "distance" between the vectors **x** and **y**, we define a function $d(\mathbf{x}, \mathbf{y})$ by the equation

$$d(\mathbf{x}, \mathbf{y}) \equiv \|\mathbf{x}-\mathbf{y}\| \tag{1.9.6}$$

which is consistent with the Euclidean distance in two and three real dimensions. For, let $n = 2$ and $\mathbf{x}, \mathbf{y} \in \mathscr{R}^2$, then

$$\|\mathbf{x}-\mathbf{y}\| = [(x_1-y_1)^2 + (x_2-y_2)^2]^{1/2}$$

which is the distance between the points (x_1, x_2) and (y_1, y_2). The distance defined by (1.9.6) has certain properties one expects of a distance measure, namely,

$$\begin{aligned}&\text{(a)}\ d(\mathbf{x}, \mathbf{y}) = d(\mathbf{y}, \mathbf{x}) \\ &\text{(b)}\ d(\mathbf{x}, \mathbf{y}) \geqslant 0 \qquad d(\mathbf{x}, \mathbf{y}) = 0 \quad \text{if and only if} \quad \mathbf{x} = \mathbf{y} \\ &\text{(c)}\ d(\mathbf{x}, \mathbf{y}) \leqslant d(\mathbf{x}, \mathbf{z}) + d(\mathbf{z}, \mathbf{y})\end{aligned} \tag{1.9.7}$$

That (a) and (b) hold is a direct consequence of the definition. To prove (c) we use (1.9.5) for the vectors **s** and **t** where $\mathbf{s} = \mathbf{x}-\mathbf{z}$, $\mathbf{t} = \mathbf{z}-\mathbf{y}$. That is,

$$\|\mathbf{s}+\mathbf{t}\| = \|\mathbf{x}-\mathbf{y}\| \leqslant \|\mathbf{s}\| + \|\mathbf{t}\| = \|\mathbf{x}-\mathbf{z}\| + \|\mathbf{z}-\mathbf{y}\|$$

Remarks

The inequality (c) is known as the "triangle inequality." To understand the motivation behind this name consider Figure 1.9.1. Inequality (c) is thus a statement that each side of a triangle is shorter than the sum of the lengths of the other two sides.

Figure 1.9.1

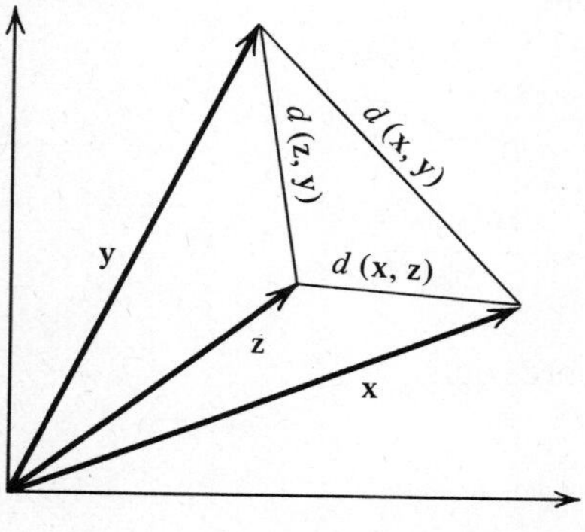

It will be important for us to determine to what extent the norm of $\mathbf{y} = \mathrm{A}\mathbf{x}$ depends upon the norm of $\mathbf{x}$. To examine this question, suppose we write A as

$$\mathrm{A} = \begin{bmatrix} a_{11} & a_{12} \cdots a_{1n} \\ a_{21} & a_{22} \cdots a_{2n} \\ \cdot & \cdot & \cdot \\ \cdot & \cdot & \cdot \\ \cdot & \cdot & \cdot \\ a_{n1} & a_{n2} \cdots a_{nn} \end{bmatrix} = \begin{bmatrix} \mathbf{a}_1 \\ \mathbf{a}_2 \\ \cdot \\ \cdot \\ \cdot \\ \mathbf{a}_n \end{bmatrix} \tag{1.9.8}$$

where $\mathbf{a}_i = [a_{i1}, a_{i2}, \ldots, a_{in}]$, $i = 1, 2, \ldots, n$.
Then if

$$\mathbf{x} = \begin{bmatrix} x_1 \\ x_2 \\ \cdot \\ \cdot \\ \cdot \\ x_n \end{bmatrix} \quad \text{and} \quad \mathbf{y} = \begin{bmatrix} y_1 \\ y_2 \\ \cdot \\ \cdot \\ \cdot \\ y_n \end{bmatrix}$$

we have

$$\mathrm{A}\mathbf{x} = \begin{bmatrix} \mathbf{a}_1 \cdot \mathbf{x} \\ \mathbf{a}_2 \cdot \mathbf{x} \\ \cdot \\ \cdot \\ \cdot \\ \mathbf{a}_n \cdot \mathbf{x} \end{bmatrix} = \begin{bmatrix} y_1 \\ y_2 \\ \cdot \\ \cdot \\ \cdot \\ y_n \end{bmatrix}$$

Therefore,

$$\begin{aligned} \|\mathbf{y}\|^2 = |\mathrm{A}\mathbf{x}\|^2 &= (\mathbf{a}_1 \cdot \mathbf{x})^2 + (\mathbf{a}_2 \cdot \mathbf{x})^2 + \cdots + (\mathbf{a}_n \cdot \mathbf{x})^2 \\ &\leqslant \|\mathbf{a}_1\|^2 \, \|\mathbf{x}\|^2 + \|\mathbf{a}_2\|^2 \, \|\mathbf{x}\|^2 + \cdots + \|\mathbf{a}_n\|^2 \, \|\mathbf{x}\|^2 \\ &= (\|\mathbf{a}_1\|^2 + \|\mathbf{a}_2\|^2 + \cdots + \|\mathbf{a}_n\|^2) \, \|\mathbf{x}\|^2 \end{aligned}$$

But $\|\mathbf{a}_i\|^2 = a_{i1}^2 + a_{i2}^2 + \cdots + a_{in}^2$ and hence

$$\|\mathbf{y}\| \leqslant \Big(\sum_{i,j=1}^{n} a_{ij}^2 \Big)^{1/2} \|\mathbf{x}\| \tag{1.9.9}$$

Let us define $\|\mathrm{A}\|$ by

$$\Big(\sum_{i,j=1}^{n} a_{ij}^2 \Big)^{1/2} = \|\mathrm{A}\| \tag{1.9.10}$$

and call $\|\mathrm{A}\|$ the norm of A. We can now write (1.9.9) as

$$\|\mathbf{y}\| = \|\mathrm{A}\mathbf{x}\| \leqslant \|\mathrm{A}\| \, \|\mathbf{x}\| \tag{1.9.11}$$

1.9.3. THEOREM *If* $\|A\|$ *is defined as in* (1.9.10) *and*

$$\|\mathbf{a}_i\| \leqslant \alpha \qquad i = 1, 2, \ldots, n$$

then

$$\|A\| \leqslant n^{1/2}\alpha \tag{1.9.12}$$

PROOF We have $\|A\|^2 = \|\mathbf{a}_1\|^2 + \|\mathbf{a}_2\|^2 + \cdots + \|\mathbf{a}_n\|^2 \leqslant n\alpha^2$, hence $\|A\| \leqslant n^{1/2}\alpha$.

Exercise 1.9

Part 1

1. Compute the following inner products:
   (a) $[x_1, x_2, x_3] \cdot [x_1, x_2, x_3]$
   (b) $[1, i, 0, 0] \cdot [-1, -i, 0, 1]$
   (c) $[\cos\theta, \sin\theta] \cdot [\cos\theta, \sin\theta]$
2. Find $\|\mathbf{x}\|$ for
   (a) $\mathbf{x} = [0, 0, 0]$
   (b) $\mathbf{x} = [1, 0, 0]$
   (c) $\mathbf{x} = [-1, 1, -1]$
3. Verify the Schwarz Inequality for
   (a) $\mathbf{x} = [1, 0, 0] \qquad \mathbf{y} = [2, 0, 0]$
   (b) $\mathbf{x} = [1, 1, 1] \qquad \mathbf{y} = [1, 0, 1]$
4. Verify $\|\mathbf{x}+\mathbf{y}\| \leqslant \|\mathbf{x}\| + \|\mathbf{y}\|$ for the vectors in Problem 3, above.
5. Let

$$A = \begin{bmatrix} 1 & -1 \\ 2 & 1 \end{bmatrix}$$

   For the vectors $\mathbf{x}$, given below, compute $\mathbf{y} = A\mathbf{x}$ and verify that

$$\|\mathbf{y}\| \leqslant \|A\|\,\|\mathbf{x}\|$$

   (a) $\mathbf{x} = [0, 0]$ (b) $\mathbf{x} = [1, 0]$ (c) $\mathbf{x} = [-1, 1]$ (d) $\mathbf{x} = [\alpha, \beta]$
6. Let

$$A = \begin{bmatrix} a & 0 \\ 0 & b \end{bmatrix}$$

   and redo Problem 5, above.

Part 2

1. Prove Theorem 1.9.1.
2. Prove Theorem 1.9.2 parts (a) and (b).
3. Show that $\|\mathbf{x}\|$ may be interpreted as the distance from $(0, 0, 0)$ to (x_1, x_2, x_3) if $\mathbf{x} \in \mathscr{R}^3$, $\mathbf{x} = [x_1, x_2, x_3]$.
4. Suppose $\mathbf{x}$ and $\mathbf{y} \in \mathscr{R}^2$. Then Theorem 1.9.2 (c) states

$$(x_1y_1 + x_2y_2)^2 \leqslant (x_1^2 + x_2^2)(y_1^2 + y_2^2)$$

   Prove this inequality directly without reference to the proof given in the text.

1.10. MATRIX AND VECTOR CALCULUS

A natural generalization of a matrix of constants is a matrix whose elements are functions. By the expression

$$A(t) = \begin{bmatrix} a_{11}(t) & a_{12}(t) & \cdots & a_{1n}(t) \\ a_{21}(t) & a_{22}(t) & \cdots & a_{2n}(t) \\ \cdot & \cdot & & \cdot \\ \cdot & \cdot & & \cdot \\ \cdot & \cdot & & \cdot \\ a_{m1}(t) & a_{m2}(t) & \cdots & a_{mn}(t) \end{bmatrix} \tag{1.10.1}$$

we mean that for each fixed t_0 in some interval, say $-r < t_0 < r$, $A(t_0)$ is a matrix of constants. Under this interpretation a square matrix may be singular for some t and nonsingular for others. Witness

$$A(t) = \begin{bmatrix} t & t^2 \\ 3 & 2 \end{bmatrix}$$

Then $\det A(t) = 2t - 3t^2 = 0$ if and only if $t = 0$ or $t = \frac{2}{3}$. Hence $A(t)$ is singular when $t = 0$ or when $t = \frac{2}{3}$ and only then.

We can define continuity and differentiability in a natural manner: Set

$$A(t, \Delta t) = A(t + \Delta t) - A(t)$$

$$= \begin{bmatrix} a_{11}(t+\Delta t) - a_{11}(t) & \cdots & a_{1n}(t+\Delta t) - a_{1n}(t) \\ a_{21}(t+\Delta t) - a_{21}(t) & \cdots & a_{2n}(t+\Delta t) - a_{2n}(t) \\ \cdot & & \cdot \\ \cdot & & \cdot \\ \cdot & & \cdot \\ a_{m1}(t+\Delta t) - a_{m1}(t) & \cdots & a_{mn}(t+\Delta t) - a_{mn}(t) \end{bmatrix}$$

1.10.1. DEFINITION†

(a) *$A(t)$ is continuous at $t = t_0$ if*

$$\lim_{\Delta t \to 0} \Delta A(t_0, \Delta t) = O$$

(b) $A'(t_0) = \lim_{\Delta t \to 0} (1/\Delta t) \Delta A(t_0, \Delta t)$ *if this limit exists.*

It should be apparent from (a) that $A(t)$ is continuous at t_0 if and only if

†In general, $\lim_{t \to t_0} A(t)$ is defined as the matrix whose entries are the limits of the entries of $A(t)$ as $t \to t_0$. For $\lim_{t \to t_0} A(t)$ to exist it is necessary and sufficient that each entry of $A(t)$ has a limit as $t \to t_0$.

$a_{ij}(t)$ is continuous at t_0 for every i and j. The definition of $A'(t_0)$ given in (b) can be written as

$$A'(t_0) = \begin{bmatrix} a'_{11}(t_0) & a'_{12}(t_0) & \cdots & a'_{1n}(t_0) \\ a'_{21}(t_0) & a'_{22}(t_0) & \cdots & a'_{2n}(t_0) \\ \cdot & \cdot & & \cdot \\ \cdot & \cdot & & \cdot \\ \cdot & \cdot & & \cdot \\ \cdot & \cdot & & \cdot \\ a'_{m1}(t_0) & a'_{m2}(t_0) & \cdots & a'_{mn}(t_0) \end{bmatrix}$$

Indeed, $A'(t_0)$ exists if and only if $a'_{ij}(t_0)$ exists for every i and j. From these observations we deduce the following rules for differentiation:

$$\begin{aligned} &(1)\ [A(t)B(t)]' = A'(t)B(t) + A(t)B'(t) \\ &(2)\ [A(t)\mathbf{x}(t)]' = A'(t)\mathbf{x}(t) + A(t)\mathbf{x}'(t) \\ &(3)\ [A(t)+B(t)]' = A'(t) + B'(t) \\ &(4)\ [cA(t)]' = cA'(t), \text{ where } c \text{ is constant} \\ &(5)\ (\mathbf{x}(t)\cdot\mathbf{y}(t))' = \mathbf{x}'(t)\cdot\mathbf{y}(t) + \mathbf{x}(t)\cdot\mathbf{y}'(t) \\ &(6)\ \frac{d}{dt}A^{-1}(t) = -A^{-1}(t)A'(t)A^{-1}(t) \quad \text{if A is nonsingular} \end{aligned} \tag{1.10.2}$$

Rule (6) follows from the identity $A(t)A^{-1}(t) = I$ which yields

$$A'(t)A^{-1}(t) + A(t)A^{-1}(t)' = O \tag{1.10.3}$$

since $I' = O$. From (1.10.3) we can solve for $A^{-1}(t)'$ to obtain Rule (6).

The definition of $A'(t)$ is so constructed that if we interpret

$$\int_0^t A(s)\,ds = \begin{bmatrix} \int_0^t a_{11}(s)\,ds & \int_0^t a_{12}(s)\,ds & \cdots & \int_0^t a_{1n}(s)\,ds \\ \int_0^t a_{21}(s)\,ds & \int_0^t a_{22}(s)\,ds & \cdots & \int_0^t a_{2n}(s)\,ds \\ \cdot & \cdot & & \cdot \\ \cdot & \cdot & & \cdot \\ \cdot & \cdot & & \cdot \\ \int_0^t a_{m1}(s)\,ds & \int_0^t a_{m2}(s)\,ds & \cdots & \int_0^t a_{mn}(s)\,ds \end{bmatrix} \tag{1.10.4}$$

then the rules for integration become:

(1) $\int_0^t A'(s)\,ds = A(t) - A(0)$

(2) $\frac{d}{dt}\int_0^t A(s)\,ds = A(t)$ if $A(s)$ is continuous

(3) $\int_0^t (A(s)+B(s))\,ds = \int_0^t A(s)\,ds + \int_0^t B(s)\,ds$

(4) $c \int_0^t A(s)\, ds = \int_0^t cA(s)\, ds$

(5) $B \int_0^t A(s)\, ds = \int_0^t BA(s)\, ds \qquad \int_0^t A(s)C\, ds = \left(\int_0^t A(s)\, ds\right)C$ (1.10.5)

when B and C are constant matrices.

Our last result for integrals mimics the classical inequality for functions,

$$\left|\int_0^t f(s)\, ds\right| \leq \int_0^t |f(s)|\, ds \tag{1.10.6}$$

We need a preliminary result. Let

$$\mathbf{x}(t) = [x_1(t), x_2(t), \ldots, x_n(t)] \qquad \text{and} \qquad \mathbf{f}(t) = [f_1(t), f_2(t), \ldots, f_n(t)]$$

Then,

$$\begin{aligned} \mathbf{x}(t) \cdot \int_0^t \mathbf{f}(s)\, ds &= x_1(t) \overline{\int_0^t f_1(s)\, ds} + x_2(t) \overline{\int_0^t f_2(s)\, ds} \\ &\quad + \cdots + x_n(t) \overline{\int_0^t f_n(s)\, ds} \\ &= \int_0^t [x_1(t)\overline{f_1(s)} + x_2(t)\overline{f_2(s)} + \cdots + x_n(t)\overline{f_n(s)}]\, ds \\ &= \int_0^t \mathbf{x}(t) \cdot \mathbf{f}(s)\, ds \end{aligned} \tag{1.10.7}$$

1.10.2. THEOREM *For any vector function* **f** *for which* $\int_0^t \mathbf{f}(s)\, ds$ *is defined,*

$$\left\|\int_0^t \mathbf{f}(s)\, ds\right\| \leq \int_0^t \|\mathbf{f}(s)\|\, ds \tag{1.10.8}$$

PROOF Compare (1.10.8) with (1.10.6). Notice that the vertical lines represent absolute values in (1.10.6) and represent norms in (1.10.8). There should be no confusion between the different usages because (1.10.8) is clearly a vector expression for which norms make sense and absolute values do not. The proof is simple but ingenious. Assume that $\| \int_0^t \mathbf{f}(s)\, ds\| \neq 0$. Let $\mathbf{F}(t)$ and $\mathbf{u}(t)$ be defined by the equations

$$\mathbf{F}(t) = \int_0^t \mathbf{f}(s)\, ds$$

$$\mathbf{u}(t) = \frac{\mathbf{F}(t)}{\|\mathbf{F}(t)\|} = \frac{\int_0^t \mathbf{f}(s)\, ds}{\|\mathbf{F}(t)\|}$$

Clearly, $\|\mathbf{u}(t)\| = 1$. Now,

$$\|\mathbf{F}(t)\|^2 = \int_0^t \mathbf{f}(s)\, ds \cdot \int_0^t \mathbf{f}(s)\, ds$$

so that

$$\|\mathbf{F}(t)\| = \frac{\int_0^t \mathbf{f}(s)\,ds}{\|\mathbf{F}(t)\|} \cdot \int_0^t \mathbf{f}(s)\,ds$$

$$= \mathbf{u}(t) \cdot \int_0^t \mathbf{f}(s)\,ds$$

$$= \int_0^t \mathbf{u}(t) \cdot \mathbf{f}(s)\,ds$$

from (1.10.7). The Schwarz Inequality, Theorem 1.9.2 (c), yields $\mathbf{u} \cdot \mathbf{f}(s) \leqslant \|\mathbf{u}\|\,\|\mathbf{f}(s)\| = \|\mathbf{f}(s)\|$, since $\|\mathbf{u}\| = 1$. Hence

$$\|\mathbf{F}(t)\| = \int_0^t \mathbf{u}(t) \cdot \mathbf{f}(s)\,ds \leqslant \int_0^t \|\mathbf{f}(s)\|\,ds$$

which completes the proof in the case $\|\mathbf{F}(t)\| \neq 0$. If $\|\mathbf{F}(t)\| = 0$, the inequality is obviously valid.

Exercise 1.10

Part 1

1. Find $A^{-1}(t)$ for $A(t) = \begin{bmatrix} t & t^2 \\ 3 & 2 \end{bmatrix}$ by row reduction.

   Note why $A^{-1}(0)$ and $A^{-1}(\frac{2}{3})$ do not exist.
2. Show, for the constant matrix A,

   (a) $\frac{d}{dt} A = 0$

   (b) $\frac{d}{dt}\{f(t)A\} = f'(t)A$

   (c) $\frac{d}{dt}(A\mathbf{x}) = A\mathbf{x}'$
3. Prove that $\{A^2(t)\}' = A(t)A'(t) + A'(t)A(t)$. By example illustrate why $A(t)A'(t) + A'(t)A(t) \neq 2A(t)A'(t)$, in general.
4. Compute $A'(t)$ and $A^{-1}(t)'$ for $A(t)$ of Problem 1 of this set. Then verify (6) of (1.10.2).
5. Prove the rules of differentiation in (1.10.2).
6. Compute $\int_0^t A'(s)\,ds$ for $A(t)$ of Problem 1, this set. Then verify (1) of (1.10.5).
7. Compute $\int_0^t A(s)\,ds$ for $A(t)$ of Problem 1, this set. Then verify (2) of (1.10.5).
8. Prove: $(\mathbf{x} \cdot \mathbf{x})' = 2\mathbf{x} \cdot \mathbf{x}'$.
9. For the various $\mathbf{f}$ given below verify Theorem 1.10.2 by integration and comparison of the two sides of (1.10.8).

   (a) $\mathbf{f}(s) = \begin{bmatrix} \alpha \\ \beta \end{bmatrix}$ α and β are constants

   (b) $\mathbf{f}(s) = \begin{bmatrix} f(s) \\ 0 \end{bmatrix}$

   (c) $\mathbf{f}(s) = \begin{bmatrix} \sqrt{s} \\ 1 \end{bmatrix}$

   (d) $\mathbf{f}(s) = f(s)\begin{bmatrix} \alpha \\ \beta \end{bmatrix}$ α and β are constants

Part 2

1. Prove the rules of integration given by (1.10.5). Use (1.10.4) and the corresponding rules from the integral calculus.
2. Use (1.10.8) to prove (1.10.6) by choosing $\mathbf{f}(a) = \begin{bmatrix} 1 \\ 0 \end{bmatrix} f(a)$.

Chapter 2
LINEAR HOMOGENEOUS SYSTEMS AND EIGENVECTORS

2.1. THE FIRST-ORDER EQUATION IN ONE DIMENSION

In many experimental and theoretical situations it is observed that the rate of change of some variable (e.g., temperature, mass, population) is proportional to the value of the variable. A plant grows at a rate proportional to its size; a radioactive substance decays proportionately to its mass; a body loses heat at a rate proportional to its temperature. Expressed mathematically, let $f(t)$ be the value of such a physical variable at time t, $f'(t)$ the rate of change of that variable at time t, then $f'(t) = \alpha f(t)$ states the proportionality. The constant α is experimentally determined. Having observed this proportionality the question arises—what is f?

This situation is quite common. Theory or experiment supplies the scientist with a relationship between a function and its derivatives but what is desired is the function itself. Such is the problem of differential equations: to find functions satisfying a prescribed relationship between itself and its derivatives. In this section we study the equation suggested by the phenomenon described above. Specifically, the differential equation

$$x' = \alpha x$$

will be our way of expressing the problem: find a differentiable function $x(t)$ such that

$$x'(t) = \alpha x(t) \qquad \text{for all } t \text{ in some interval}$$

Any function with this property is said to be a *solution* or to *satisfy* the equation. Frequently, besides satisfying the differential equation, we shall ask that $x(t)$ take a prescribed value x_0 at some specified time $t = t_0$. Together,

$$\begin{aligned} x' &= \alpha x \\ x(t_0) &= x_0 \end{aligned} \tag{2.1.1}$$

asks for a function $x(t)$ such that $x'(t) = \alpha x(t)$ for all t and $x(t_0) = x_0$. The problem stated by (2.1.1) is known as an *initial-value problem*; the expression $x(t_0) = x_0$ as the *initial condition*; x_0 as the *initial value.* For simplicity of notation we shall often choose $t_0 = 0$.

2.1.1. THEOREM *The initial-value problem* (2.1.1) *has the unique solution*

$$x(t) = x_0 e^{\alpha(t-t_0)} \tag{2.1.2}$$

PROOF It is clear that $x(t_0) = x_0$ and that $(d/dt)x(t) = x_0 \alpha e^{\alpha(t-t_0)} = \alpha x(t)$ for all t. The only analysis of any consequence is the demonstration that there are no other solutions. To see that this is so, set $y(t) = e^{\alpha(t-t_0)}$ so that $y'(t) = \alpha y(t)$ for all t and $y(t_0) = 1$. Now suppose $z(t)$ is a solution of (2.1.1). Then

$$\frac{d}{dt}\left\{\frac{z(t)}{y(t)}\right\} = \frac{z'(t)y(t) - y'(t)z(t)}{y^2(t)} \qquad \text{for all } t \tag{2.1.3}$$

since $y(t) > 0$ for all t. But $z'(t) = \alpha z(t)$ and $y'(t) = \alpha y(t)$ so the numerator of (2.1.3) is zero for all t. That is,

$$\frac{d}{dt}\left\{\frac{z(t)}{y(t)}\right\} = 0 \qquad \text{for all } t$$

From the mean-value theorem† of the calculus $z(t)/y(t) =$ constant. But $y(t_0) = 1$, $z(t_0) = x_0$. Therefore $z(t)/y(t) = z(t_0)/y(t_0) = x_0$ and in view of the definition of $y(t)$, $z(t) = x_0 y(t) = x_0 e^{\alpha(t-t_0)}$.

Having thus discovered the unique solution of the initial-value problem, it is possible to phrase an analogous theorem for the differential equation $x' = \alpha x$ without initial conditions.

2.1.2. COROLLARY *Each element in the set of functions* $x(t) = ce^{\alpha t}$, c *an arbitrary constant, satisfies*

$$x' = \alpha x \qquad -\infty < t < \infty \tag{2.1.4}$$

and any solution of (2.1.4) is a member of this set.

PROOF The first part is clear; just differentiate and substitute into (2.1.4). If $z(t)$ satisfies (2.1.4), then at $t = 0$, $z(0)$ is some number, say z_0. Then $z'(t) = \alpha z(t)$ and $z(0) = z_0$. From Theorem 2.1.1, $z(t) = z_0 e^{\alpha t}$ concluding the proof.

†See Exercise 2.1, Part 2, Problem 1.

Technically, (2.1.4) is known as a *first-order, homogeneous linear differential equation.* The relevance of this name is only apparent after we study more complicated equations in later sections.

Exercise 2.1

Part 1

1. A cell has a fixed volume v. It is immersed in a fluid whose volume is large enough to be considered infinite. The cell has a substance S dissolved in its interior in the amount $s(t)$ at time t. S leaves the cell through its outer membrane at a rate proportional to its concentration, $s(t)/v$. Write a differential equation describing this phenomenon.
2. A body, B, has temperature $T(t)$ at time t. The surrounding atmosphere has fixed temperature T_a. B cools at a rate proportional to the difference $T(t)-T_a$. Write a differential equation satisfied by the difference between body and atmospheric temperature, $\delta(t)=T(t)-T_a$. What happens if $T(t_0)=T_a$ for some t_0?
3. A radioactive substance, R, has mass $m(t)$ at time t. It decays, that is, its mass diminishes, at a rate proportional to $m(t)$. Write a differential equation satisfied by $m(t)$. Show that the constant of proportionality is $\frac{1}{10}\ln 2$ if its half-life (that is, the time required for the mass to diminish to one-half the initial amount) is 10 seconds.
4. If the population of bacteria satisfies (2.1.1) and the population doubles every 5 minutes, find the constant of proportionality, α. If $P(0)=0$, find $P(t_1)$, $t_1=1$ hour.

Part 2

1. The mean-value theorem may be phrased:
   If f is continuous in $a \leqq t \leqq b$ and f is differentiable in $a<t<b$, then there exists a point t^*, $a<t^*<b$ such that
   $$f(b)-f(a)=f'(t^*)(b-a)$$
   Suppose f is continuous for all t and $f'(t)=0$ for all t. Use the mean-value theorem to prove that $f(t)=c$, c constant, for all t.

2.2. THE ORIGIN OF LINEAR SYSTEMS OF LINEAR DIFFERENTIAL EQUATIONS

Many important physical and biological systems involve several interrelated variables, for example, the magnitude of currents and voltages in various branches of an electrical circuit, the position of masses in a spring network, and the concentration of solutes in various cells of an organism. In some cases, one may very closely approximate the laws that govern the behavior of these systems by a linear differential model. That is, a model which assumes that the rate of growth, x_i', of each variable, x_i, is a linear combination of the other variables, say

$$x_i' = a_{i1}x_1 + \cdots + a_{in}x_n$$

where a_{ij}, $i, j = 1, \ldots, n$ are constants determined by physical or other considerations, and n is the number of variables in the system modeled by the equations.

As a specific example, let us consider the radioactive decay of bismuth 214, a branching process in the radioactive uranium–radium series.

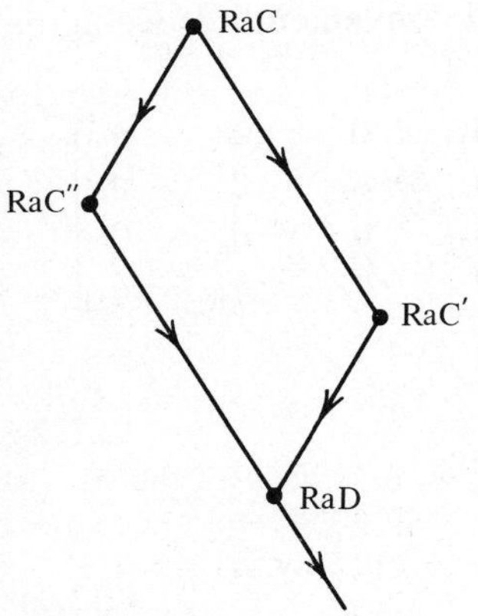

Radium C (bismuth 214), by emitting β-particles, forms radium C′ (polonium 214) and, by emitting α-particles, forms radium C″ (thalium 210). Radium C′ emits α-particles and forms radium D (lead 210) while radium C″ forms radium D by emission of β-particles. Radium D is also radioactive.

Let $x_1(t)$, $x_2(t)$, $x_3(t)$, and $x_4(t)$ represent the amounts of radium C, C′, C″, and D, respectively, present at time t. Assuming there is no replacement of radium C, and the rate of decay of radium C is proportional to the amount of radium C present, we have

$$x_1'(t) = -a_{11}x_1(t)$$

Radium C′ and C″ also decay but are replaced by the decay products of radium C. Since the rates of decay and production are proportional to the masses present,

$$x_2'(t) = a_{21}x_1(t) - a_{22}x_2(t)$$

and

$$x_3'(t) = a_{31}x_1(t) - a_{33}x_3(t)$$

Radium D decays but is replaced by the products of both radium C′ and C″, thus

$$x_4'(t) = a_{42}x_2(t) + a_{43}x_3(t) - a_{44}x_4(t)$$

The a_{ij} are constants determining the rates of decay and the rates of product formation and are discovered experimentally.

We collect these equations to form the *system*

$$
\begin{aligned}
x_1'(t) &= -a_{11}x_1(t) \\
x_2'(t) &= a_{21}x_1(t) - a_{22}x_2(t) \\
x_3'(t) &= a_{31}x_1(t) \qquad\qquad\qquad - a_{33}x_3(t) \\
x_4'(t) &= \qquad\qquad\qquad a_{42}x_2(t) + a_{43}x_3(t) - a_{44}x_4(t)
\end{aligned}
\tag{2.2.1}
$$

This system is dealt with far more efficiently and conveniently if we agree to use the matrix–vector notation:

$$
\mathbf{x}'(t) = \begin{bmatrix} x_1'(t) \\ x_2'(t) \\ x_3'(t) \\ x_4'(t) \end{bmatrix} \qquad
\mathbf{x}(t) = \begin{bmatrix} x_1(t) \\ x_2(t) \\ x_3(t) \\ x_4(t) \end{bmatrix} \qquad
\mathrm{A} = \begin{bmatrix} -a_{11} & 0 & 0 & 0 \\ a_{21} & -a_{22} & 0 & 0 \\ a_{31} & 0 & -a_{33} & 0 \\ 0 & a_{42} & a_{43} & -a_{44} \end{bmatrix}
$$

Then (2.2.1) becomes

$$
\mathbf{x}'(t) = \mathbf{A}\mathbf{x}(t) \tag{2.2.2}
$$

The resemblance between (2.2.2) and

$$
x'(t) = ax(t) \tag{2.2.3}
$$

is so striking that without the bold face type they are indistinguishable. The similarities between (2.2.2) and (2.2.3) go deeper than their superficial resemblance. A theory paralleling the theory of (2.2.3) as developed in Section 2.1 is valid for (2.2.2) and is the subject of this chapter and Chapter 3.

One illustration of the usefulness of the notation in (2.2.2) is its suggestiveness. For $x(t) = e^{at}$ satisfies (2.2.3), as we know, and this suggests that exponentials may also satisfy (2.2.2). Let us study this possibility in detail. Suppose we search for a constant vector $\mathbf{u}$ and a number λ such that $\mathbf{x}(t) = e^{\lambda t}\mathbf{u}$ satisfies (2.2.2). Since $\mathbf{x}'(t) = \lambda e^{\lambda t}\mathbf{u}$, for $\mathbf{x}(t)$ to be a solution it is necessary and sufficient that

$$
\lambda e^{\lambda t}\mathbf{u} = \mathbf{A}e^{\lambda t}\mathbf{u} = e^{\lambda t}\mathbf{A}\mathbf{u}
$$

or equivalently, since $e^{\lambda t} \neq 0$,

$$
\mathbf{A}\mathbf{u} - \lambda\mathbf{u} = (\mathbf{A} - \lambda\mathbf{I})\mathbf{u} = \mathbf{0} \tag{2.2.4}
$$

Since the numerical coefficients of the uranium–radium system are fairly nasty we shall invent another system of equations (which represent nothing in particular so far as we know) with conveniently chosen integral coefficients.

Consider the system of differential equations

$$
\begin{aligned}
x_1'(t) &= 7x_1(t) + 2x_2(t) - 3x_3(t) \\
x_2'(t) &= 4x_1(t) + 6x_2(t) - 4x_3(t) \\
x_3'(t) &= 5x_1(t) + 2x_2(t) - x_3(t)
\end{aligned}
\tag{2.2.5}
$$

Here,

$$\mathbf{x}'(t) = \begin{bmatrix} x_1'(t) \\ x_2'(t) \\ x_3'(t) \end{bmatrix} \qquad \mathbf{x}(t) = \begin{bmatrix} x_1(t) \\ x_2(t) \\ x_3(t) \end{bmatrix} \qquad \mathrm{A} = \begin{bmatrix} 7 & 2 & -3 \\ 4 & 6 & -4 \\ 5 & 2 & -1 \end{bmatrix}$$

satisfy Equation 2.2.2. Thus (2.2.4) becomes, in expanded form,

$$\begin{aligned} (7-\lambda)u_1 + 2u_2 - 3u_3 &= 0 \\ 4u_1 + (6-\lambda)u_2 - 4u_3 &= 0 \\ 5u_1 + 2u_2 + (-1-\lambda)u_3 &= 0 \end{aligned} \tag{2.2.6}$$

reducing (2.2.5), a problem in differential equations, to (2.2.6), a problem of algebra. The determinant of the coefficients, $C(\lambda)$, must vanish in order that there be solutions u_1, u_2, u_3 of (2.2.6) not all zero, as we know from Corollary 1.5.6. But

$$\begin{aligned} C(\lambda) \equiv \det(\mathrm{A} - \lambda \mathrm{I}) &\equiv \det \begin{bmatrix} 7-\lambda & 2 & -3 \\ 4 & 6-\lambda & -4 \\ 5 & 2 & -1-\lambda \end{bmatrix} \\ &= -(\lambda-2)(\lambda-4)(\lambda-6) = 0 \end{aligned}$$

if and only if $\lambda = 2, \lambda = 4$, or $\lambda = 6$.

Let us choose $\lambda = 2$. Then (2.2.6) reduces to

$$\begin{aligned} 5u_1 + 2u_2 - 3u_3 &= 0 \\ 4u_1 + 4u_2 - 4u_3 &= 0 \\ 5u_1 + 2u_2 - 3u_3 &= 0 \end{aligned}$$

which is equivalent, after row operations, to

$$\begin{aligned} 5u_1 + 2u_2 - 3u_3 &= 0 \\ -3u_2 + 2u_3 &= 0 \end{aligned}$$

and hence $u_3 = 3k_1$, $u_2 = 2k_1$ and then $u_1 = k_1$ are solutions of (2.2.6) when $\lambda = 2$. In vector form,

$$\begin{bmatrix} x_1(t) \\ x_2(t) \\ x_3(t) \end{bmatrix} \equiv \mathbf{x}(t) = \begin{bmatrix} k_1 \\ 2k_1 \\ 3k_1 \end{bmatrix} e^{2t} = k_1 \begin{bmatrix} 1 \\ 2 \\ 3 \end{bmatrix} e^{2t} \tag{2.2.7}$$

satisfies (2.2.5), as one can verify without difficulty. Equation (2.2.7) is interpreted as meaning:

$$x_1(t) = k_1 e^{2t} \qquad x_2(t) = 2k_1 e^{2t} \qquad x_3(t) = 3k_1 e^{2t}$$

are simultaneous solutions of (2.2.5). The other two eigenvalues, $\lambda_2 = 4$ and $\lambda_3 = 6$, lead to the solutions

$$\mathbf{x}(t) = k_2 \begin{bmatrix} 1 \\ 0 \\ 1 \end{bmatrix} e^{4t} \qquad \text{and} \qquad \mathbf{x}(t) = k_3 \begin{bmatrix} 1 \\ 1 \\ 1 \end{bmatrix} e^{6t} \tag{2.2.8}$$

by exactly the same analysis.

Section 2.3 will be devoted to a general treatment of the equation $\mathbf{x}' = \mathrm{A}\mathbf{x}$ where A is an arbitrary $n \times n$ matrix of constants. In it we shall generalize the methods of this section. The fundamental question, Does every initial value problem $\mathbf{x}' = \mathrm{A}\mathbf{x}$, $\mathbf{x}(0) = \mathbf{x}_0$ have exactly one solution?, will be treated in Chapters 3 and 4.

Exercise 2.2

1. Find $x_1(t)$, $x_2(t)$, $x_3(t)$ by solving (2.2.5) using the second and third solutions given in (2.2.8).
2. Call

$$\mathbf{x}_1(t) = k_1\begin{bmatrix}1\\2\\3\end{bmatrix}e^{2t} \qquad \mathbf{x}_2(t) = k_2\begin{bmatrix}1\\0\\1\end{bmatrix}e^{4t} \qquad \text{and} \qquad \mathbf{x}_3(t) = \begin{bmatrix}1\\1\\1\end{bmatrix}e^{6t}$$

Show by substitution that $\mathbf{x}_1(t) + \mathbf{x}_2(t) + \mathbf{x}_3(t)$ solves (2.2.5).

3. Referring to Problem 2, show that $\alpha\mathbf{x}_1(t) + \beta\mathbf{x}_2(t) + \gamma\mathbf{x}_3(t)$ solves (2.2.2) for any choice of α, β, γ. Hint: For any matrix A,

$$\mathrm{A}\{\alpha\mathbf{x}_1(t) + \beta\mathbf{x}_2(t) + \gamma\mathbf{x}_3(t)\} = \alpha\mathrm{A}\mathbf{x}_1(t) + \beta\mathrm{A}\mathbf{x}_2(t) + \gamma\mathrm{A}\mathbf{x}_3(t)$$

2.3. EIGENVALUES, EIGENVECTORS, AND SYSTEMS OF EQUATIONS

Let us suppose then that A is an arbitrary $n \times n$ matrix of *real* (from here on unless explicitly stated to the contrary, all matrices have real entries) constants and ask for a vector (a set of n functions) satisfying

$$\mathbf{x}' = \mathrm{A}\mathbf{x} \tag{2.3.1}$$

Such a vector is called a solution of this system. This system of equations is known as a *linear, homogeneous, first-order* system of differential equations with constant coefficients. If we assume a solution of the form $\mathbf{x} = e^{\lambda t}\mathbf{u}$, we find $\mathbf{u}$ and $\boldsymbol{\lambda}$ must satisfy

$$(\mathrm{A} - \lambda\mathrm{I})\mathbf{u} = \mathbf{0} \tag{2.3.2}$$

Now (2.3.2) represents n homogeneous algebraic equations in the n unknown variables,

$$\mathbf{u} = \begin{bmatrix}u_1\\u_2\\ \cdot\\ \cdot\\ \cdot\\ u_n\end{bmatrix}$$

Such a system has a solution $\mathbf{u} \neq \mathbf{0}$ if and only if the determinant of the coefficients is zero, that is,

$$\det(\mathrm{A} - \lambda\mathrm{I}) = \mathrm{C}(\lambda) = 0$$

Since $\det(A-\lambda I)$ is a polynomial in λ of degree n, $C(\lambda)$ has n roots, the *characteristic values* or *eigenvalues* of A. For the case $n=3$,

$$C(\lambda)=\det\begin{bmatrix} a_{11}-\lambda & a_{12} & a_{13} \\ a_{21} & a_{22}-\lambda & a_{23} \\ a_{31} & a_{32} & a_{33}-\lambda \end{bmatrix}$$

when

$$A=\begin{bmatrix} a_{11} & a_{12} & a_{13} \\ a_{21} & a_{22} & a_{23} \\ a_{31} & a_{32} & a_{33} \end{bmatrix}$$

If $\mathbf{u}\neq\mathbf{0}$ satisfies (2.3.2) for some value of λ, we call $\mathbf{u}$ an *eigenvector* corresponding to the eigenvalue λ. Furthermore, if λ is real, $\mathbf{u}$ is a solution of a homogeneous system of algebraic equations (linear) with real coefficients and therefore may be selected as a real vector. That is, $\mathbf{u}\in\mathscr{R}^3$ in this case.

2.3.1. THEOREM *If $\mathbf{u}_1$ is a real eigenvector of A corresponding to the real eigenvalue λ_1, then $\mathbf{x}_1=e^{\lambda_1 t}\mathbf{u}_1$ is a solution of $\mathbf{x}'=A\mathbf{x}$, for all t.*

PROOF We compute

$$\mathbf{x}'(t)=\frac{d}{dt}e^{\lambda_1 t}\mathbf{u}_1=\lambda_1 e^{\lambda_1 t}\mathbf{u}_1$$

while

$$\begin{aligned} A\mathbf{x}_1(t) &= Ae^{\lambda_1 t}\mathbf{u}_1=e^{\lambda_1 t}A\mathbf{u}_1 \\ &= \lambda_1 e^{\lambda_1 t}\mathbf{u}_1 \end{aligned}$$

Hence $\mathbf{x}'(t)=A\mathbf{x}_1(t)$ and $\mathbf{x}_1(t)$ is a solution of $\mathbf{x}'=A\mathbf{x}$.

Theorem 2.3.1 shows that solutions of $\mathbf{x}'=A\mathbf{x}$ can be found by studying the eigenvalues and eigenvectors of A.

2.3.2. DEFINITION *If for a given $n\times n$ matrix A, there exists a vector $\mathbf{u}\neq\mathbf{0}$ and a number λ such that*

$$A\mathbf{u}=\lambda\mathbf{u} \tag{2.3.3}$$

then λ is called an eigenvalue of A and $\mathbf{u}$ the corresponding eigenvector.

Equation (2.3.3) is equivalent to the homogeneous system

$$(A-\lambda I)\mathbf{u}=\mathbf{0} \tag{2.3.4}$$

The solution of equation (2.3.4) will be nontrivial ($\mathbf{u}\neq\mathbf{0}$) if and only if $\det(A-\lambda I)=0$. This proves the following theorem.

2.3.3. THEOREM *The number λ is an eigenvalue of A if and only if*

$$\det(A-\lambda I)=0 \tag{2.3.5}$$

2.3.4. DEFINITION *Let* A *be a given* $n \times n$ *matrix. The characteristic polynomial of* A, C(λ) *is then defined by*

$$C(\lambda) = \det \begin{bmatrix} a_{11}-\lambda & a_{12} \cdots\cdots & a_{1n} \\ a_{21} & a_{22}-\lambda \cdots & a_{2n} \\ \cdot & \cdot & \cdot \\ \cdot & \cdot & \cdot \\ \cdot & \cdot & \cdot \\ a_{n1} & a_{n2} \cdots\cdots & a_{nn}-\lambda \end{bmatrix}$$

$$= (-1)^n \lambda^n + c_{n-1}\lambda^{n-1} + \cdots + c_1\lambda + c_0 \tag{2.3.6}$$

2.3.1. Example

Find the eigenvalues of

$$A = \begin{bmatrix} 1 & 1 \\ 0 & 1 \end{bmatrix}$$

Solution

$$C(\lambda) = \det(A - \lambda I) = \det \begin{bmatrix} 1-\lambda & 1 \\ 0 & 1-\lambda \end{bmatrix} = (1-\lambda)^2$$

Hence $\lambda = 1$ is the only eigenvalue.

2.3.2. Example

Find the eigenvalues of

$$A = \begin{bmatrix} -1 & 2 & 2 \\ 2 & 2 & 2 \\ -3 & -6 & -6 \end{bmatrix}$$

Solution

Here

$$C(\lambda) = \det \begin{bmatrix} -1-\lambda & 2 & 2 \\ 2 & 2-\lambda & 2 \\ -3 & -6 & -6-\lambda \end{bmatrix}$$

$$= -\lambda(\lambda+2)(\lambda+3)$$

so that $\lambda = 0, -2, -3$ are the eigenvalues.

In determining the eigenvectors we need to solve the homogeneous system (2.3.4).

2.3.3. Example

For A of Example 2.3.1 find the eigenvectors.

Solution

We set

$$\mathbf{u} = \begin{bmatrix} u_1 \\ u_2 \end{bmatrix}$$

and setting $\lambda = 1$ in $(\mathrm{A} - \lambda\mathrm{I})\mathbf{u} = \mathbf{0}$, we solve

$$(\mathrm{A} - \mathrm{I})\mathbf{u} = \begin{bmatrix} 0 & 1 \\ 0 & 0 \end{bmatrix}\begin{bmatrix} u_1 \\ u_2 \end{bmatrix} = \begin{bmatrix} u_2 \\ 0 \end{bmatrix} = \begin{bmatrix} 0 \\ 0 \end{bmatrix}$$

Thus $u_2 = 0$ and $\mathbf{u} = \begin{bmatrix} u_1 \\ u_2 \end{bmatrix} = u_1\begin{bmatrix} 1 \\ 0 \end{bmatrix}$. Thus $\begin{bmatrix} 1 \\ 0 \end{bmatrix}$ and any scalar multiple of $\begin{bmatrix} 1 \\ 0 \end{bmatrix}$ is an eigenvector.

2.3.4. Example

Find the eigenvectors for the A of Example 2.3.2.

Solution

For $\lambda = 0$,

$$\mathbf{Au} = \begin{bmatrix} -1 & 2 & 2 \\ 2 & 2 & 2 \\ -3 & -6 & -6 \end{bmatrix}\begin{bmatrix} u_1 \\ u_2 \\ u_3 \end{bmatrix} = \begin{bmatrix} 0 \\ 0 \\ 0 \end{bmatrix}$$

$$\begin{bmatrix} -1 & 2 & 2 \\ 0 & 6 & 6 \\ 0 & 0 & 0 \end{bmatrix}\begin{bmatrix} u_1 \\ u_2 \\ u_3 \end{bmatrix} = \begin{bmatrix} 0 \\ 0 \\ 0 \end{bmatrix}$$

$$\begin{bmatrix} -1 & 2 & 2 \\ 0 & 1 & 1 \\ 0 & 0 & 0 \end{bmatrix}\begin{bmatrix} u_1 \\ u_2 \\ u_3 \end{bmatrix} = \begin{bmatrix} 0 \\ 0 \\ 0 \end{bmatrix}$$

$$\begin{bmatrix} -1 & 0 & 0 \\ 0 & 1 & 1 \\ 0 & 0 & 0 \end{bmatrix}\begin{bmatrix} u_1 \\ u_2 \\ u_3 \end{bmatrix} = \begin{bmatrix} 0 \\ 0 \\ 0 \end{bmatrix}$$

Thus $u_1 = 0$ and $u_2 + u_3 = 0$. Set $u_3 = -k$ then $u_2 = k$, and

$$\mathbf{u} = k\begin{bmatrix} 0 \\ 1 \\ -1 \end{bmatrix} \qquad \text{for} \quad \lambda = 0$$

For $\lambda = -2$, we solve

$$(\mathrm{A} + 2\mathrm{I})\mathbf{u} = \begin{bmatrix} 1 & 2 & 2 \\ 2 & 4 & 2 \\ -3 & -6 & -4 \end{bmatrix}\begin{bmatrix} u_1 \\ u_2 \\ u_3 \end{bmatrix} = \begin{bmatrix} 0 \\ 0 \\ 0 \end{bmatrix}$$

It turns out that $u_3 = 0$ and $u_1 + 2u_2 = 0$. Set $u_2 = k$, then

$$\mathbf{u} = k\begin{bmatrix} -2 \\ 1 \\ 0 \end{bmatrix} \quad \text{for} \quad \lambda = -2$$

Finally, for $\lambda = -3$, we solve

$$(\mathrm{A} + 3\mathrm{I})\mathbf{u} = \begin{bmatrix} 2 & 2 & 2 \\ 2 & 5 & 2 \\ -3 & -6 & -3 \end{bmatrix}\begin{bmatrix} u_1 \\ u_2 \\ u_3 \end{bmatrix} = \begin{bmatrix} 0 \\ 0 \\ 0 \end{bmatrix}$$

and obtain $u_2 = 0$ and $u_1 + u_3 = 0$. Hence

$$\mathbf{u} = k\begin{bmatrix} 1 \\ 0 \\ -1 \end{bmatrix} \quad \text{for} \quad \lambda = -3$$

2.3.5. Example

Find the eigenvalues and the corresponding eigenvectors for the matrix

$$\mathrm{A} = \begin{bmatrix} 1 & 0 & 0 \\ 0 & 1 & 0 \\ 0 & 0 & 1 \end{bmatrix}$$

Solution

$\mathrm{C}(\lambda) = (1 - \lambda)^3$ and hence $\lambda = 1$ is the only eigenvalue. But

$$(\mathrm{A} - \mathrm{I})\mathbf{u} = \begin{bmatrix} 0 & 0 & 0 \\ 0 & 0 & 0 \\ 0 & 0 & 0 \end{bmatrix}\begin{bmatrix} u_1 \\ u_2 \\ u_3 \end{bmatrix} = \begin{bmatrix} 0 \\ 0 \\ 0 \end{bmatrix}$$

for any choice of u_1, u_2, u_3. That is, $\lambda = 1$ and

$$\mathbf{u} = u_1\begin{bmatrix} 1 \\ 0 \\ 0 \end{bmatrix} + u_2\begin{bmatrix} 0 \\ 1 \\ 0 \end{bmatrix} + u_3\begin{bmatrix} 0 \\ 0 \\ 1 \end{bmatrix}$$

2.3.6. Example

Find solutions of $\mathbf{x}' = \mathrm{A}\mathbf{x}$ where A is the matrix of Example 2.3.5.

Solution

We use Theorem 2.3.1. We have

$$\begin{aligned} \mathbf{x}(t) &= e^t\mathbf{u} \\ &= \begin{bmatrix} u_1 \\ u_2 \\ u_3 \end{bmatrix} e^t \end{aligned}$$

for any choice of u_1, u_2, u_3.

Exercise 2.3

Part 1

1. Find the characteristic polynomial and eigenvalues of each of the following matrices:

(a) $\begin{bmatrix} 1 & 4 \\ 2 & 3 \end{bmatrix}$ (b) $[1]$ (c) $\begin{bmatrix} 2 & 2 & 0 \\ 1 & 2 & 1 \\ 1 & 2 & 1 \end{bmatrix}$ (d) $\begin{bmatrix} 1 & 2 & 4 \\ 0 & 1 & 0 \\ 0 & 2 & 1 \end{bmatrix}$

(e) $\begin{bmatrix} 1 & 1 & 0 \\ 0 & 1 & 1 \\ 0 & 0 & 1 \end{bmatrix}$ (f) $\begin{bmatrix} 1 & 1 & -1 & 2 \\ 0 & 2 & 0 & 1 \\ 0 & 0 & -1 & 1 \\ 0 & 0 & 0 & 0 \end{bmatrix}$

2. In each part of Problem 1, compute the eigenvectors.
3. Find the characteristic polynomials for the matrices

(a) $\begin{bmatrix} 0 & 1 \\ -b & -a \end{bmatrix}$ (b) $\begin{bmatrix} 0 & 1 & 0 \\ 0 & 0 & 1 \\ -c & -b & -a \end{bmatrix}$

4. Find the inverse of $\begin{bmatrix} 1 & 4 \\ 2 & 3 \end{bmatrix}$ and compute the eigenvalues of the inverse. Relate your answers to the answers given for Problem 1(a) above.
5. Use your answers to Problems 1 and 2 of this set and Theorem 2.3.1 to find solutions of $\mathbf{x}' = \mathrm{A}\mathbf{x}$ for the A of each part of Problem 1.

Part 2

1. Prove: If A is singular then $\lambda = 0$ is an eigenvalue and conversely.
2. Prove that eigenvalues of A are the same as those of $\mathrm{P}^{-1}\mathrm{AP}$.
3. Prove that λ^{-1} is an eigenvalue of A^{-1} if λ is an eigenvalue of A. Hint: $\mathrm{A}^{-1} - \mu\mathrm{I} = \mathrm{A}^{-1}(\mathrm{I} - \mu\mathrm{A})$.
4.* The matrix

$$\mathrm{C} = \begin{bmatrix} 0 & 1 & 0 \\ 0 & 0 & 0 \\ \cdot & \cdot & \cdot \\ \cdot & \cdot & \cdot \\ \cdot & \cdot & \cdot \\ 0 & 0 & 1 \\ -a_0 & -a_1 \cdots\cdots & -a_n \end{bmatrix}$$

is the companion matrix to $\lambda^n + a_{n-1}\lambda^{n-1} + \cdots + a_0$. Prove that $(-1)^n(\lambda^n + a_{n-1}\lambda^{n-1} + a_{n-2}\lambda^{n-2} + \cdots + a_0)$ is the characteristic polynomial of C.

5.* Let the $n \times n$ matrix J_n be defined by

$$\mathrm{J}_n = \begin{bmatrix} 1 & 1 \cdots & 1 \\ 1 & 1 \cdots & 1 \\ \cdot & \cdot & \cdot \\ \cdot & \cdot & \cdot \\ \cdot & \cdot & \cdot \\ 1 & 1 \cdots & 1 \end{bmatrix}$$

Show that $\det(\mathrm{J}_n - \lambda \mathrm{I}) = (-\lambda)^{n-1}(n-\lambda)$ and prove that the eigenvectors corresponding to $\lambda = 0$ are representable as

$$\mathbf{u}_0 = k_1 \begin{bmatrix} -1 \\ 0 \\ \cdot \\ \cdot \\ \cdot \\ 0 \\ 0 \\ 1 \end{bmatrix} + k_2 \begin{bmatrix} -1 \\ 0 \\ \cdot \\ \cdot \\ \cdot \\ 0 \\ 1 \\ 0 \end{bmatrix} + \cdots k_{n-1} \begin{bmatrix} -1 \\ 1 \\ 0 \\ \cdot \\ \cdot \\ \cdot \\ 0 \\ 0 \end{bmatrix}$$

where $k_1, k_2, \ldots, k_{n-1}$ are not all zero and corresponding to $\lambda = n$ is

$$\mathbf{u}_1 = k \begin{bmatrix} 1 \\ \cdot \\ \cdot \\ \cdot \\ 1 \end{bmatrix} \qquad k \neq 0$$

6. Use the results of Problem 5 above to solve

$$\mathbf{x}' = \mathrm{J}_n \mathbf{x} \qquad \mathbf{x}(0) = \begin{bmatrix} a_1 \\ a_2 \\ \cdot \\ \cdot \\ a_n \end{bmatrix}$$

2.4. INITIAL-VALUE PROBLEMS

If we supplement the system of equations $\mathbf{x}' = \mathrm{A}\mathbf{x}$ with a condition such as $\mathbf{x}(t_0) = \mathbf{x}_0$, then together

$$\mathbf{x}' = \mathrm{A}\mathbf{x} \qquad \mathbf{x}(t_0) = \mathbf{x}_0 \tag{2.4.1}$$

is known as an *initial-value problem.* The vector $\mathbf{x}_0$ is called the *initial value.* By a *solution* of this problem we mean a differentiable function $\mathbf{x}(t)$ which is a solution of $\mathbf{x}' = \mathrm{A}\mathbf{x}$ and whose value at t_0 is the given vector $\mathbf{x}_0$. Often $t_0 = 0$ for convenience. If we return for a moment to the problem given in (2.2.5), that is,

$$\mathbf{x}' = \begin{bmatrix} 7 & 2 & -3 \\ 4 & 6 & -4 \\ 5 & 2 & -1 \end{bmatrix} \mathbf{x} \tag{2.4.2}$$

and ask for a solution meeting

$$\mathbf{x}(0) = \begin{bmatrix} 2 \\ 4 \\ 6 \end{bmatrix} = \mathbf{x}_0$$

then (2.2.7) supplies the answer with $k_1 = 2$. None of the three solutions given, however, solves (2.4.2) subject to the initial condition

$$\mathbf{x}(0) = \begin{bmatrix} 0 \\ 1 \\ 1 \end{bmatrix} = \mathbf{x}_0$$

What we need are more solutions! An important feature of linear equations is that new solutions may be obtained by forming linear combinations of old solutions.

2.4.1. THEOREM *Let $\mathbf{x}_1(t)$ and $\mathbf{x}_2(t)$ be any solutions of $\mathbf{x}' = \mathbf{A}\mathbf{x}$. Then for any constants k_1 and k_2, $\mathbf{x}(t) = k_1\mathbf{x}_1(t) + k_2\mathbf{x}_2(t)$ is also a solution.*

PROOF

$$\begin{aligned} \mathbf{x}'(t) &= \frac{d}{dt}[k_1\mathbf{x}_1(t) + k_2\mathbf{x}_2(t)] \\ &= k_1\mathbf{x}_1'(t) + k_2\mathbf{x}_2'(t) \\ &= k_1\mathbf{A}\mathbf{x}_1(t) + k_2\mathbf{A}\mathbf{x}_2(t) \\ &= \mathbf{A}[k_1\mathbf{x}_1(t) + k_2\mathbf{x}_2(t)] \\ &= \mathbf{A}\mathbf{x}(t) \end{aligned}$$

Hence $\mathbf{x}(t)$ is a solution of $\mathbf{x}' = \mathbf{A}\mathbf{x}$.

We return to the problem at hand—the solution of the initial-value problem. Some examples follow which are chosen to illustrate the kinds of difficulties we may encounter. First a corollary.

2.4.2. COROLLARY *Let $\mathbf{x}_1(t), \ldots, \mathbf{x}_r(t)$ be solutions of $\mathbf{x}' = \mathbf{A}\mathbf{x}$. Then for any constants $k_1, \ldots, k_r$*

$$\mathbf{x}(t) = k_1\mathbf{x}_1(t) + \cdots + k_r\mathbf{x}_r(t) \tag{2.4.3}$$

is also a solution of $\mathbf{x}' = \mathbf{A}\mathbf{x}$.

Corollary 2.4.2 is known as the *first principle of superposition.* One role of this corollary is to expand the power of Theorem 2.3.1 and hence aid in our attempts to find a solution of the initial-value problem. Clearly, the more roots of $C(\lambda)$ which are different, the greater the number of constants in (2.4.3) and hence the more likely our success will be.

2.4.1. Example

Solve

$$\mathbf{x}' = \begin{bmatrix} 1 & 1 \\ 1 & 1 \end{bmatrix}\mathbf{x} \qquad \mathbf{x}(0) = \mathbf{x}_0$$

Solution

We have $C(\lambda) = (1-\lambda)^2 - 1 = \lambda(-2+\lambda)$. Thus $\lambda = 0$ and $\lambda = 2$.

(a) Take $\lambda = 0$. Then

$$\begin{bmatrix} 1 & 1 \\ 1 & 1 \end{bmatrix}\begin{bmatrix} u_1 \\ u_2 \end{bmatrix} = \begin{bmatrix} u_1 + u_2 \\ u_1 + u_2 \end{bmatrix} = \begin{bmatrix} 0 \\ 0 \end{bmatrix}$$

and

$$\mathbf{u}_1 = k_1\begin{bmatrix} 1 \\ -1 \end{bmatrix}$$

From this we have the solutions,

$$\mathbf{x}_1(t) = k_1 e^{0t}\begin{bmatrix} 1 \\ -1 \end{bmatrix} = k_1\begin{bmatrix} 1 \\ -1 \end{bmatrix}$$

(b) Take $\lambda = 2$. Then

$$\begin{bmatrix} -1 & 1 \\ 1 & -1 \end{bmatrix}\begin{bmatrix} u_1 \\ u_2 \end{bmatrix} = \begin{bmatrix} -u_1 + u_2 \\ u_1 - u_2 \end{bmatrix} = \begin{bmatrix} 0 \\ 0 \end{bmatrix}$$

and hence $\mathbf{u}_2 = k_2[1, 1]^T$. This eigenvalue and eigenvector enable us to construct the solution

$$\mathbf{x}_2(t) = k_2 e^{2t}\begin{bmatrix} 1 \\ 1 \end{bmatrix}$$

The corollary asserts that

$$\mathbf{x}(t) = k_1\begin{bmatrix} 1 \\ -1 \end{bmatrix} + k_2 e^{2t}\begin{bmatrix} 1 \\ 1 \end{bmatrix} \tag{2.4.4}$$

is a solution of the given system for each choice of k_1 and k_2. When $t = 0$, $\mathbf{x}(t)$ has value

$$\mathbf{x}(0) = k_1\begin{bmatrix} 1 \\ -1 \end{bmatrix} + k_2\begin{bmatrix} 1 \\ 1 \end{bmatrix} = \begin{bmatrix} 1 & 1 \\ -1 & 1 \end{bmatrix}\begin{bmatrix} k_1 \\ k_2 \end{bmatrix}$$

Therefore, k_1 and k_2 are determined by the equation

$$\begin{bmatrix} 1 & 1 \\ -1 & 1 \end{bmatrix}\begin{bmatrix} k_1 \\ k_2 \end{bmatrix} = \mathbf{x}_0$$

This has the unique solution

$$\begin{bmatrix} k_1 \\ k_2 \end{bmatrix} = \begin{bmatrix} 1 & 1 \\ -1 & 1 \end{bmatrix}^{-1}\mathbf{x}_0 = \begin{bmatrix} \frac{1}{2} & -\frac{1}{2} \\ \frac{1}{2} & \frac{1}{2} \end{bmatrix}\mathbf{x}_0$$

Remarks

The expression (2.4.4) could reasonably be called the "general" solution of

$$\mathbf{x}' = \begin{bmatrix} 1 & 1 \\ 1 & 1 \end{bmatrix}\mathbf{x}$$

because if $\mathbf{x}_0$ is any initial vector the computation just given shows that k_1 and k_2 can always be found so that $\mathbf{x}(0) = \mathbf{x}_0$.

Now the n roots of $C(\lambda)$ may not all be distinct. In fact there may be as few as one root repeated n times as is the case of $C(\lambda) = (\lambda_1 - \lambda)^n$. In this instance it may happen that the "variety" of solutions in (2.4.3) is drastically reduced. Suppose, to illustrate this point, that $\mathbf{u}_1 = \mathbf{u}_2 = \cdots = \mathbf{u}_n$ and $\lambda_1 = \lambda_2 = \cdots = \lambda_n$. Then

$$\begin{aligned} k_1 e^{\lambda_1 t}\mathbf{u}_1 + k_2 e^{\lambda_2 t}\mathbf{u}_2 + \cdots + k_n e^{\lambda_n t}\mathbf{u}_n &= (k_1 + k_2 + \cdots k_n) e^{\lambda_1 t}\mathbf{u}_1 \\ &= k e^{\lambda_1 t}\mathbf{u}_1 \end{aligned}$$

Thus the n parameters are illusory. There is but one. This is an important observation because without sufficiently many "essentially different" solutions (n in this case) we are unable to solve all initial-value problems. For instance;

2.4.2. Example

Solve

$$\mathbf{x}' = \begin{bmatrix} 1 & 1 \\ 0 & 1 \end{bmatrix}\mathbf{x}, \qquad \mathbf{x}(0) = \begin{bmatrix} 1 \\ 1 \end{bmatrix}$$

Solution

We find $(1-\lambda)^2 = C(\lambda)$ and hence $\lambda = 1$. Thus

$$\begin{bmatrix} 0 & 1 \\ 0 & 0 \end{bmatrix}\mathbf{u} = \begin{bmatrix} 0 & 1 \\ 0 & 0 \end{bmatrix}\begin{bmatrix} u_1 \\ u_2 \end{bmatrix} = \begin{bmatrix} u_2 \\ 0 \end{bmatrix} = \begin{bmatrix} 0 \\ 0 \end{bmatrix}$$

Hence $\mathbf{u} = k\begin{bmatrix} 1 \\ 0 \end{bmatrix}$ and correspondingly we have the solutions $\mathbf{x}(t) = ke^t\begin{bmatrix} 1 \\ 0 \end{bmatrix}$;

but $\mathbf{x}(0) = \begin{bmatrix} k \\ 0 \end{bmatrix} \neq \begin{bmatrix} 1 \\ 1 \end{bmatrix}$. *We are unable to find a solution meeting the required initial condition with the knowledge we have so far.* This difficulty is not due so much to a lack of eigenvalues but a lack of eigenvectors. We can see this in the next example.

2.4.3. Example

Solve

$$\mathbf{x}' = \mathbf{I}\mathbf{x}$$
$$\mathbf{x}(0) = \mathbf{x}_0 \neq \mathbf{0}$$

Solution

Immediately, $C(\lambda) = (1-\lambda)^n$ and $\lambda = 1$ is the only eigenvalue. However, every nonzero vector is an eigenvector of I. Thus, $\mathbf{x}_0$ is an eigenvector and $\mathbf{x}(t) = e^t\mathbf{x}_0$ is our solution. Our success in Example 2.4.3 was due to the structure of I which permits the existence of n linearly independent eigenvectors corresponding to the single eigenvalue $\lambda = 1$. Contrast this with Example 2.4.2.

The following general theorem is easy to prove.

2.4.3. THEOREM *If there exist n real, linearly independent eigenvectors, corresponding to real (not necessarily distinct) eigenvalues of the $n \times n$ matrix, A, then $\mathbf{x}' = A\mathbf{x}$, $\mathbf{x}(t_0) = \mathbf{x}_0$ has a solution for any $\mathbf{x}_0$.*

PROOF Corresponding to the eigenvector $\mathbf{u}$ is the solution $\mathbf{u}e^{\lambda t}$. At $t = 0$, this solution has the value $\mathbf{u}$. Let $\mathbf{u}_1, \mathbf{u}_2, \ldots, \mathbf{u}_n$ be linearly independent eigenvectors corresponding to the eigenvalues (not necessarily distinct) $\lambda_1, \lambda_2, \ldots, \lambda_n$. Then by Corollary 2.4.2

$$\mathbf{x}(t) = c_1 e^{\lambda_1 t}\mathbf{u}_1 + c_2 e^{\lambda_2 t}\mathbf{u}_2 + \cdots + c_n e^{\lambda_n t}\mathbf{u}_n$$

is a solution of $\mathbf{x}' = A\mathbf{x}$ and

$$\mathbf{x}(t_0) = c_1 \mathbf{u}_1 e^{\lambda_1 t_0} + \cdots + c_n \mathbf{u}_n e^{\lambda_n t_0}$$

Since $\mathbf{u}_1, \ldots, \mathbf{u}_n$ are linearly independent so are $e^{\lambda_1 t_0}\mathbf{u}_1, \ldots, e^{\lambda_n t_0}\mathbf{u}_n$, and they form a basis in $\mathscr{C}^n$. If $\mathbf{x}_0 \in \mathscr{C}^n$ then the choice $c_1 = c_1^*, c_2 = c_2^*, \ldots, c_n = c_n^*$ can be made so that

$$\mathbf{x}_0 = c_1^* \mathbf{u}_1 e^{\lambda_1 t_0} + c_2^* \mathbf{u}_2 e^{\lambda_2 t_0} + \cdots + c_n^* \mathbf{u}_n e^{\lambda_n t_0}$$

From what has been said about $\mathbf{x}(t)$ at $t = 0$, we see that

$$\mathbf{x}(t) = c_1^* e^{\lambda_1 t}\mathbf{u}_1 + c_2^* e^{\lambda_2 t}\mathbf{u}_2 + \cdots + c_n^* e^{\lambda_n t}\mathbf{u}_n \tag{2.4.5}$$

is a solution of the initial-value problem.

The hypothesis of Theorem 2.4.3 will be satisfied when A has n distinct eigenvalues. To prove this we must first prove a lemma. We prove the lemma in the case $k = 3$; the general case of arbitrary k can be stated and proved in a totally analogous manner.

2.4.4. LEMMA *If A has eigenvalues $\lambda_1, \lambda_2, \lambda_3$ which are distinct and $\mathbf{u}_1, \mathbf{u}_2, \mathbf{u}_3$ are any three eigenvectors corresponding to λ_1, λ_2, and λ_3, respectively, then $\mathbf{u}_1, \mathbf{u}_2, \mathbf{u}_3$ are linearly independent, that is, $k_1\mathbf{u}_1 + k_2\mathbf{u}_2 + k_3\mathbf{u}_3 = \mathbf{0}$ only if $k_1 = k_2 = k_3 = 0$.*

PROOF Define the matrix $A_1 \equiv (A - \lambda_2 I)(A - \lambda_3 I)$. We examine the effects of A_1 on $\mathbf{u}_1, \mathbf{u}_2$, and $\mathbf{u}_3$. Specifically,

$$\begin{aligned} A_1\mathbf{u}_1 &= (A - \lambda_2 I)(A - \lambda_3 I)\mathbf{u}_1 = (A - \lambda_2 I)(A\mathbf{u}_1 - \lambda_3 I\mathbf{u}_1) \\ &= (A - \lambda_2 I)(\lambda_1\mathbf{u}_1 - \lambda_3\mathbf{u}_1) = (A - \lambda_2 I)\mathbf{u}_1(\lambda_1 - \lambda_3) \\ &= \mathbf{u}_1(\lambda_1 - \lambda_2)(\lambda_1 - \lambda_3) \neq \mathbf{0} \end{aligned}$$

since $\mathbf{u}_1 \neq \mathbf{0}$ and $\lambda_3 \neq \lambda_1 \neq \lambda_2$. But

$$A_1\mathbf{u}_2 = (A-\lambda_2 I)[(A-\lambda_3 I)\mathbf{u}_2] = (\lambda_2-\lambda_3)[(A-\lambda_2 I)\mathbf{u}_2] = \mathbf{0}$$

and

$$A_1\mathbf{u}_3 = (A-\lambda_2 I)[(A-\lambda_3 I)\mathbf{u}_3] = \mathbf{0}$$

That is, $A_1\mathbf{u}_1 \neq 0$, $A_1\mathbf{u}_2 = \mathbf{0}$, and $A_1\mathbf{u}_3 = \mathbf{0}$. Now suppose

$$k_1\mathbf{u}_1 + k_2\mathbf{u}_2 + k_3\mathbf{u}_3 \equiv \mathbf{0}$$

Then

$$\begin{aligned}\mathbf{0} = A_1\mathbf{0} &= A_1[k_1\mathbf{u}_1 + k_2\mathbf{u}_2 + k_3\mathbf{u}_3] \\ &= k_1A_1\mathbf{u}_1 + k_2A_1\mathbf{u}_2 + k_3A_1\mathbf{u}_3 \\ &= k_1(\lambda_1-\lambda_2)(\lambda_1-\lambda_3)\mathbf{u}_1\end{aligned}$$

which can happen only if $k_1 = 0$.

Replacing A_1 in the above argument by $A_2 = (A-\lambda_1 I)(A-\lambda_3 I)$, we see that $k_2 = 0$. Similarly, replacing A_1 by $A_3 = (A-\lambda_1 I)(A-\lambda_2 I)$, we find $k_3 = 0$.

Thus the only vanishing linear combination of u_1, u_2, and u_3 is the trivial combination, from which it follows that $\mathbf{u}_1$, $\mathbf{u}_2$, and $\mathbf{u}_3$ are linearly independent.

It now follows that n real, distinct eigenvalues lead to a set of n linearly independent, real eigenvectors. From Theorem 2.4.3 we deduce

2.4.5. THEOREM *If* A *has* n *real, distinct eigenvalues then* $\mathbf{x}' = A\mathbf{x}$, $\mathbf{x}(t_0) = \mathbf{x}_0$ *has a solution for any* $\mathbf{x} \in \mathcal{R}^n$.

We study two more examples.

2.4.4. Example

Solve

$$\mathbf{x}' = \begin{bmatrix} 6 & 8 \\ -3 & -4 \end{bmatrix}\mathbf{x} \qquad \mathbf{x}(1) = \mathbf{x}_0$$

Solution

Here, $C(\lambda) = (6-\lambda)(-4-\lambda)+24 = -24-2\lambda+\lambda^2+24 = 0$ so that $\lambda_1 = 0$ and $\lambda_2 = 2$. Although the eigenvalues are the same in this example as they were in Example 2.4.1, we find here:

(a) For $\lambda_1 = 0$,

$$\begin{bmatrix} 6 & 8 \\ -3 & -4 \end{bmatrix}\begin{bmatrix} u_1 \\ u_2 \end{bmatrix} = \begin{bmatrix} 0 \\ 0 \end{bmatrix}$$

implies

$$\mathbf{u}_1 = k_1\begin{bmatrix} 4 \\ -3 \end{bmatrix} \qquad \text{and} \qquad \mathbf{x}_1 = k_1\begin{bmatrix} 4 \\ -3 \end{bmatrix}$$

(b) For $\lambda_2 = 2$,

$$\begin{bmatrix} 4 & 8 \\ -3 & -6 \end{bmatrix}\begin{bmatrix} u_1 \\ u_2 \end{bmatrix} = \begin{bmatrix} 0 \\ 0 \end{bmatrix}$$

implies

$$\mathbf{u}_2 = k_2\begin{bmatrix} -2 \\ 1 \end{bmatrix} \quad \text{and} \quad \mathbf{x}_2 = k_2\begin{bmatrix} -2 \\ 1 \end{bmatrix}e^{2t}$$

From these results and the principle of superposition it follows that

$$\begin{aligned} \mathbf{x}(t) &= k_1 e^{0t}\begin{bmatrix} 4 \\ -3 \end{bmatrix} + k_2 e^{2t}\begin{bmatrix} -2 \\ 1 \end{bmatrix} \\ &= k_1\begin{bmatrix} 4 \\ -3 \end{bmatrix} + k_2 e^{2t}\begin{bmatrix} -2 \\ 1 \end{bmatrix} \end{aligned}$$

Finally,

$$\mathbf{x}(1) = k_1\begin{bmatrix} 4 \\ -3 \end{bmatrix} + k_2 e^2\begin{bmatrix} -2 \\ 1 \end{bmatrix} = \mathbf{x}_0$$

yields

$$\begin{bmatrix} 4 & -2e^2 \\ -3 & e^2 \end{bmatrix}\begin{bmatrix} k_1 \\ k_2 \end{bmatrix} = \mathbf{x}_0$$

or

$$\begin{bmatrix} k_1 \\ k_2 \end{bmatrix} = \begin{bmatrix} 4 & -2e^2 \\ -3 & e^2 \end{bmatrix}^{-1}\mathbf{x}_0 = \begin{bmatrix} -\frac{1}{2} & -1 \\ (-\frac{3}{2})e^{-2} & -2e^{-2} \end{bmatrix}\mathbf{x}_0$$

Remarks

Theorem 2.4.5 assures us of the existence of a solution because the eigenvalues are distinct.

2.4.5. Example

Solve

$$\mathbf{x}' = \begin{bmatrix} 0 & 1 & 0 \\ 0 & 0 & 1 \\ 2 & 1 & -2 \end{bmatrix}\mathbf{x}, \qquad \mathbf{x}(0) = \begin{bmatrix} 1 \\ 0 \\ 1 \end{bmatrix}$$

Solution

Here $C(\lambda) = -(\lambda - 1)(\lambda + 1)(\lambda + 2)$ and therefore $\lambda_1 = -1$, $\lambda_2 = 1$, and $\lambda_3 = -2$. We compute

$$\mathbf{u}_1 = \begin{bmatrix} 1 \\ -1 \\ 1 \end{bmatrix} \qquad \mathbf{u}_2 = \begin{bmatrix} 1 \\ 1 \\ 1 \end{bmatrix} \qquad \mathbf{u}_3 = \begin{bmatrix} 1 \\ -2 \\ 4 \end{bmatrix}$$

corresponding to λ_1, λ_2, λ_3, respectively. We know that k_1, k_2, k_3 exist such that

$$\mathbf{x}(0) = \begin{bmatrix} 1 \\ 0 \\ 1 \end{bmatrix} = k_1\mathbf{u}_1 + k_2\mathbf{u}_2 + k_3\mathbf{u}_3 = \begin{bmatrix} 1 & 1 & 1 \\ -1 & 1 & -2 \\ 1 & 1 & 4 \end{bmatrix} \begin{bmatrix} k_1 \\ k_2 \\ k_3 \end{bmatrix}$$

Specifically,

$$\begin{aligned} k_1 + k_2 + k_3 &= 1 \\ -k_1 + k_2 - 2k_3 &= 0 \\ k_1 + k_2 + 4k_3 &= 1 \end{aligned} \tag{2.4.6}$$

from which we conclude: $k_1 = k_2 = \frac{1}{2}$, $k_3 = 0$ and

$$\mathbf{x}(t) = \frac{e^t}{2} \begin{bmatrix} 1 \\ -1 \\ 1 \end{bmatrix} + \frac{e^{-t}}{2} \begin{bmatrix} 1 \\ 1 \\ 1 \end{bmatrix} = \begin{bmatrix} \dfrac{e^t + e^{-t}}{2} \\ \dfrac{e^t - e^{-t}}{-2} \\ \dfrac{e^t + e^{-t}}{2} \end{bmatrix} = \begin{bmatrix} \cosh t \\ -\sinh t \\ \cosh t \end{bmatrix}$$

and $\mathbf{x}(t)$ solves the posed initial-value problem.

We note for emphasis that (2.4.6) has a unique solution because the determinant of the coefficients does not vanish, that is,

$$\det\,[\mathbf{u}_1, \mathbf{u}_2, \mathbf{u}_3] = \det \begin{bmatrix} 1 & 1 & 1 \\ -1 & 1 & -2 \\ 1 & 1 & 4 \end{bmatrix} \neq 0$$

Exercise 2.4

Part 1

The A in each of the following problems has n distinct, real eigenvalues, where A is $n \times n$. Find a solution of $\mathbf{x}' = \mathrm{A}\mathbf{x}$ containing n arbitrary constants.

1. $\mathrm{A} = \begin{bmatrix} 1 & 3 \\ 1 & -1 \end{bmatrix}$

2. $\mathrm{A} = \begin{bmatrix} 1 & 1 \\ 3 & -1 \end{bmatrix}$

3. $\mathrm{A} = \begin{bmatrix} 1 & \alpha^2 - 1 \\ 1 & -1 \end{bmatrix} \quad \alpha \neq 0$

4. $\mathrm{A} = \begin{bmatrix} 2 & 1 \\ 2 & 3 \end{bmatrix}$

5. For $0 < \theta < \pi/2$, $\mathrm{A} = \begin{bmatrix} \cos\theta & \sin\theta \\ \sin\theta & -\cos\theta \end{bmatrix}$

6. $\mathrm{A} = \begin{bmatrix} 4 & -3 & -2 \\ 2 & -1 & -2 \\ 3 & -3 & -1 \end{bmatrix}$

7. $\mathrm{A} = \begin{bmatrix} 1 & 1 & 1 \\ 0 & 2 & 1 \\ 0 & 0 & 3 \end{bmatrix}$

8. $\mathrm{A} = \begin{bmatrix} 1 & 1 & 1 \\ 1 & -1 & 1 \\ 0 & 0 & 0 \end{bmatrix}$

9. $A = \begin{bmatrix} 1 & 1 & -1 \\ 0 & 0 & 1 \\ 0 & -2 & -3 \end{bmatrix}$

10. $A = \begin{bmatrix} 2 & -2 & 2 & 1 \\ -1 & 3 & 0 & 3 \\ 0 & 0 & 4 & -2 \\ 0 & 0 & 2 & -1 \end{bmatrix}$

In each of the following problems, solve $\mathbf{x}' = A\mathbf{x}$, $\mathbf{x}(t_0) = \mathbf{x}_0$ for the A, t_0, and $\mathbf{x}_0$ as given.

11. A of Problem 1

$\mathbf{x}(0) = \begin{bmatrix} 1 \\ 0 \end{bmatrix}$

12. A of Problem 1

$\mathbf{x}(0) = \begin{bmatrix} 0 \\ 1 \end{bmatrix}$

13. A of Problem 2

$\mathbf{x}(-1) = \begin{bmatrix} 1 \\ 2 \end{bmatrix}$

14. A of Problem 3

$\mathbf{x}(1) = \begin{bmatrix} -2 \\ 0 \end{bmatrix}$

15. A of Problem 1

$\mathbf{x}(0) = \begin{bmatrix} 1 \\ 1 \end{bmatrix}$

16. A of Problem 4

$\mathbf{x}(0) = \begin{bmatrix} 1 \\ -1 \end{bmatrix}$

17. A of Problem 6

$\mathbf{x}(1) = \begin{bmatrix} 0 \\ 0 \\ 1 \end{bmatrix}$

18. A of Problem 6

$\mathbf{x}(1) = \begin{bmatrix} 0 \\ 0 \\ -1 \end{bmatrix}$

19. A of Problem 7

$\mathbf{x}(-1) = \begin{bmatrix} 1 \\ 1 \\ 1 \end{bmatrix}$

20. A of Problem 8

$\mathbf{x}(0) = \begin{bmatrix} 1 \\ 0 \\ 1 \end{bmatrix}$

21. A of Problem 9

$\mathbf{x}(0) = \begin{bmatrix} 0 \\ 1 \\ 0 \end{bmatrix}$

22. A of Problem 9

$\mathbf{x}(0) = \begin{bmatrix} 1 \\ 0 \\ 0 \end{bmatrix}$

23. A of Problem 9

$\mathbf{x}(0) = \begin{bmatrix} 0 \\ 0 \\ 1 \end{bmatrix}$

24. A of Problem 9

$\mathbf{x}(0) = \begin{bmatrix} \alpha \\ \rho \\ \gamma \end{bmatrix}$

Part 2

1. For $A = \begin{bmatrix} 0 & 1 & 0 \\ 0 & 0 & 1 \\ 2 & 1 & -2 \end{bmatrix}$, $\lambda_1 = -1$, $\lambda_2 = 1$, $\lambda_3 = -2$ and

$$\mathbf{u}_1 = \begin{bmatrix} 1 \\ -1 \\ 1 \end{bmatrix}, \quad \mathbf{u}_2 = \begin{bmatrix} 1 \\ 1 \\ 1 \end{bmatrix}, \quad \mathbf{u}_3 = \begin{bmatrix} 1 \\ -2 \\ 4 \end{bmatrix}$$

(a) Compute

$$A_1 = (A - \lambda_2 I)(A - \lambda_3 I)$$

and verify

$$A_1\mathbf{u}_1 = \mathbf{u}_1(\lambda_1 - \lambda_2)(\lambda_1 - \lambda_3) \neq \mathbf{0}$$
$$A_1\mathbf{u}_2 = \mathbf{0}$$
$$A_1\mathbf{u}_3 = \mathbf{0}$$

(b) Compute

$$A_2 = (A - \lambda_1 I)(A - \lambda_3 I)$$

and verify

$$A_2\mathbf{u}_1 = \mathbf{0}$$
$$A_2\mathbf{u}_2 = \mathbf{u}_2(\lambda_2 - \lambda_3)(\lambda_2 - \lambda_1) \neq \mathbf{0}$$
$$A_2\mathbf{u}_3 = \mathbf{0}$$

2. If a nonsingular matrix P exists such that $P^{-1}AP =$ diagonal matrix D, then $\mathbf{x}' = A\mathbf{x}$ can be simplified by the following two steps:
(a) $P^{-1}\mathbf{x}' = P^{-1}A\mathbf{x}$
(b) Let $P\mathbf{y} = \mathbf{x}$, then $P\mathbf{y}' = \mathbf{x}'$ and $P^{-1}P\mathbf{y}' = \mathbf{y}' = P^{-1}AP\mathbf{y} = D\mathbf{y}$. After $\mathbf{y}' = D\mathbf{y}$ is solved for $\mathbf{y}$, $P\mathbf{y} = \mathbf{x}$ gives $\mathbf{x}$. Use this method for problems 1–5, Part 1 1 of this set. Hint: Verify that P can be picked in these problems so that its columns are the eigenvectors of A.

3. As with $x' = ax$ we might expect $\mathbf{x} = e^{At}\mathbf{k}$ to be a solution of $x' = Ax$ if we could make reasonable sense out of e^{At}. Let

$$e^{At} \equiv I + At + A^2\frac{t^2}{2!} + \cdots + A^n\frac{t^n}{n!} + \cdots$$

Choose $A = \begin{bmatrix} 0 & 1 \\ 1 & 0 \end{bmatrix}$ and compute A^n.

Thus find a matrix equal to

$$\exp\left(\begin{bmatrix} 0 & 1 \\ 1 & 0 \end{bmatrix} t\right)$$

Is $\exp\left(\begin{bmatrix} 0 & 1 \\ 1 & 0 \end{bmatrix} t\right)\mathbf{k}$ a solution of $\mathbf{x}' = \begin{bmatrix} 0 & 1 \\ 1 & 0 \end{bmatrix}\mathbf{x}$?

2.5. SYMMETRIC MATRICES

In view of Theorem 2.4.3 it is natural to ask which matrices have linearly independent eigenvectors. A particularly important class enjoying this property are the symmetric matrices. Fortunately, symmetric matrices occur with great frequency in applications. This section is devoted to the study of symmetric matrices. The main theorem is stated but not proved until Chapter 3.

We recall from Chapter 1 that
(1) A is symmetric if $A^T = A$, where A^T is the transpose of A
(2) $(A^T)^T = A$
(3) $(AB)^T = B^T A^T$

Items (1) and (2) are immediate. Item (3) is easily proved by comparing the entry in the ith row jth column on both sides.

Let us continue our convention that A has only real entries. If $\mathbf{u} = [u_1, u_2, \ldots, u_n]^T$ is given with possibly complex entries, the vector $\bar{\mathbf{u}}$ is that vector whose entries are the complex conjugates of the entries of $\mathbf{u}$, i.e., $\bar{\mathbf{u}} = [\bar{u}_1, \bar{u}_2, \ldots, \bar{u}_n]^T$. Since for real numbers, $\bar{\alpha} = \alpha$, if $\mathbf{u}$ has only real entries, then $\bar{\mathbf{u}} = \mathbf{u}$.

2.5.1. LEMMA *For any vectors* $\mathbf{u}$ *and* $\mathbf{v}$ *in* $\mathscr{C}^n$ *and any real symmetric matrix* A

$$\mathbf{u}^T A \bar{\mathbf{v}} = \bar{\mathbf{v}}^T A \mathbf{u} \tag{2.5.1}$$

PROOF The product $A\bar{\mathbf{v}}$ is a column matrix. Since $\mathbf{u}^T$ is a row matrix $\mathbf{u}^T A \bar{\mathbf{v}}$ is a matrix with one entry. Therefore $(\mathbf{u}^T A \bar{\mathbf{v}})^T = \mathbf{u}^T A \bar{\mathbf{v}}$. But $(\mathbf{u}^T A \bar{\mathbf{v}})^T = \bar{\mathbf{v}}^T A^T (\mathbf{u}^T)^T = \bar{\mathbf{v}}^T A \mathbf{u}$, since $A^T = A$. Therefore, (2.5.1) follows.

2.5.2. LEMMA *If* λ *is an eigenvalue of* A *with corresponding eigenvector* $\mathbf{u}$ *then* $\bar{\mathbf{u}}$ *is an eigenvector of* A *corresponding to the eigenvalue* $\bar{\lambda}$.

PROOF By hypothesis, $A\mathbf{u} = \lambda\mathbf{u}$. Therefore the complex conjugates of both sides are equal, that is $\overline{A\mathbf{u}} = A\bar{\mathbf{u}} = \bar{\lambda}\bar{\mathbf{u}}$ (recall A has real entries so $\bar{A} = A$). But $A\bar{\mathbf{u}} = \bar{\lambda}\bar{\mathbf{u}}$ is precisely the asserted conclusion.

The next and last lemma is a result of some interest in its own right.

2.5.3. LEMMA *A symmetric matrix has only real eigenvalues.*

PROOF Note that Equation (2.5.1) is valid for any $\mathbf{v}$ and therefore for $\mathbf{v} = \mathbf{u}$. Assume A symmetric and $\mathbf{u}$ eigenvector of A and λ its corresponding eigenvalue. Then

$$\mathbf{u}^T A \bar{\mathbf{u}} = \mathbf{u}^T(\bar{\lambda}\bar{\mathbf{u}}) = \bar{\lambda}\mathbf{u}^T\bar{\mathbf{u}}$$

by Lemma 2.5.2. Also,

$$\bar{\mathbf{u}}^T A \mathbf{u} = \bar{\mathbf{u}}^T(\lambda\mathbf{u}) = \lambda\bar{\mathbf{u}}^T\mathbf{u}$$

by Lemma 2.5.1. Then

$$\lambda\bar{\mathbf{u}}^T\mathbf{u} = \bar{\lambda}\mathbf{u}^T\bar{\mathbf{u}}$$

Note that $\bar{\mathbf{u}}^T\mathbf{u}$ and $\mathbf{u}^T\bar{\mathbf{u}}$ are matrices with but one entry, namely, $|u_1|^2 + |u_2|^2 + \cdots + |u_n|^2$. Since the eigenvector $\mathbf{u} \neq \mathbf{0}$, this entry is positive and hence $\lambda\bar{\mathbf{u}}^T\mathbf{u} = \bar{\lambda}\mathbf{u}^T\bar{\mathbf{u}}$ implies $\bar{\lambda} = \lambda$ which means λ is real, as asserted.

The next theorem is the highlight of this section. It is a deep result whose proof must be deferred until Section 3.6.

2.5.4. THEOREM *If* A *is a real* $n \times n$ *symmetric matrix then there exists* n *linearly independent eigenvectors each belonging to* $\mathscr{R}^n$ *(and hence spanning* $\mathscr{R}^n$*).*

As a consequence of Theorem 2.5.4 we have the following theorem.

2.5.5. THEOREM *If* A *is symmetric then the initial-value problem*

$$\mathbf{x}' = A\mathbf{x} \qquad \mathbf{x}(t_0) = \mathbf{x}_0$$

has a solution for every $\mathbf{x}_0 \in \mathscr{R}^n$

Exercise 2.5

1. For matrix A find three linearly independent eigenvectors.

$$\text{(a) } A = \begin{bmatrix} 2 & 0 & 0 \\ 0 & 1 & 0 \\ 0 & 0 & 1 \end{bmatrix} \qquad \text{(b) } A = \begin{bmatrix} 1 & 0 & 0 \\ 0 & 3 & \sqrt{2} \\ 0 & \sqrt{2} & 2 \end{bmatrix} \qquad \text{(c) } A = \begin{bmatrix} 1 & 1 & 1 \\ 1 & 1 & 1 \\ 1 & 1 & 1 \end{bmatrix}$$

2. Write out the details which prove $\overline{A\mathbf{u}} = A\bar{\mathbf{u}}$ if A is real.
3. Solve $\mathbf{x}' = A\mathbf{x}$, $\mathbf{x}(0) = \mathbf{x}_0$ for the A of Problem 1(c) above, when

$$\text{(a) } \mathbf{x}_0 = \begin{bmatrix} 1 \\ 1 \\ 1 \end{bmatrix} \qquad \text{(b) } \mathbf{x}_0 = \begin{bmatrix} 1 \\ 0 \\ 0 \end{bmatrix} \qquad \text{(c) } \mathbf{x}_0 = \begin{bmatrix} 0 \\ 1 \\ 0 \end{bmatrix} \qquad \text{(d) } \mathbf{x}_0 = \begin{bmatrix} 0 \\ 0 \\ 1 \end{bmatrix}$$

4. If A and B are symmetric and AB = BA prove that AB is symmetric.
5.* Recall, A is antisymmetric if $A^T = -A$. Show $\mathbf{u}^T A \bar{\mathbf{v}} = -\bar{\mathbf{v}}^T A \mathbf{u}$ in this case.
6.* Show that the eigenvalues of a real, antisymmetric matrix are pure imaginary (0 may be considered a pure imaginary number).
7.* If A is a real, $(2n+1)\times(2n+1)$, antisymmetric matrix, then $\lambda = 0$ is an eigenvalue.

2.6. DILUTE SOLUTIONS: A SYMMETRIC EXAMPLE

We consider three cells, c_1, c_2, and c_3, each of which contains the solute S dissolved in a unit volume of solvent. These cells are separated from each other by semipermeable membranes through which the solute may flow but not the solvent. Figure 2.6.1 illustrates the configuration under study.

Figure 2.6.1

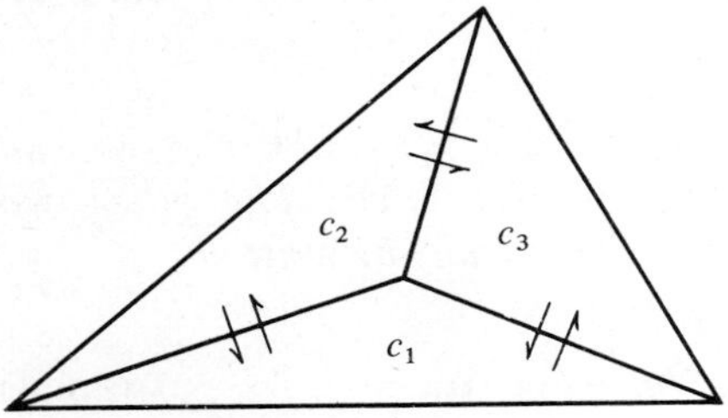

Let us denote by $x_1(t)$, $x_2(t)$, and $x_3(t)$, the amount of solute dissolved in c_1, c_2, c_3, respectively, at time t. Since each cell contains a unit volume of solvent, $x_1(t)$, $x_2(t)$, and $x_3(t)$ also represent the concentrations of solute in the respective cells.

It is a plausible physical assumption that for weak (dilute) solutions the rate of change of solute due to the passage of solute from c_i to c_j is proportional to the difference in concentrations in these cells. Furthermore, the amount of solute entering c_j through the membrane separating c_i and c_j is the negative of the amount entering c_i from c_j during any time interval. We also assume that the net change in concentration (per unit time) in c_i is the sum of the changes in concentrations due to passage of solute through the two membranes bounding c_i. This leads to the following mathematical formulation of this diffusion problem:

$$\begin{aligned} x_1'(t) &= k_{12}\{x_2(t)-x_1(t)\}+k_{13}\{x_3(t)-x_1(t)\} \\ x_2'(t) &= k_{12}\{x_1(t)-x_2(t)\}+k_{23}\{x_3(t)-x_2(t)\} \\ x_3'(t) &= k_{13}\{x_1(t)-x_3(t)\}+k_{23}\{x_2(t)-x_3(t)\} \end{aligned} \tag{2.6.1}$$

where the constants k_{12}, k_{13}, k_{23} are positive constants dependent upon the physical properties of the membranes. We may collect terms and write in vector–matrix form,

$$\mathbf{x}' = \mathrm{K}\mathbf{x} \tag{2.6.2}$$

where

$$\mathrm{K} = \begin{bmatrix} -(k_{12}+k_{13}) & k_{12} & k_{13} \\ k_{12} & -(k_{12}+k_{23}) & k_{23} \\ k_{13} & k_{23} & -(k_{13}+k_{23}) \end{bmatrix} \tag{2.6.3}$$

If (2.6.2) is a good model for this diffusion problem, it should be possible to derive mathematically various conclusions which are experimentally verifiable. A little study of (2.6.2) and (2.6.3) leads us to Theorem 2.6.1.

2.6.1. THEOREM *Equation (2.6.1) has only real eigenvalues and three linearly independent, real eigenvectors.*

PROOF Note that K is symmetric and hence the theorem is just a restatement of Theorem 2.5.4.

2.6.2. THEOREM *There is an eigenvalue, say λ_1, which is zero and corresponding to $\lambda_1 = 0$ is an eigenvector $\mathbf{u}_1 = [1, 1, 1]^{\mathrm{T}}$. The other two eigenvalues are negative and the solution of (2.6.1) may be written*

$$\mathbf{x}(t) = a_1 \begin{bmatrix} 1 \\ 1 \\ 1 \end{bmatrix} + a_2 e^{\lambda_2 t}\mathbf{u}_2 + a_3 e^{\lambda_3 t}\mathbf{u}_3 \tag{2.6.4}$$

PROOF There are three assertions in this theorem. The first is that $\lambda = 0$ is an eigenvalue with $[1, 1, 1]^{\mathrm{T}}$ as a corresponding eigenvector. The second is that $\lambda = 0$ has no other linearly independent eigenvector and the last is that the remaining eigenvalues are negative. To prove these facts observe that taking m_{ij} to be the element in the ith row and jth column of K, we find

$$\text{(a)}\ m_{ij} > 0 \qquad \text{if} \quad i \neq j$$

$$\text{(b)}\ \sum_{j=1}^{3} m_{ij} = 0 \qquad \text{for} \quad i = 1, 2, 3$$

$$\text{(c)}\ \sum_{j=1}^{3} m_{ij}u_j = \lambda u_i \qquad \text{for} \quad i = 1, 2, 3$$

Item (c) is a statement of the fact that $\mathrm{K}\mathbf{u} = \lambda \mathbf{u}$ if λ is an eigenvalue and $\mathbf{u}$ a corresponding eigenvector.

But (c) is satisfied by $\mathbf{u} = [1, 1, 1]^{\mathrm{T}}$, $\lambda = 0$, since (c) reduces to (b) in this case. This proves the first assertion.

From (b) and (c) we deduce

$$\begin{aligned}\text{(d)}\ \lambda u_i &= \sum_{j=1}^{3} m_{ij}u_j - u_i\Big(\sum_{j=1}^{3} m_{ij}\Big)\\ &= \sum_{j=1}^{3} m_{ij}(u_j - u_i) \qquad i = 1, 2, 3\end{aligned}$$

If $\lambda = 0$ then (d) reduces to

$$0 = \sum_{j=1}^{3} m_{ij}(u_j - u_i)$$

If $u_1 = u_2 = u_3$ is false, we contradict this latter result by choosing u_i the smallest entry in $\mathbf{u}$. Since $m_{ij} > 0, i \neq j$, and $u_j - u_i \geqslant 0$

$$0 < \sum_{j=1}^{3} m_{ij}(u_j - u_i)$$

Hence, the only eigenvectors corresponding to $\lambda = 0$ are of the form $\mathbf{u} = [c, c, c]^{\mathrm{T}}$.

The last assertion follows also from (d). Suppose $\lambda > 0$ is an eigenvalue and $\mathbf{v}$ is a corresponding eigenvector. If $\mathbf{v}$ has no negative entries set, $\mathbf{u} = -\mathbf{v}$; if $\mathbf{v}$ has even one negative entry set, $\mathbf{u} = \mathbf{v}$. In either case $\mathbf{u}$ is an eigenvector corresponding to $\lambda > 0$ and there is a smallest entry in $\mathbf{u}$ *which is negative*. Call this entry u_i. From the fact that $u_j - u_i \geqslant 0$, $\lambda > 0$, and $u_i < 0$, we deduce from (d) that

$$0 > \lambda u_i = \sum_{j=1}^{3} m_{ij}(u_j - u_i) \geqslant 0$$

a contradiction. Hence, all the nonzero eigenvalues are negative.

Now for the experimental implications.

(1) If the solute is distributed equally in each of the cells, there is no observable net change in the distribution. It remains in an equilibrium state. Mathematically, suppose $\mathbf{x}(0) = a_1[1, 1, 1]^{\mathrm{T}}$. Then, from (2.6.4), $\mathbf{x}(t) = a_1 [1, 1, 1]^{\mathrm{T}}$ and theory and experiment agree. (We are assuming that no

solutions of $\mathbf{x}' = \mathrm{K}\mathbf{x}$, $\mathbf{x}(0) = a_1[1, 1, 1]^{\mathrm{T}}$ exist other than the one given. We prove this in Section 3.5.)

(2) Experiments show that any distribution of solute, other than equidistribution, does not remain in that state. To derive this from our model, assume $\mathbf{x}(0) \neq a_1[1, 1, 1]^{\mathrm{T}}$. Then, from (2.6.4), one of the constants a_2 or a_3 or both are not zero and hence $\mathbf{x}(t)$ varies with t.

(3) Experimentally, and rather obviously, the total amount of solute does not change with time—there is conservation of mass. In our model

$$\begin{aligned} x_1'(t) + x_2'(t) + x_3'(t) &= [-(k_{12}+k_{13}) + k_{12} + k_{13}]x_1(t) \\ &\quad + [k_{12} - (k_{12}+k_{23}) + k_{23}]\, x_2(t) \\ &\quad + [k_{13} + k_{23} - (k_{13}+k_{23})]\, x_3(t) \\ &= 0 \end{aligned}$$

The mean-value theorem of the calculus then yields $x_1(t) + x_2(t) + x_3(t) =$ total mass of solute = a constant.

(4) It is observed that the distribution of solute tends to the equilibrium state (an equidistribution) as t tends to infinity. Since $\lambda_2 < 0$ and $\lambda_3 < 0$, we see that (2.6.4) implies this conclusion mathematically. Moreover, if λ_2 and λ_3 are known, we can actually estimate the time needed for $\mathbf{x}(t)$ to be "essentially" constant.

Exercise 2.6

1.* Let K be an $n \times n$ symmetric matrix with positive "off-diagonal" entries. Suppose the sum of the entries in each row is zero. Does Theorem 2.6.2 generalize? If so state and prove the generalization; if not, give counter-examples.

2.* Which, if any, of the four concluding statements of this section hold for the K given in Problem 1, above?

2.7. COMPLEX EIGENVALUES AND EIGENVECTORS

Even though the elements of A are real it often happens that the eigenvalues of A are complex. For example, the eigenvalues of

$$\mathrm{A} = \begin{bmatrix} 0 & 1 \\ -1 & 0 \end{bmatrix}$$

are $\lambda = \pm i$.

We would like to extend the theory of the preceding sections to this case as well. To this end we shall define $e^{\lambda t}$ for λ complex. Following Euler we set, for $\lambda = a + ib$,

$$e^{(a+ib)t} = e^{at}(\cos bt + i \sin bt) \tag{2.7.1}$$

For each real value of t, (2.7.1) defines a complex number as the value of $e^{(a+ib)t}$.

The reader is asked to verify in Exercise 2.7, Part 2, Problem 1 that

$$\frac{d}{dt}e^{(a+ib)t} = (a+ib)e^{(a+ib)t}$$

Furthermore,†

$$\begin{aligned}|e^{(a+ib)t}|^2 &= |e^{at}|^2(\cos^2 bt + \sin^2 bt)\\ &= |e^{at}|^2 > 0\end{aligned}$$

for any value of t.

We apply this extension of $e^{\lambda t}$ to the differential equation $\mathbf{x}' = A\mathbf{x}$: suppose λ is a complex eigenvalue of A and $\mathbf{u}$ is a corresponding eigenvector. Setting $\mathbf{x}(t) = e^{\lambda t}\mathbf{u}$, we see that

$$\mathbf{x}'(t) = \lambda e^{\lambda t}\mathbf{u} = \lambda\mathbf{x}(t)$$

and

$$A\mathbf{x}(t) = e^{\lambda t}A\mathbf{u} = \lambda e^{\lambda t}\mathbf{u} = \lambda\mathbf{x}(t)$$

so that $\mathbf{x}(t)$ is a solution of $\mathbf{x}' = A\mathbf{x}$ as in the real case.

Let us consider $e^{\lambda t}\mathbf{u}$ in detail. If $\lambda = a+ib$ and $\mathbf{u} = \mathbf{v}+i\mathbf{w}$ where a and b are real numbers and $\mathbf{v}$ and $\mathbf{w}$ belong to $\mathcal{R}^n$ then

$$\begin{aligned}e^{\lambda t}\mathbf{u} &= e^{at}(\cos bt + i\sin bt)(\mathbf{v}+i\mathbf{w})\\ &= e^{at}\{\cos bt\mathbf{v} - \sin bt\mathbf{w}\} + ie^{at}\{\sin bt\mathbf{v} + \cos bt\mathbf{w}\}\end{aligned}$$

The real part of $e^{\lambda t}\mathbf{u}$ is written Re $\{e^{\lambda t}\mathbf{u}\}$ and is defined by

$$\text{Re}\,\{e^{\lambda t}\mathbf{u}\} = e^{at}\{\cos bt\mathbf{v} - \sin bt\mathbf{w}\} \tag{2.7.2}$$

and the imaginary part of $e^{\lambda t}\mathbf{u}$ is written Im $(e^{\lambda t}\mathbf{u})$ and is defined

$$\text{Im}\,\{e^{\lambda t}\mathbf{u}\} = e^{at}\{\sin bt\mathbf{v} + \cos bt\mathbf{w}\} \tag{2.7.3}$$

We assert the following theorem.

2.7.1. THEOREM *If* $\mathbf{x}(t)$ *is a solution of* $\mathbf{x}' = A\mathbf{x}$, A real, *then* Re $[\mathbf{x}(t)]$ *and* Im $[\mathbf{x}(t)]$ *are solutions.*

PROOF

$$\begin{aligned}\frac{d}{dt}\{\text{Re}\,\mathbf{x}(t)\} &= \text{Re}\left\{\frac{d}{dt}\mathbf{x}(t)\right\}\\ &= \text{Re}\,(A\mathbf{x}(t))\end{aligned}$$

since $\mathbf{x}(t)$ is a solution $\mathbf{x}' = A\mathbf{x}$. But

$$\text{Re}[A\,\mathbf{x}(t)] = A\{\text{Re}\,\mathbf{x}(t)\}$$

since A has real elements. Therefore,

$$\begin{aligned}\frac{d}{dt}\{\text{Re}\,\mathbf{x}(t)\} &= \text{Re}\{A(\text{Re}\,\mathbf{x}(t) + i\,\text{Im}\,\mathbf{x}(t)\}\\ &= A\{\text{Re}\,\mathbf{x}(t)\}\end{aligned}$$

A similar argument is valid for Im $\{\mathbf{x}(t)\}$.

†Recall that $|u+iv|^2 = u^2+v^2$.

Because of Equations (2.7.2) and (2.7.3) we can be more explicit about the nature of Re $\{\mathbf{x}(t)\}$ and Im $\{\mathbf{x}(t)\}$.

2.7.2. COROLLARY *If $\lambda = a + ib$ is an eigenvalue of A and $\mathbf{u} = \mathbf{v} + i\mathbf{w}$ is a corresponding eigenvector then*

$$\mathbf{x}_1(t) = e^{at}\{\cos bt\,\mathbf{v} - \sin bt\,\mathbf{w}\} \tag{2.7.4}$$

and

$$\mathbf{x}_2(t) = e^{at}\{\sin bt\,\mathbf{v} + \cos bt\,\mathbf{w}\} \tag{2.7.5}$$

are two solutions of $\mathbf{x}' = \mathrm{A}\mathbf{x}$.

We illustrate this with an example.

2.7.1. Example

Solve $\mathbf{x}' = \begin{bmatrix} 0 & 1 \\ -1 & 0 \end{bmatrix}\mathbf{x}$

Solution

Here $\lambda = i, -i$. We find, using $\lambda = i$,

$$\begin{bmatrix} -i & 1 \\ -1 & -i \end{bmatrix}\begin{bmatrix} u_1 \\ u_2 \end{bmatrix} = \begin{bmatrix} 0 \\ 0 \end{bmatrix} \quad \text{leads to} \quad \begin{bmatrix} u_1 \\ u_2 \end{bmatrix} = \begin{bmatrix} 1 \\ i \end{bmatrix}$$

Hence

$$\mathbf{x}_1(t) = \mathrm{Re}\left\{e^{it}\begin{bmatrix} 1 \\ i \end{bmatrix}\right\} \qquad \mathbf{x}_2(t) = \mathrm{Im}\left\{e^{it}\begin{bmatrix} 1 \\ i \end{bmatrix}\right\}$$

But

$$\begin{aligned} e^{it}\begin{bmatrix} 1 \\ i \end{bmatrix} &= (\cos t + i\sin t)\left\{\begin{bmatrix} 1 \\ 0 \end{bmatrix} + i\begin{bmatrix} 0 \\ 1 \end{bmatrix}\right\} \\ &= \left\{\cos t\begin{bmatrix} 1 \\ 0 \end{bmatrix} - \sin t\begin{bmatrix} 0 \\ 1 \end{bmatrix}\right\} + i\left\{\cos t\begin{bmatrix} 0 \\ 1 \end{bmatrix} + \sin t\begin{bmatrix} 1 \\ 0 \end{bmatrix}\right\} \\ &= \begin{bmatrix} \cos t \\ -\sin t \end{bmatrix} + i\begin{bmatrix} \sin t \\ \cos t \end{bmatrix} \end{aligned}$$

Hence

$$\mathbf{x}_1(t) = \begin{bmatrix} \cos t \\ -\sin t \end{bmatrix} \qquad \mathbf{x}_2(t) = \begin{bmatrix} \sin t \\ \cos t \end{bmatrix}$$

as is asserted in (2.7.4) and (2.7.5)

We can check these functions directly:

$$\mathbf{x}_1'(t) = \begin{bmatrix} -\sin t \\ -\cos t \end{bmatrix} \qquad \mathrm{A}\mathbf{x}_1(t) = \begin{bmatrix} -\sin t \\ -\cos t \end{bmatrix}$$

$$\mathbf{x}_2'(t) = \begin{bmatrix} \cos t \\ -\sin t \end{bmatrix} \qquad \mathrm{A}\mathbf{x}_2(t) = \begin{bmatrix} \cos t \\ -\sin t \end{bmatrix}$$

Exercise 2.7

Part 1

1. In Example 2.7.1, use $\lambda = -i$ and thus find two solutions of

$$\mathbf{x}' = \begin{bmatrix} 0 & 1 \\ -1 & 0 \end{bmatrix} \mathbf{x}$$

How are these solutions related to those computed in the text?

2. Solve

$$\mathbf{x}' = \begin{bmatrix} 0 & 1 \\ -1 & 0 \end{bmatrix} \mathbf{x} \qquad \mathbf{x}(0) = \begin{bmatrix} 1 \\ 1 \end{bmatrix}$$

3. Solve $\mathbf{x}' = \mathrm{A}\mathbf{x}$, where

$$\mathrm{A} = \begin{bmatrix} 1 & 0 & 0 \\ 0 & 0 & 2 \\ 0 & -2 & 0 \end{bmatrix}$$

4. Solve $\mathbf{x}' = \mathrm{A}\mathbf{x}$, where

$$\mathrm{A} = \begin{bmatrix} 0 & 1 & 0 & 0 \\ 0 & 0 & 1 & 0 \\ 0 & 0 & 0 & 1 \\ -1 & 0 & 0 & 0 \end{bmatrix}$$

5. Solve $\mathbf{x}' = \mathrm{A}\mathbf{x}$, for

$$\mathrm{A} = \begin{bmatrix} 0 & 1 & 1 \\ -1 & 0 & 1 \\ -1 & -1 & 0 \end{bmatrix}$$

Part 2

1. Verify, using (2.7.1) that $(d/dt)\exp^{[(a+ib)t]} = (a+ib)\exp^{[(a+ib)t]}$.
2.* Verify by direction substitution of (2.7.4) and (2.7.5) into $\mathbf{x}' = \mathrm{A}\mathbf{x}$ that Corollary 2.7.2 is valid.

2.8. A SUMMARY

We are given an $n \times n$ matrix with real entries and a constant vector $\mathbf{x}_0$. We are asked to find vector functions $\mathbf{x}(t)$ which are solutions of

$$\mathbf{x}' = \mathrm{A}\mathbf{x} \qquad \mathbf{x}(t_0) = \mathbf{x}_0 \tag{2.8.1}$$

The function $\mathbf{x}(t) = \mathbf{u}e^{\lambda t}$ is a solution of $\mathbf{x}' = \mathrm{A}\mathbf{x}$ if and only if λ is an eigenvalue of A and $\mathbf{u}$ is one of its corresponding eigenvectors. Three cases occur according to the nature of the roots of the characteristic equation, $\det(\mathrm{A} - \lambda\mathrm{I}) = 0$. Namely,

(1) There are n distinct eigenvalues of A. In this case we find n solutions, one for each root; that linear combination of the eigenvectors which equals $\mathbf{x}_0$ is precisely the linear combination of solutions which form the solution of (2.8.1).

Specifically, if

$$\mathbf{x}_0 = c_1\mathbf{u}_1 + c_2\mathbf{u}_2 + \cdots + c_n\mathbf{u}_n \tag{2.8.2}$$

then

$$\mathbf{x}(t) = c_1\mathbf{u}_1 e^{\lambda_1 t} + c_2\mathbf{u}_2 e^{\lambda_2 t} + \cdots + c_n\mathbf{u}_n e^{\lambda_n t} \tag{2.8.3}$$

is the required solution of $\mathbf{x}' = \mathrm{A}\mathbf{x}$, $\mathbf{x}_0 = \mathbf{x}(0)$.

(2) There may be fewer than n distinct eigenvalues but we can find n linearly independent eigenvectors. Since (2.8.2) still holds, $\mathbf{x}(t)$ defined in (2.8.3) is the required solution with the understanding that some (or even all) of the eigenvalues may be repeated. This case subsumes (1) and either (1) or (2) occurs when A is symmetric.

(3) There are fewer than n linearly independent eigenvectors (and therefore by necessity fewer than n distinct eigenvalues).

If, however, $\mathbf{x}_0$ is in the span of the eigenvectors then a solution of (2.8.1) can be constructed. For instance, if $\mathbf{x}(0) = \mathbf{x}_0$,

$$\mathbf{x}_0 = c_1\mathbf{u}_1 + c_2\mathbf{u}_2 + \cdots c_k\mathbf{u}_k \qquad (k \leqslant n) \tag{2.8.4}$$

then

$$\mathbf{x}(t) = c_1\mathbf{u}_1 e^{\lambda_1 t} + c_2\mathbf{u}_2 e^{\lambda_2 t} + \cdots + c_k\mathbf{u}_k e^{\lambda_k t} \tag{2.8.5}$$

is a solution. If an expression like (2.8.4) is impossible, which is to say, if $\mathbf{x}_0, \mathbf{u}_1, \ldots, \mathbf{u}_k$ are linearly independent, then no solution of the form (2.8.5) is possible.

The determination of the constants in (2.8.2) or (2.8.4) is an algebraic problem. Specifically, we solve the system

$$\mathrm{U}\mathbf{c} = \mathbf{x}_0 \qquad k \leqslant n \tag{2.8.6}$$

where U is the matrix whose columns are $\mathbf{u}_1, \mathbf{u}_2, \ldots,$ respectively, and $\mathbf{c}$ is the vector whose entries are to be found.

In the event that $\lambda = a + ib$ is complex, we know that $\bar{\lambda} = a - ib$ is a second eigenvalue. Furthermore, $\mathrm{Re}\,\{\mathbf{u}e^{\lambda t}\}$ and $\mathrm{Im}\,\{\mathbf{u}e^{\lambda t}\}$ are real solutions which may replace $\mathbf{u}e^{\lambda t}$ and $\bar{\mathbf{u}}e^{\bar{\lambda} t}$ in (2.8.3) or (2.8.5). Indeed,

$$\begin{aligned} \mathrm{Re}\,\{ue^{\lambda t}\} &= e^{at}(\cos bt\,\mathbf{v} - \sin bt\,\mathbf{w}) \\ \mathrm{Im}\,\{ue^{\lambda t}\} &= e^{at}(\sin bt\,\mathbf{v} + \cos bt\,\mathbf{w}) \end{aligned}$$

where $\mathbf{u} = \mathbf{v} + i\mathbf{w}$ and $\lambda = a + ib$.

Exercise 2.8

In each of the next four problems you are given A and some $\mathbf{x}_0$'s for the initial-value problem $\mathbf{x}' = \mathrm{A}\mathbf{x}$, $\mathbf{x}(0) = \mathbf{x}_0$. Find a solution where possible.

1. $A = \begin{bmatrix} 1 & 1 \\ 0 & 1 \end{bmatrix}$ (a) $\mathbf{x}_0 = \begin{bmatrix} 1 \\ -1 \end{bmatrix}$ (b) $\mathbf{x}_0 = \begin{bmatrix} 1 \\ 0 \end{bmatrix}$

2. $A = \begin{bmatrix} 1 & 0 \\ 1 & 1 \end{bmatrix}$ (a) $\mathbf{x}_0 = \begin{bmatrix} 2 \\ 3 \end{bmatrix}$ (b) $\mathbf{x}_0 = \begin{bmatrix} 0 \\ 1 \end{bmatrix}$

3. $A = \begin{bmatrix} 1 & 1 & 0 \\ 0 & 1 & 0 \\ 0 & 0 & 2 \end{bmatrix}$ (a) $\mathbf{x}_0 = \begin{bmatrix} 2 \\ 0 \\ -1 \end{bmatrix}$ (b) $\mathbf{x}_0 = \begin{bmatrix} 1 \\ 1 \\ 1 \end{bmatrix}$

4. $A = \begin{bmatrix} 1 & 1 \\ -1 & -1 \end{bmatrix}$ (a) $\mathbf{x}_0 = \begin{bmatrix} 1 \\ 1 \end{bmatrix}$ (b) $\mathbf{x}_0 = \begin{bmatrix} 0 \\ 0 \end{bmatrix}$ (c) $\mathbf{x}_0 = \begin{bmatrix} 1 \\ 0 \end{bmatrix}$

Chapter 3
LINEAR SYSTEMS AND ROOT VECTORS

3.1. NEW SOLUTIONS OF $\mathbf{x}' = \mathrm{A}\mathbf{x}$

In the last chapter we explored solutions of $\mathbf{x}' = \mathrm{A}\mathbf{x}$ which were combinations of functions of the form $\mathbf{u}e^{at}$. We learned that if the span of the eigenvectors of A did not include the vector $\mathbf{x}_0$, then we could not solve $\mathbf{x}' = \mathrm{A}\mathbf{x}$, $\mathbf{x}(t_0) = \mathbf{x}_0$ with the functions at our disposal. We attempt to remedy this situation by searching for new solutions of $\mathbf{x}' = \mathrm{A}\mathbf{x}$.

We discover such solutions by studying simple systems whose solutions can be obtained by inspection. From the knowledge thus gained we guess at the form of the solutions of more complicated systems. We shall see that this leads us to an interesting generalization of the eigenvector and finally to the complete solution of the initial-value problem for both $\mathbf{x}' = \mathrm{A}\mathbf{x}$ and the more general $\mathbf{x}' = \mathrm{A}\mathbf{x} + \mathbf{f}(t)$.

The system

$$\mathbf{x}' = \begin{bmatrix} 0 & 1 \\ 0 & 0 \end{bmatrix} \mathbf{x}$$

provides our first example amenable to a direct approach. We find $\lambda = 0$ is a twofold root of the characteristic equation and $\mathbf{u} = [1, 0]^{\mathrm{T}}$ is a corresponding eigenvector. Hence one solution is $\mathbf{x}(t) = [1, 0]^{\mathrm{T}}$. To find a second solution set $\hat{\mathbf{x}} = [x_1, x_2]^{\mathrm{T}}$, then in expanded form this system is

$$x_1' = x_2 \qquad \text{and} \qquad x_2' = 0$$

A solution, $\hat{\mathbf{x}}$, distinct from $\mathbf{x}$ can be obtained by choosing $x_2 = 1$ and $x_1 = t$. Then

$$\hat{\mathbf{x}}(t) = \begin{bmatrix} t \\ 1 \end{bmatrix} = t \begin{bmatrix} 1 \\ 0 \end{bmatrix} + \begin{bmatrix} 0 \\ 1 \end{bmatrix}$$

As a second illustration, we study

$$\mathbf{x}' = \begin{bmatrix} 1 & 1 \\ 0 & 1 \end{bmatrix} \mathbf{x}$$

in which $\lambda = 1$ is a twofold eigenvalue and $\mathbf{u} = [1, 0]$ is a corresponding eigenvector. Hence one solution is

$$\mathbf{x}(t) = \begin{bmatrix} 1 \\ 0 \end{bmatrix} e^t$$

This same system can be described by setting $\hat{\mathbf{x}} = [x_1, x_2]$ and obtaining

$$x_1' = x_1 + x_2 \qquad \text{and} \qquad x_2' = x_2$$

Then $x_2(t) = e^t$ and $x_1'(t) = x_1(t) + e^t$. The latter equation is first-order but not homogeneous. We can find a solution by multiplying this equation through by e^{-t} obtaining

$$\frac{d}{dt}[e^{-t}x_1(t)] = e^{-t}x_1'(t) - e^{-t}x_1(t) = 1$$

which is satisfied by $e^{-t}x_1(t) = t$. In vector form the solution is

$$\mathbf{x}_2(t) = \begin{bmatrix} te^t \\ e^t \end{bmatrix} = \begin{bmatrix} t \\ 1 \end{bmatrix} e^t = \left\{ t\begin{bmatrix} 1 \\ 0 \end{bmatrix} + \begin{bmatrix} 0 \\ 1 \end{bmatrix} \right\} e^t$$

That this is indeed a solution can easily be checked by substitution.

A number of examples of this kind suggest that functions of the form $(\mathbf{a}t + \mathbf{b})e^{\lambda t}$, $(\mathbf{a}t^2 + \mathbf{b}t + \mathbf{c})e^{\lambda t}$, etc. may be solutions when there is a deficiency in the number of linearly independent eigenvectors.

3.1.1. Example

Find solutions of

$$\mathbf{x}' = \begin{bmatrix} 0 & 1 & 0 \\ 0 & 0 & 1 \\ 0 & 0 & 0 \end{bmatrix} \mathbf{x} \tag{3.1.1}$$

Solution

Set $\mathbf{x}_1(t) = \mathbf{u}e^{\lambda t}$. Then $\lambda = 0$ and we find

$$\mathbf{u}_1 = \begin{bmatrix} 1 \\ 0 \\ 0 \end{bmatrix} \qquad \mathbf{x}_1(t) = \begin{bmatrix} 1 \\ 0 \\ 0 \end{bmatrix} e^{0t} = \begin{bmatrix} 1 \\ 0 \\ 0 \end{bmatrix}$$

This exhausts the possibilities of solutions of the form $\mathbf{u}_1 e^{\lambda t}$. We try $\mathbf{x}_2(t) = (\mathbf{a}t + \mathbf{b})e^{\lambda t}$ and find by substitution

$$\begin{aligned} \mathbf{x}_2'(t) &= \mathbf{a}e^{\lambda t} + \lambda(\mathbf{a}t + \mathbf{b})e^{\lambda t} \\ &= e^{\lambda t}\{\lambda \mathbf{a}t + \lambda \mathbf{b} + \mathbf{a}\} \end{aligned}$$

while

$$\mathbf{A}\mathbf{x}_2(t) = e^{\lambda t}\{t\mathbf{A}\mathbf{a} + \mathbf{A}\mathbf{b}\}$$

Now for $\mathbf{x}_2(t)$ to be a solution it is necessary and sufficient that $\mathbf{x}_2'(t) \neq \mathbf{A}_2\mathbf{x}(t)$. Therefore,

$$e^{\lambda t}\{\lambda \mathbf{a}t + \lambda \mathbf{b} + \mathbf{a}\} \equiv e^{\lambda t}\{t\mathbf{A}\mathbf{a} + \mathbf{A}\mathbf{b}\}$$

Since $e^{\lambda t} \neq 0$,

$$t(\lambda \mathbf{a}) + (\lambda \mathbf{b} + \mathbf{a}) \equiv t(\mathbf{A}\mathbf{a}) + (\mathbf{A}\mathbf{b})$$

But two polynomials are identical if and only if their coefficients of equal powers agree. Applying this to each component in the latter equation, we infer

$$\mathbf{A}\mathbf{a} = \lambda \mathbf{a} \qquad \text{and} \qquad \mathbf{A}\mathbf{b} = \lambda \mathbf{b} + \mathbf{a} \tag{3.1.2}$$

The first relationship in (3.1.2) shows that $\lambda = 0$ and $\mathbf{a} = \mathbf{u}_1 = [l, 0, 0]^T$. This is now used in the second relationship given in (3.1.2). That is,

$$\begin{bmatrix} 0 & 1 & 0 \\ 0 & 0 & 1 \\ 0 & 0 & 0 \end{bmatrix} \mathbf{b} = \begin{bmatrix} l \\ 0 \\ 0 \end{bmatrix} \tag{3.1.3}$$

so that $\mathbf{b} = [k, l, 0]^T$. The constants k and l are arbitrary so we set $k = 0$, $l = 1$, giving $\mathbf{a} = [1, 0, 0]$ and $\mathbf{b} = [0, 1, 0]^T$. (Retaining the arbitrary constants gains us nothing, as it turns out. See, for instance, Exercise 3.1, Problem 12.) Thus

$$\mathbf{x}_2(t) = \begin{bmatrix} 1 \\ 0 \\ 0 \end{bmatrix} t + \begin{bmatrix} 0 \\ 1 \\ 0 \end{bmatrix}$$

We next set $\mathbf{x}_3(t) = (\mathbf{d}t^2 + \mathbf{e}t + \mathbf{f})e^{\lambda t}$ and learn by analysis similar to that carried out above,

$$\mathbf{A}\mathbf{d} = \lambda \mathbf{d} \qquad \mathbf{A}\mathbf{e} = \lambda \mathbf{e} + 2\mathbf{d} \qquad \mathbf{A}\mathbf{f} = \lambda \mathbf{f} + \mathbf{e}$$

Once again $\mathbf{d} = \mathbf{u} = [1, 0, 0]^T$ and $\lambda = 0$. From the second equation $\mathbf{e} = [0, 2, 0]^T$. The last equation is

$$\begin{bmatrix} 0 & 1 & 0 \\ 0 & 0 & 1 \\ 0 & 0 & 0 \end{bmatrix} \mathbf{f} = \begin{bmatrix} 0 \\ 2 \\ 0 \end{bmatrix}$$

which yields $\mathbf{f} = [0, 0, 2]^T$. Again we have chosen specific values for various constants. Having determined the vectors **d**, **e**, and **f** we have

$$\mathbf{x}_3(t) = \begin{bmatrix} 1 \\ 0 \\ 0 \end{bmatrix} t^2 + \begin{bmatrix} 0 \\ 2 \\ 0 \end{bmatrix} t + \begin{bmatrix} 0 \\ 0 \\ 2 \end{bmatrix}$$

as a third solution. By superposition

$$\begin{aligned}\mathbf{x}(t) &= k_1\mathbf{x}_1(t) + k_2\mathbf{x}_2(t) + k_3\mathbf{x}_3(t)\\ &= k_1\begin{bmatrix}1\\0\\0\end{bmatrix} + k_2\left\{\begin{bmatrix}1\\0\\0\end{bmatrix}t + \begin{bmatrix}0\\1\\0\end{bmatrix}\right\}\\ &\quad + k_3\left\{\begin{bmatrix}1\\0\\0\end{bmatrix}t^2 + \begin{bmatrix}0\\2\\0\end{bmatrix}t + \begin{bmatrix}0\\0\\2\end{bmatrix}\right\}\end{aligned}$$

is a solution for each choice of k_1, k_2, and k_3. We conclude with one last example.

3.1.2. Example

Solve the initial-value problem

$$\mathbf{x}' = \begin{bmatrix}1 & 1 & 0\\0 & 1 & 0\\0 & 1 & 1\end{bmatrix}\mathbf{x} \qquad \mathbf{x}(0) = \mathbf{x}_0 = \begin{bmatrix}1\\1\\1\end{bmatrix}$$

Solution

The characteristic equation for this problem is $(1-\lambda)^3 = 0$ and hence $\lambda = 1$ is a threefold eigenvalue. The eigenvectors for this eigenvalue form a two-parameter family which we may write as $\mathbf{u} = [k_1, 0, k_2]^T$. Therefore

$$\mathbf{x}_1(t) = \begin{bmatrix}k_1\\0\\k_2\end{bmatrix}e^t$$

is a solution of the differential system. But for no choice of k_1 and k_2 will $\mathbf{x}_1(0) = [1, 1, 1]^T$ be possible. We attempt a solution of the form

$$\mathbf{x}_2(t) = (\mathbf{a}t + \mathbf{b})e^{\lambda t}$$

and find, as above, $\mathbf{A}\mathbf{a} = \lambda\mathbf{a}$ and $\mathbf{A}\mathbf{b} = \lambda\mathbf{b} + \mathbf{a}$. Thus λ is an eigenvalue, hence $\lambda = 1$. Also $\mathbf{a}$ is an eigenvector. But not all eigenvectors will lead to a consistent set in $\mathbf{A}\mathbf{b} = \lambda\mathbf{b} + \mathbf{a}$. For instance, the choice $\mathbf{a} = [1, 0, 0]^T$ leads to

$$\begin{bmatrix}0 & 1 & 0\\0 & 0 & 0\\0 & 1 & 0\end{bmatrix}\begin{bmatrix}b_1\\b_2\\b_3\end{bmatrix} = \begin{bmatrix}b_2\\0\\b_2\end{bmatrix} = \begin{bmatrix}1\\0\\0\end{bmatrix} \tag{3.1.4}$$

an impossibility. In fact (3.1.4) indicates that $\mathbf{a}$ must be selected with equal entries in the two nonzero positions. We therefore choose $\mathbf{a} = [1, 0, 1]^T$ and now

$$\begin{bmatrix}0 & 1 & 0\\0 & 0 & 0\\0 & 1 & 0\end{bmatrix}\begin{bmatrix}b_1\\b_2\\b_3\end{bmatrix} = \begin{bmatrix}1\\0\\1\end{bmatrix}$$

yields $\mathbf{b} = [0, 1, 0]^T$ as one of many possible solutions. With $\mathbf{a} = [1, 0, 1]^T$ and $\mathbf{b} = [0, 1, 0]^T$ we have

$$\mathbf{x}_2(t) = \left\{ \begin{bmatrix} 1 \\ 0 \\ 1 \end{bmatrix} t + \begin{bmatrix} 0 \\ 1 \\ 0 \end{bmatrix} \right\} e^t$$

and hence, by superposition, $\mathbf{x}(t) = \mathbf{x}_1(t) + \mathbf{x}_2(t)$, i.e.,

$$\mathbf{x}(t) = \left[k_1 \begin{bmatrix} 0 \\ 0 \\ 1 \end{bmatrix} + k_2 \begin{bmatrix} 1 \\ 0 \\ 0 \end{bmatrix} + k_3 \left\{ \begin{bmatrix} 0 \\ 1 \\ 0 \end{bmatrix} + \begin{bmatrix} 1 \\ 0 \\ 1 \end{bmatrix} t \right\} \right] e^t$$

From $\mathbf{x}(0) = [1, 1, 1]^T$ we find $k_1 = k_2 = k_3 = 1$ and then combining in the obvious way,

$$\mathbf{x}(t) = \left\{ t \begin{bmatrix} 1 \\ 0 \\ 1 \end{bmatrix} + \begin{bmatrix} 1 \\ 1 \\ 1 \end{bmatrix} \right\} e^t$$

which is the solution to the given problem.

Remarks

Although both examples have three equal eigenvalues, their solutions take different forms; in the first a "quadratic" coefficient of e^t is necessary while in the second a "linear" term appears as the coefficient of e^t. An attempt, incidentally, to find a solution of the form $(\mathbf{a}t^2 + \mathbf{b}t + \mathbf{c})e^{\lambda t}$ for Example 3.1.2 will lead to $\mathbf{a} = \mathbf{0}$ and from there back to a previously considered solution. (See Exercise 3.1, Problem 14.)

Exercise 3.1

In the following six problems you are given A and $\mathbf{x}_0$. Find a solution to the initial-value problem $\mathbf{x}' = \mathbf{A}\mathbf{x}$, $\mathbf{x}(0) = \mathbf{x}_0$.

1. $\begin{bmatrix} 0 & 1 \\ -4 & 4 \end{bmatrix}$, $\quad \mathbf{x}_0 = \begin{bmatrix} 1 \\ 2 \end{bmatrix}$

2. $\begin{bmatrix} 0 & 1 \\ -4 & 4 \end{bmatrix}$, $\quad \mathbf{x}_0 = \begin{bmatrix} 1 \\ 1 \end{bmatrix}$

3. $\begin{bmatrix} 1 & 1 & 0 \\ 0 & 1 & 0 \\ 0 & 0 & -2 \end{bmatrix}$, $\quad \mathbf{x}_0 = \begin{bmatrix} 1 \\ 0 \\ 1 \end{bmatrix}$

4. $\begin{bmatrix} 1 & 1 & 0 \\ 0 & 1 & 0 \\ 0 & 0 & -2 \end{bmatrix}$, $\quad \mathbf{x}_0 = \begin{bmatrix} 1 \\ 1 \\ 1 \end{bmatrix}$

5. $\begin{bmatrix} 0 & 1 \\ -\alpha^2 & 2\alpha \end{bmatrix}$, $\quad \mathbf{x}_0 = \begin{bmatrix} 1 \\ \alpha \end{bmatrix}$

6. $\begin{bmatrix} 0 & 1 \\ -\alpha^2 & 2\alpha \end{bmatrix}$, $\quad \mathbf{x}_0 = \begin{bmatrix} 0 \\ 1 \end{bmatrix}$

7. Given

$$\mathbf{x}' = \begin{bmatrix} 0 & 1 & 0 \\ 0 & 0 & 1 \\ 4 & -8 & 5 \end{bmatrix} \mathbf{x}$$

choose $\mathbf{x}_0$ so that $\mathbf{u}_1 e^t + \mathbf{u}_2 e^{2t}$ is a solution of $\mathbf{x}' = \mathrm{A}\mathbf{x}$, $\mathbf{x}(0) = \mathbf{x}_0$, where

$$\mathbf{v} = \begin{bmatrix} 0 \\ 1 \\ 4 \end{bmatrix}$$

8. Same as Problem 7 except that $(\mathbf{u}_2 t + \mathbf{v})e^{2t}$ is a solution; here

$$\mathbf{u}_1 = \begin{bmatrix} 1 \\ 1 \\ 1 \end{bmatrix} \quad \text{and} \quad \mathbf{u}_2 = \begin{bmatrix} 1 \\ 2 \\ 4 \end{bmatrix}$$

9. Solve

$$\mathbf{x}' = \begin{bmatrix} -1 & 1 & 0 \\ 0 & -1 & 1 \\ 0 & 0 & -1 \end{bmatrix} \mathbf{x} \qquad \mathbf{x}_0 = \begin{bmatrix} -3 \\ 1 \\ -2 \end{bmatrix}$$

10. Solve

$$\mathbf{x}' = \begin{bmatrix} 0 & 1 & 0 \\ 0 & 0 & 1 \\ 1 & -3 & 3 \end{bmatrix} \mathbf{x} \qquad \mathbf{x}_0 = \begin{bmatrix} 1 \\ 1 \\ 1 \end{bmatrix}$$

11. Same as Problem 9 except $\mathbf{x}_0 = [0, 1, 2]^{\mathrm{T}}$.
12. Same as Problem 9 except $\mathbf{x}_0 = [0, 0, 1]^{\mathrm{T}}$.
13. In Example 3.1.1, Equation (3.1.3), show that $\mathbf{b} = [k, 1, 0]^{\mathrm{T}}$ leads to a solution which is a linear combination of $\mathbf{x}_1(t)$ and $\mathbf{x}_2(t)$ and therefore the choice $k = 0$ loses no solutions.
14. In Example 3.1.1 show that $\mathbf{v} = [k, 2, 0]^{\mathrm{T}}$ leads to a solution which is a combination of $\mathbf{x}_1(t)$ and $\mathbf{x}_3(t)$ and therefore the choice $k = 0$ loses no solution.
15. Show that $(\mathbf{a}t^2 + \mathbf{b}t + \mathbf{c})e^{\lambda t} \neq \mathbf{0}$ is a solution of the system of Example 3.1.2 if and only if $\mathbf{a} = \mathbf{0}$ and $(\mathrm{A} - \lambda\mathrm{I})\mathbf{b} = \mathbf{0}$, $(\mathrm{A} - \lambda\mathrm{I})\mathbf{c} = \mathbf{b}$.

3.2. ROOT VECTORS

If $(\mathbf{a}t + \mathbf{b})e^{\lambda t}$, $\mathbf{a} \neq \mathbf{0}$ is a solution of $\mathbf{x}' = \mathrm{A}\mathbf{x}$ then, as we have seen in Section 3.1, λ is an eigenvalue of A and $\mathbf{a}$ is a corresponding eigenvector. Also $\mathbf{b}$ is a solution of

$$(\mathrm{A} - \lambda\mathrm{I})\mathbf{b} = \mathbf{a} \neq \mathbf{0} \tag{3.2.1}$$

for some eigenvector $\mathbf{a}$. The matrix $(\mathrm{A} - \lambda\mathrm{I})$ is singular so that Equation (3.2.1) may not have any solution for some (perhaps all) choices of $\mathbf{a}$. Examples 3.1.1 and 3.1.2 illustrate some of the possibilities.

A simpler form of Equation (3.2.1) may be obtained by multiplying both sides through by $(\mathrm{A} - \lambda\mathrm{I})$. This leads to

$$(\mathrm{A} - \lambda\mathrm{I})^2\mathbf{b} = (\mathrm{A} - \lambda\mathrm{I})\mathbf{a} = \mathbf{0} \tag{3.2.2}$$

in which $\mathbf{a}$ does not explicitly appear.

The assumption that $(\mathbf{a}t^2+\mathbf{b}t+\mathbf{c})e^{\lambda t}$, $\mathbf{a} \neq \mathbf{0}$, be a solution of $\mathbf{x}' = A\mathbf{x}$ requires similar restraints on **a**, **b**, and **c**. We find specifically,

$$\begin{aligned} (A-\lambda I)\mathbf{a} &= \mathbf{0} \qquad \mathbf{a} \neq \mathbf{0} \\ (A-\lambda I)\mathbf{b} &= 2\mathbf{a} \\ (A-\lambda I)\mathbf{c} &= \mathbf{b} \end{aligned} \tag{3.2.3}$$

For the same reasons given before, we multiply the second and third equations through by $(A-\lambda I)$ and $(A-\lambda I)^2$, respectively. We obtain

$$\begin{aligned} (A-\lambda I)\ \mathbf{a} &= \mathbf{0} \qquad \mathbf{a} \neq \mathbf{0} \\ (A-\lambda I)^2\mathbf{b} &= \mathbf{0} \\ (A-\lambda I)^3\mathbf{c} &= \mathbf{0} \end{aligned} \tag{3.2.4}$$

All this suggests that we study solutions of systems of the form

$$(A-\lambda^* I)^k\mathbf{u} = \mathbf{0}$$

when λ^* is an eigenvalue. The vectors **u** which satisfy such systems are sometimes called *generalized eigenvectors* and sometimes *root vectors*. Their importance in constructing solutions of $\mathbf{x}' = A\mathbf{x}$ when A has repeated eigenvalues has been amply demonstrated in the previous section. For this reason we devote the rest of this section to the study of root vectors.

Suppose $\lambda = \lambda^*$ is an eigenvalue of A. Then an eigenvector of A corresponding to λ^* may be thought of as a nonzero vector "annihilated" by the singular matrix $A-\lambda^* I$, that is, $(A-\lambda^* I)\mathbf{u} = \mathbf{0}$, $\mathbf{u} \neq \mathbf{0}$. A root vector **v** of order k corresponding to the eigenvalue λ^* of A is a nonzero vector annihilated by $(A-\lambda^* I)^k$ and not by $(A-\lambda^* I)^{k-1}$.

3.2.1. DEFINITION *The vector* **v** *is a root vector of order* k, $k \geq 1$, *of* A *if there is a number* λ^* *such that*

$$(A-\lambda^* I)^{k-1}\mathbf{v} \neq \mathbf{0} \tag{3.2.5}$$

$$(A-\lambda^* I)^{k}\mathbf{v} = \mathbf{0} \tag{3.2.6}$$

The point of Equation (3.2.5) is to insure that k is the smallest integer for which a power of $(A-\lambda^* I)$ annihilates **v**. Of course, if $(A-\lambda^* I)^k\mathbf{v} = \mathbf{0}$, then $(A-\lambda^* I)^n\mathbf{v} = \mathbf{0}$ for all integers $n \geq k$, also. Thus Equation (3.2.5) is necessary if we wish a unique definition of the "order" of a root vector. It might also be noted that unless $A-\lambda^* I$ is singular, $(A-\lambda^* I)^k\mathbf{v}$ would be a nonzero vector for all k. Hence the existence of k, $\mathbf{v} \neq \mathbf{0}$ and λ^* satisfying Equation (3.2.5) implies $A-\lambda^* I$ is singular which means that λ^* must be an eigenvalue (see Exercise 3.2, Part 2, Problem 6).

In terms of the notion of root vector, an eigenvector is simply a root vector of order 1 and the vectors **b** and **c** in Equations (3.2.3) and (3.2.4) are root vectors of order 2 and 3, respectively.

Suppose, again, that $\mathbf{v}_k$ is a root vector of order k corresponding to the eigenvalue λ^* of A. From $\mathbf{v}_k$ we can readily construct a sequence of vectors,

$\mathbf{v}_k, \mathbf{v}_{k-1}, \ldots, \mathbf{v}_2, \mathbf{v}_1$ which, as is suggested by the notation, are root vectors of decreasing order. Consider the defining equations

$$\begin{aligned}(\mathbf{A}-\lambda^*\mathbf{I})\mathbf{v}_k &\equiv \mathbf{v}_{k-1}\\ (\mathbf{A}-\lambda^*\mathbf{I})\mathbf{v}_{k-1} &\equiv \mathbf{v}_{k-2}\\ &\cdot\\ &\cdot\\ &\cdot\\ (\mathbf{A}-\lambda^*\mathbf{I})\mathbf{v}_2 &\equiv \mathbf{v}_1\end{aligned} \tag{3.2.7}$$

The case $k=4$ illustrates the general technique. Equations (3.2.7) reduce to

$$\begin{aligned}(\mathbf{A}-\lambda^*\mathbf{I})\mathbf{v}_4 &\equiv \mathbf{v}_3\\ (\mathbf{A}-\lambda^*\mathbf{I})\mathbf{v}_3 &\equiv \mathbf{v}_2\\ (\mathbf{A}-\lambda^*\mathbf{I})\mathbf{v}_2 &\equiv \mathbf{v}_1\end{aligned} \tag{3.2.8}$$

From these relationships we deduce

$$\begin{aligned}(\mathbf{A}-\lambda^*\mathbf{I})^4\mathbf{v}_4 &= (\mathbf{A}-\lambda^*\mathbf{I})^3\{(\mathbf{A}-\lambda^*\mathbf{I})\mathbf{v}_4\} = (\mathbf{A}-\lambda^*\mathbf{I})^3\mathbf{v}_3\\ &= (\mathbf{A}-\lambda^*\mathbf{I})^2\{(\mathbf{A}-\lambda^*\mathbf{I})\mathbf{v}_3\} = (\mathbf{A}-\lambda^*\mathbf{I})^2\mathbf{v}_2\\ &= (\mathbf{A}-\lambda^*\mathbf{I})\{(A-\lambda^*\mathbf{I})\mathbf{v}_2\} = (\mathbf{A}-\lambda^*\mathbf{I})\mathbf{v}_1\end{aligned} \tag{3.2.9}$$

and similarly,

$$\begin{aligned}(\mathbf{A}-\lambda^*\mathbf{I})^3\mathbf{v}_4 &= (\mathbf{A}-\lambda^*\mathbf{I})^2\{(\mathbf{A}-\lambda^*\mathbf{I})\mathbf{v}_4\} = (\mathbf{A}-\lambda^*\mathbf{I})^2\mathbf{v}_3\\ &= (\mathbf{A}-\lambda^*\mathbf{I})\ \{(\mathbf{A}-\lambda^*\mathbf{I})\mathbf{v}_3\} = (\mathbf{A}-\lambda^*\mathbf{I})\mathbf{v}_2 = \mathbf{v}_1\end{aligned} \tag{3.2.10}$$

If we now assume $\mathbf{v}_1$ is an eigenvector corresponding to the eigenvalue λ^*, then Equations (3.2.9) imply

$$\begin{aligned}(\mathbf{A}-\lambda^*\mathbf{I})^4\mathbf{v}_4 &= (\mathbf{A}-\lambda^*\mathbf{I})^3\mathbf{v}_3 = (\mathbf{A}-\lambda^*\mathbf{I})^2\mathbf{v}_2\\ &= (\mathbf{A}-\lambda^*\mathbf{I})\mathbf{v}_1 = \mathbf{0}\end{aligned} \tag{3.2.11}$$

and Equations (3.2.10) imply

$$(\mathbf{A}-\lambda^*\mathbf{I})^3\mathbf{v}_4 = (\mathbf{A}-\lambda^*\mathbf{I})^2\mathbf{v}_3 = (\mathbf{A}-\lambda^*\mathbf{I})\mathbf{v}_2 = \mathbf{v}_1 \neq \mathbf{0} \tag{3.2.12}$$

These are the required conditions that $\mathbf{v}_4$ be a root vector of order 4, $\mathbf{v}_3$ a root vector of order 3, and $\mathbf{v}_2$ a root vector of order 2.

On the other hand, if we assume $\mathbf{v}_4$ is a root vector of order 4 corresponding to the eigenvalue λ^* we see that Equations (3.2.9) and (3.2.10) again imply Equations (3.2.11) and (3.2.12). Thus $\mathbf{v}_3$, $\mathbf{v}_2$, $\mathbf{v}_1$ are root vectors of order 3, 2, and 1, respectively, if $\mathbf{v}_4$ is a root vector of order 4. What we have done for $k=4$ may be extended in an analogous manner to any k. All this means we have proved a theorem which (1) provides a criterion for judging when we have a root vector of order k and, simultaneously, (2) yields root vectors of each successively lower order. Before formally presenting this theorem, we study an example.

3.2.1. Example

Verify that $[2, 0, 1]^T$ is a root vector of order 3 corresponding to the eigenvalue $\lambda = 0$ of

$$A = \begin{bmatrix} 0 & 1 & 0 \\ 0 & 0 & 1 \\ 0 & 0 & 0 \end{bmatrix}$$

Solution

If we label $\mathbf{v}_3 = [2, 0, 1]$, then $A - \lambda I = A$ and

$$\begin{bmatrix} 0 & 1 & 0 \\ 0 & 0 & 1 \\ 0 & 0 & 0 \end{bmatrix} \mathbf{v}_3 = \begin{bmatrix} 0 \\ 1 \\ 0 \end{bmatrix} \equiv \mathbf{v}_2$$

defines $\mathbf{v}_2$

$$\begin{bmatrix} 0 & 1 & 0 \\ 0 & 0 & 1 \\ 0 & 0 & 0 \end{bmatrix} \mathbf{v}_2 = \begin{bmatrix} 1 \\ 0 \\ 0 \end{bmatrix} \equiv \mathbf{v}_1$$

defines $\mathbf{v}_1$. Finally,

$$\begin{bmatrix} 0 & 1 & 0 \\ 0 & 0 & 1 \\ 0 & 0 & 0 \end{bmatrix} \mathbf{v}_1 = \begin{bmatrix} 0 \\ 0 \\ 0 \end{bmatrix}$$

means that $\mathbf{v}_3, \mathbf{v}_2, \mathbf{v}_1$ are, respectively, root vectors of order 3, 2, and 1.

3.2.2. THEOREM *Suppose λ^* is an eigenvalue of A and the sequence of vectors $\mathbf{v}_k, \mathbf{v}_{k-1}, \ldots, \mathbf{v}_1$ are defined recursively by Equation (3.2.7). Then $\mathbf{v}_k, \mathbf{v}_{k-1}, \ldots, \mathbf{v}_1$ are root vectors of order $k, k-1, \ldots, 2, 1$, respectively, if and only if $\mathbf{v}_1$ is an eigenvector of A corresponding to λ^*.*

The sequence of vectors $\mathbf{v}_k, \mathbf{v}_{k-1}, \ldots, \mathbf{v}_1$ play a central role in the solution of $\mathbf{x}' = A\mathbf{x}$. We refer to such sequences sufficiently often to warrant a special name for them. Since these sequences depend upon the choice of eigenvalue of A, we need to distinguish between sequences originating from different eigenvalues. This is most conveniently done by double subscripting the vectors. Study the following definition.

3.2.3. DEFINITION *If $\mathbf{u}_{i,k}$ is a root vector of order k corresponding to the eigenvalue λ_i of A then the sequence of root vectors*

$$\mathbf{u}_{i,k}, \mathbf{u}_{i,k-1}, \ldots, \mathbf{u}_{i,1} \tag{3.2.13}$$

defined by

$$\begin{aligned} (A - \lambda_i I)\mathbf{u}_{i,k} &= \mathbf{u}_{i,k-1} \\ (A - \lambda_i I)\mathbf{u}_{i,k-1} &= \mathbf{u}_{i,k-2} \\ &\vdots \\ (A - \lambda_i I)\mathbf{u}_{i,2} &= \mathbf{u}_{i,1} \end{aligned} \tag{3.2.14}$$

is called a chain of length k.

The question of finding $\mathbf{u}_{i,k}$ in the first place is unanswered by anything we have presented here. Theorem 3.2.2 begs the question. Furthermore, Equations (3.2.7) have singular coefficient matrices and an attempt to proceed from $\mathbf{v}_1$ backward to $\mathbf{v}_k$ will only succeed if $\mathbf{v}_1$ is an eigenvector for which each of the succeeding equations is consistent. Some vectors lead to consistent equations and others do not. In terms of Definition 3.2.3, some eigenvectors are terminal vectors in a chain and others are not. How do we recognize which are which? If A is sufficiently simple, that is, of low order or of special type, the latter question can be readily answered; otherwise we have a problem of some complexity.

3.2.2. Example

Find a chain of length 3 for the matrix

$$\begin{bmatrix} 0 & 1 & 0 \\ 0 & 0 & 1 \\ 8 & -12 & 6 \end{bmatrix}$$

Solution

We find that $\lambda = 2$ is the sole eigenvalue and $[1, 2, 4]^T$ is an eigenvector. These facts are readily verified. We attempt to find "$\mathbf{v}_2$" from

$$\begin{bmatrix} -2 & 1 & 0 \\ 0 & -2 & 1 \\ 8 & -12 & 4 \end{bmatrix} \begin{bmatrix} v_1 \\ v_2 \\ v_3 \end{bmatrix} = \begin{bmatrix} 1 \\ 2 \\ 4 \end{bmatrix}$$

Row reductions lead to

$$\begin{bmatrix} -2 & 1 & 0 \\ 0 & -2 & 1 \\ 0 & 0 & 0 \end{bmatrix} \quad \text{and} \quad \begin{bmatrix} v_1 \\ v_2 \\ v_3 \end{bmatrix} = \begin{bmatrix} 1 \\ 2 \\ 0 \end{bmatrix}$$

from which we deduce $\mathbf{v}_2 = [-1, -1, 0]^T$. Finally,

$$\begin{bmatrix} -2 & 1 & 0 \\ 0 & -2 & 1 \\ 8 & -12 & 4 \end{bmatrix} \begin{bmatrix} w_1 \\ w_2 \\ w_3 \end{bmatrix} = \begin{bmatrix} -1 \\ -1 \\ 0 \end{bmatrix}$$

from which $\mathbf{v}_3 = [\frac{3}{4}, \frac{1}{2}, 0]^T$ follows. Proceeding backward, $(A - 2I)\mathbf{v}_3 = [-1, -1, 0]^T = \mathbf{v}_2$; $(A - 2I)\mathbf{v}_2 = [1, 2, 4]^T = \mathbf{v}_1$, provides a check on our arithmetic.

3.2.3 Example

Find all the chains for the matrix

$$\begin{bmatrix} 1 & 1 & 0 & 0 \\ 0 & 1 & 1 & 0 \\ 0 & 0 & 1 & 0 \\ 0 & 6 & 0 & 1 \end{bmatrix}$$

Solution

The eigenvalues are $\lambda = 1$ repeated four times and, clearly, the set of all eigenvectors may be represented via

$$\mathbf{u} = \begin{bmatrix} k_1 \\ 0 \\ 0 \\ k_2 \end{bmatrix}$$

Let $\mathbf{v}$ be a root vector of order 2. Then

$$\begin{bmatrix} 0 & 1 & 0 & 0 \\ 0 & 0 & 1 & 0 \\ 0 & 0 & 0 & 0 \\ 0 & 6 & 0 & 0 \end{bmatrix} \begin{bmatrix} v_1 \\ v_2 \\ v_3 \\ v_4 \end{bmatrix} = \begin{bmatrix} k_1 \\ 0 \\ 0 \\ k_2 \end{bmatrix}$$

which implies

$$\begin{bmatrix} v_2 \\ v_3 \\ 0 \\ 6v_2 \end{bmatrix} = \begin{bmatrix} k_1 \\ 0 \\ 0 \\ k_2 \end{bmatrix}$$

Hence $\mathbf{v}$ is a root vector of order 2 and $\mathbf{v}$, $\mathbf{u}$ is a chain if and only if $v_2 = k_1$, $v_3 = 0$, $k_2 = 6v_2 = 6k_1$. We are free to choose v_1 and v_4 arbitrarily. Let $v_1 = k_3$, $v_4 = k_4$. Hence for every choice of k_1, k_3, and k_4

$$\mathbf{v} = \begin{bmatrix} k_3 \\ k_1 \\ 0 \\ k_4 \end{bmatrix} \qquad \mathbf{u} = \begin{bmatrix} k_1 \\ 0 \\ 0 \\ 6k_1 \end{bmatrix}$$

is a chain of length 2. *Note that only a selected set of eigenvectors terminate chains of length 2.* We now search for root vectors of order 3 and in the process discover that not all root vectors of order 2 are part of chains of length 3! To wit, let $\mathbf{w} = [w_1, w_2, w_3, w_4]^T$. We solve

$$\begin{bmatrix} 0 & 1 & 0 & 0 \\ 0 & 0 & 1 & 0 \\ 0 & 0 & 0 & 0 \\ 0 & 6 & 0 & 0 \end{bmatrix} \mathbf{w} = \mathbf{v} = \begin{bmatrix} k_3 \\ k_1 \\ 0 \\ k_4 \end{bmatrix}$$

This leads to $w_2 = k_3$, $w_3 = k_1$, $k_4 = 6w_2 = 6k_3$ with w_1 and w_4 arbitrary. Let $w_1 = k_5$, $w_4 = k_6$. Then for any choice of k_1, k_3, k_5, and k_6

$$\mathbf{w} = \begin{bmatrix} k_5 \\ k_3 \\ k_1 \\ k_6 \end{bmatrix} \qquad \mathbf{v} = \begin{bmatrix} k_3 \\ k_1 \\ 0 \\ 6k_3 \end{bmatrix} \qquad \mathbf{u} = \begin{bmatrix} k_1 \\ 0 \\ 0 \\ 6k_1 \end{bmatrix}$$

is a chain of length 3. Thus only root vectors of order 2 whose first and fourth components are k_3 and $6k_3$, respectively, are part of chains of length 3. There are no chains of length 4 because

$$\begin{bmatrix} 0 & 1 & 0 & 0 \\ 0 & 0 & 1 & 0 \\ 0 & 0 & 0 & 0 \\ 0 & 6 & 0 & 0 \end{bmatrix} \mathbf{z} = \begin{bmatrix} k_5 \\ k_3 \\ k_1 \\ k_6 \end{bmatrix}$$

would imply that $k_1 = 0$ contradicting the fact that **u** is an eigenvector.

Exercise 3.2

Part 1

In Problems 1–4 show there are no root vectors of order 2.

1. $\begin{bmatrix} 3 & 1 \\ 2 & 2 \end{bmatrix}$

2. $\begin{bmatrix} 1 & 0 & 0 \\ 0 & 1 & 0 \\ 0 & 0 & 1 \end{bmatrix}$

3. $\begin{bmatrix} 0 & 1 \\ 1 & 0 \end{bmatrix}$

4. $\begin{bmatrix} 0 & 0 & 0 \\ 0 & 0 & 0 \\ 0 & 0 & 0 \end{bmatrix}$

In Problems 5–8 find at least one chain of length 2.

5. $\begin{bmatrix} 1 & 1 & 0 \\ 0 & 1 & 0 \\ 1 & 2 & -2 \end{bmatrix}$

6. $\begin{bmatrix} 1 & 1 & 0 \\ 0 & 1 & 0 \\ 0 & 0 & 1 \end{bmatrix}$

7. $\begin{bmatrix} 0 & -1 \\ \dfrac{\alpha^2}{4} & \alpha \end{bmatrix}$

8. $\begin{bmatrix} 0 & 1 & 0 \\ 0 & 0 & 1 \\ 4 & -8 & 5 \end{bmatrix}$

In Problems 9 and 10 find at least one chain of length 3.

9. $\begin{bmatrix} 1 & 1 & 0 \\ 0 & 1 & 1 \\ 0 & 0 & 1 \end{bmatrix}$

10. $\begin{bmatrix} 0 & 1 & 0 \\ 0 & 0 & 1 \\ 1 & -3 & 3 \end{bmatrix}$

In Problems 11 and 12 find at least one chain of length 3 and a chain of length 2.

11. $\begin{bmatrix} 1 & 1 & 0 & 0 & 0 \\ 0 & 1 & 1 & 0 & 0 \\ 0 & 0 & 1 & 0 & 0 \\ 0 & 0 & 0 & 1 & 1 \\ 0 & 0 & 0 & 0 & 1 \end{bmatrix}$

12. $\begin{bmatrix} 1 & 1 & 0 & 0 & 0 \\ 0 & 1 & 1 & 0 & 0 \\ 0 & 0 & 1 & 0 & 0 \\ 0 & 0 & 0 & 2 & 1 \\ 0 & 0 & 0 & 0 & 2 \end{bmatrix}$

13. Find a chain of length 3, where

$$\mathrm{A} = \begin{bmatrix} 1 & 1 & 0 & 0 & 1 \\ 0 & 1 & 0 & 0 & 2 \\ 0 & 0 & 1 & 1 & 3 \\ 0 & 0 & 0 & 1 & 4 \\ 0 & 0 & 0 & 0 & 1 \end{bmatrix}$$

(Use the method outlined in Example 3.2.3.)

Part 2

1. Prove that the identity and zero matrices have no chains of length 2.
2. Can a matrix have a chain of length $k > 1$ and no chain of length $k-1$?
3. If $\mathbf{u}$ is a root vector of order k corresponding to the eigenvalue λ and $\mathbf{v}$ is a root vector of order $j < k$, corresponding to the same eigenvalue, is $\mathbf{u}+\mathbf{v}$ a root vector? How about $a\mathbf{u}+b\mathbf{v}$?
4. Prove that $(\mathrm{A}-\lambda_i\mathrm{I})^n\mathbf{u}_{i,k}$ is a root vector of order $k-n$, assuming $0 \leqslant n < k$ and $\mathbf{u}_{i,k}$ is a root vector of order k, corresponding to λ_i.
5. Prove that $\lambda = \alpha$ is the only eigenvalue and that $[0, \ldots, 0, \alpha]^{\mathrm{T}}$ is a root vector of order n for the $n \times n$ matrix,

$$\begin{bmatrix} \alpha & 1 & 0 & & \cdots & 0 \\ 0 & \alpha & 1 & 0 \cdots & \cdots & 0 \\ & & & & & \cdot \\ \cdot & 0 & & & & \cdot \\ \cdot & \cdot & & & & \cdot \\ \cdot & \cdot & & & \cdot & 0 \\ & & & & \cdot & 1 \\ 0 & 0 & \cdots & & 0 & \alpha \end{bmatrix}$$

6. Prove: If $(\mathrm{A}-\mu\mathrm{I})^k\mathbf{u} = \mathbf{0}$, then $\mathbf{u} \neq \mathbf{0}$ is an eigenvalue of A.
7. Prove: If $\mathbf{u}$ is a root vector of order k so is $c\mathbf{u}$, $c \neq 0$.

3.3. THE ROOT VECTOR AND SOLUTIONS OF $\mathbf{x}' = \mathrm{A}\mathbf{x}$

The eigenvector $\mathbf{u}$ of A leads via Theorem 2.3.1 to the solution $e^{\lambda t}\mathbf{u}$ of $\mathbf{x}' = \mathrm{A}\mathbf{x}$. What solutions derive from root vectors? A study of the examples in Section 3.1 suggests we examine $\mathbf{x}(t) = (\mathbf{a}t^n + \mathbf{b}t^{n-1} + \cdots + \mathbf{u})e^{\lambda t}$ expecting $\mathbf{a}, \mathbf{b}, \ldots$ to be root vectors. This is essentially the case. A precise statement of this result and its applications to some examples are the themes of this section.

3.3.1. THEOREM *Let $\mathbf{u}_k$ be a root vector of order k corresponding to the eigenvalue λ of* A. *Then*

$$\mathbf{x}(t) = e^{\lambda t}\left(\mathbf{u}_k + t\mathbf{u}_{k-1} + \cdots + \frac{t^{k-1}}{(k-1)!}\mathbf{u}_1\right) \tag{3.3.1}$$

is a solution of $\mathbf{x}' = \mathrm{A}\mathbf{x}$.

PROOF We first note

$$(\mathrm{A}-\lambda\mathrm{I})\mathbf{x}(t) = e^{\lambda t}\left(\mathbf{u}_{k-1} + t\mathbf{u}_{k-2} + \cdots + \frac{t^{k-2}}{(k-2)!}\mathbf{u}_2\right) \tag{3.3.2}$$

because $(\mathrm{A}-\lambda\mathrm{I})\mathbf{u}_i = \mathbf{u}_{i-1}$. Next, differentiating $\mathbf{x}(t)$ leads to

$$\begin{aligned} \mathbf{x}'(t) &= \lambda e^{\lambda t}\left(\mathbf{u}_k + \cdots + \frac{t^{k-1}}{(k-1)!}\mathbf{u}_1\right) \\ &\quad + e^{\lambda t}\left(\mathbf{u}_{k-1} + \cdots + \frac{t^{k-2}}{(k-2)!}\mathbf{u}_1\right) \\ &= \lambda\mathbf{x}(t) + (\mathrm{A}-\lambda\mathrm{I})\mathbf{x}(t) \end{aligned}$$

from (3.3.2). But $\lambda\mathbf{x}(t) + (\mathrm{A}-\lambda\mathrm{I})\mathbf{x}(t) = \mathrm{A}\mathbf{x}(t)$, proving that $\mathbf{x}'(t) \equiv \mathrm{A}\mathbf{x}(t)$.

3.3.1. Example

Solve

$$\mathbf{x}' = \begin{bmatrix} 1 & 1 & 0 & 0 \\ 0 & 1 & 0 & 0 \\ 0 & 0 & 1 & 1 \\ 0 & 0 & 0 & 1 \end{bmatrix} \mathbf{x}$$

Solution

It is easy to see that $\lambda = 1$ is the sole eigenvalue. The system

$$\begin{bmatrix} 0 & 1 & 0 & 0 \\ 0 & 0 & 0 & 0 \\ 0 & 0 & 0 & 1 \\ 0 & 0 & 0 & 0 \end{bmatrix} \begin{bmatrix} u_1 \\ u_2 \\ u_3 \\ u_4 \end{bmatrix} = \begin{bmatrix} 0 \\ 0 \\ 0 \\ 0 \end{bmatrix}$$

leads to $u_2 = u_4 = 0$, with u_1 and u_3 arbitrary. Set $u_1 = k_1$ and $u_3 = k_3$. We search for root vectors of order 2 by solving, if possible,

$$\begin{bmatrix} 0 & 1 & 0 & 0 \\ 0 & 0 & 0 & 0 \\ 0 & 0 & 0 & 1 \\ 0 & 0 & 0 & 0 \end{bmatrix} \begin{bmatrix} v_1 \\ v_2 \\ v_3 \\ v_4 \end{bmatrix} = \begin{bmatrix} k_1 \\ 0 \\ k_3 \\ 0 \end{bmatrix} \equiv \mathbf{u}$$

This requires $v_2 = k_1$, $v_4 = k_3$ and leaves v_1 and v_3 arbitrary. Hence

$$\mathbf{v} = \begin{bmatrix} k_2 \\ k_1 \\ k_4 \\ k_3 \end{bmatrix} \qquad \mathbf{u} = \begin{bmatrix} k_1 \\ 0 \\ k_3 \\ 0 \end{bmatrix}$$

form a chain of length 2. The reader is invited to show that there are no chains of length 3. (Note, as a check on the computation, that $(\mathbf{A} - \mathbf{I})\mathbf{v} = \mathbf{u}$, as required.) From Theorem 3.3.1 we may write a solution from this information in the form

$$\mathbf{x}(t) = \left\{ \begin{bmatrix} k_2 \\ k_1 \\ k_4 \\ k_3 \end{bmatrix} + \begin{bmatrix} k_1 \\ 0 \\ k_3 \\ 0 \end{bmatrix} t \right\} e^t$$

Since

$$\mathbf{x}(0) = [k_1, k_2, k_3, k_4]^{\mathrm{T}}$$

the "general" initial-value problem $\mathbf{x}' = \mathbf{A}\mathbf{x}$, $\mathbf{x}(0) = \mathbf{x}_0$ is always solvable by some choice of k_1, k_2, k_3, and k_4.

Remarks

Each selection of values for the four parameters k_1, k_2, k_3, and k_4 leads to a particular solution of the given system. The choices $k_1 = k_2 = k_3 = 0$, $k_4 = 1$, and $k_1 = k_3 = k_4 = 0$, $k_2 = 1$ yields the "eigenvalue–eigenvector" solutions

$$\mathbf{x}_1(t) = \begin{bmatrix} 0 \\ 0 \\ 1 \\ 0 \end{bmatrix} e^t \qquad \mathbf{x}_2(t) = \begin{bmatrix} 1 \\ 0 \\ 0 \\ 0 \end{bmatrix} e^t$$

respectively. On the other hand, the choices $k_2 = k_3 = k_4 = 0$, $k_1 = 1$, and $k_1 = k_2 = k_4 = 0$, $k_3 = 1$ generate

$$\mathbf{x}_3(t) = \left\{ \begin{bmatrix} 0 \\ 1 \\ 0 \\ 0 \end{bmatrix} + \begin{bmatrix} 1 \\ 0 \\ 0 \\ 0 \end{bmatrix} t \right\} e^t$$

and

$$\mathbf{x}_4(t) = \left\{ \begin{bmatrix} 0 \\ 0 \\ 0 \\ 1 \end{bmatrix} + \begin{bmatrix} 0 \\ 0 \\ 1 \\ 0 \end{bmatrix} t \right\} e^t$$

respectively. This procedure of selecting all but one of the parameters zero and the remaining parameter 1 is a convenient means for generating specific solutions from a solution given with arbitrary constants.

3.3.2 Example

Solve

$$\mathbf{x}' = \begin{bmatrix} 1 & 1 & 0 & 0 \\ 0 & 1 & 0 & 0 \\ 0 & 0 & 1 & 0 \\ 0 & 0 & 0 & 1 \end{bmatrix} \mathbf{x}$$

Solution

Here as in the previous example, $\lambda = 1$ is the sole eigenvalue. To find eigenvectors corresponding to this eigenvalue we solve

$$\begin{bmatrix} 0 & 1 & 0 & 0 \\ 0 & 0 & 0 & 0 \\ 0 & 0 & 0 & 0 \\ 0 & 0 & 0 & 0 \end{bmatrix} \begin{bmatrix} u_1 \\ u_2 \\ u_3 \\ u_4 \end{bmatrix} = \begin{bmatrix} 0 \\ 0 \\ 0 \\ 0 \end{bmatrix}$$

Therefore $u_2 = 0$ and u_1, u_3, and u_4 are arbitrary. We set $u_1 = k_1$, $u_3 = k_3$, and $u_4 = k_4$ and look for root vectors by solving

$$\begin{bmatrix} 0 & 1 & 0 & 0 \\ 0 & 0 & 0 & 0 \\ 0 & 0 & 0 & 0 \\ 0 & 0 & 0 & 0 \end{bmatrix} \begin{bmatrix} v_1 \\ v_2 \\ v_3 \\ v_4 \end{bmatrix} = \begin{bmatrix} k_1 \\ 0 \\ k_3 \\ k_4 \end{bmatrix} = \mathbf{u}$$

Here, $v_2 = k_1$, $k_3 = k_4 = 0$, and v_1, v_3, and v_4 are arbitrary. For neatness, let us set $v_1 = c_1$, $v_2 = c_2 = k_1$, $v_3 = c_3$, $v_4 = c_4$. Then

$$\mathbf{v} = \begin{bmatrix} c_1 \\ c_2 \\ c_3 \\ c_4 \end{bmatrix} \qquad \mathbf{u} = \begin{bmatrix} c_2 \\ 0 \\ 0 \\ 0 \end{bmatrix}$$

is a chain of length 2 and via Theorem 3.3.1

$$\mathbf{x}(t) = e^t \left\{ \begin{bmatrix} c_1 \\ c_2 \\ c_3 \\ c_4 \end{bmatrix} + \begin{bmatrix} c_2 \\ 0 \\ 0 \\ 0 \end{bmatrix} t \right\} \tag{3.3.3}$$

is a solution. Note that the solutions

$$\mathbf{x}_1(t) = k_1 \begin{bmatrix} 1 \\ 0 \\ 0 \\ 0 \end{bmatrix} e^t \qquad \mathbf{x}_2(t) = k_2 \begin{bmatrix} 0 \\ 0 \\ 1 \\ 0 \end{bmatrix} e^t \qquad \mathbf{x}_3 = k_3 \begin{bmatrix} 0 \\ 0 \\ 0 \\ 1 \end{bmatrix} e^t$$

are all special cases of (3.3.3). For instance, set $c_1 = c_2 = c_4 = 0$ and $c_3 = k_2$ and observe that $\mathbf{x}(t) \equiv \mathbf{x}_2(t)$ in this case.

3.3.3. Example

Solve

$$\mathbf{x}' = \begin{bmatrix} 1 & 1 & 0 & 0 \\ 0 & 1 & 1 & 0 \\ 0 & 0 & 1 & 0 \\ 0 & 6 & 0 & 1 \end{bmatrix} \mathbf{x} \qquad \mathbf{x}(1) = \begin{bmatrix} 2 \\ -1 \\ 1 \\ 1 \end{bmatrix}$$

Solution

In Example 3.2.3 we have computed the chain

$$\mathbf{w} = \begin{bmatrix} c_1 \\ c_2 \\ c_3 \\ c_4 \end{bmatrix} \qquad (\mathbf{A} - \mathbf{I})\mathbf{w} = \mathbf{v} = \begin{bmatrix} c_2 \\ c_3 \\ 0 \\ 6c_2 \end{bmatrix} \qquad (\mathbf{A} - \mathbf{I})\mathbf{v} = \mathbf{u} = \begin{bmatrix} c_3 \\ 0 \\ 0 \\ 6c_3 \end{bmatrix}$$

with a different notation for the arbitrary constants. Hence,

$$\mathbf{x}(t) = e^t \left\{ \begin{bmatrix} c_1 \\ c_2 \\ c_3 \\ c_4 \end{bmatrix} + \begin{bmatrix} c_2 \\ c_3 \\ 0 \\ 6c_2 \end{bmatrix} t + \begin{bmatrix} c_3 \\ 0 \\ 0 \\ 6c_3 \end{bmatrix} \frac{t^2}{2} \right\}$$

is a solution from Theorem 3.3.1. Set $t = 1$. Then

$$\mathbf{x}(1) = e \begin{bmatrix} c_1 + c_2 + \dfrac{c_3}{2} \\ c_2 + c_3 \\ c_3 \\ c_4 \end{bmatrix} = \begin{bmatrix} 2 \\ -1 \\ 1 \\ 1 \end{bmatrix}$$

is required by the initial condition. Thus

$$c_4 = e^{-1} \qquad c_3 = e^{-1} \qquad c_2 = -2e^{-1} \qquad c_1 = \tfrac{7}{2}e^{-1}$$

Finally,

$$\mathbf{x}(t) = e^{t-1} \left\{ \begin{bmatrix} \frac{7}{2} \\ -2 \\ 1 \\ 1 \end{bmatrix} + \begin{bmatrix} -2 \\ 1 \\ 0 \\ -12 \end{bmatrix} t + \begin{bmatrix} 1 \\ 0 \\ 0 \\ 6 \end{bmatrix} \frac{t^2}{2} \right\}$$

is a solution of the given initial-value problem.

3.3.4. Example

Solve

$$\mathbf{x}' = \begin{bmatrix} 1 & 1 & 0 & 0 \\ 0 & 1 & 1 & 0 \\ 0 & 0 & 1 & 1 \\ 0 & 0 & 0 & 1 \end{bmatrix} \mathbf{x} \qquad \mathbf{x}(0) = \begin{bmatrix} 0 \\ 0 \\ 0 \\ 1 \end{bmatrix}$$

Solution

The reader can check that $\lambda = 1$ is the only eigenvalue and that

$$\mathbf{z} = \begin{bmatrix} k_1 \\ k_2 \\ k_3 \\ k_4 \end{bmatrix} \qquad \mathbf{w} = \begin{bmatrix} k_2 \\ k_3 \\ k_4 \\ 0 \end{bmatrix} \qquad \mathbf{v} = \begin{bmatrix} k_3 \\ k_4 \\ 0 \\ 0 \end{bmatrix} \qquad \mathbf{u} = \begin{bmatrix} k_4 \\ 0 \\ 0 \\ 0 \end{bmatrix}$$

is a chain of length 4 and hence

$$\mathbf{x}(t) = e^t \left\{ \begin{bmatrix} k_1 \\ k_2 \\ k_3 \\ k_4 \end{bmatrix} + \begin{bmatrix} k_2 \\ k_3 \\ k_4 \\ 0 \end{bmatrix} t + \begin{bmatrix} k_3 \\ k_4 \\ 0 \\ 0 \end{bmatrix} \frac{t^2}{2} + \begin{bmatrix} k_4 \\ 0 \\ 0 \\ 0 \end{bmatrix} \frac{t^3}{3!} \right\}$$

is a solution which solves every initial-value problem constructed with the given matrix. To meet the initial condition, we can see by inspection that $k_1 = k_2 = k_3 = 0$ and $k_4 = 1$. This yields

$$\mathbf{x}(t) = e^t \begin{bmatrix} \frac{t^3}{3!} \\ \frac{t^2}{2!} \\ t \\ 1 \end{bmatrix}$$

3.3.5. Example

Solve

$$\mathbf{x}' = \begin{bmatrix} 0 & 1 & 0 & 0 \\ 0 & 0 & 1 & 0 \\ 0 & 0 & 0 & 1 \\ -4 & -4 & 3 & 2 \end{bmatrix} \mathbf{x}$$

Solution

We find $C(\lambda) = (\lambda - 2)^2(\lambda + 1)^2$.

Case 1

Select $\lambda = -1$ and solve

$$\begin{bmatrix} 1 & 1 & 0 & 0 \\ 0 & 1 & 1 & 0 \\ 0 & 0 & 1 & 1 \\ -4 & -4 & 3 & 3 \end{bmatrix} \begin{bmatrix} u_1 \\ u_2 \\ u_3 \\ u_4 \end{bmatrix} = \begin{bmatrix} u_1 + u_2 \\ u_2 + u_3 \\ u_3 + u_4 \\ -4u_1 - 4u_2 + 3u_3 + 3u_4 \end{bmatrix} = \begin{bmatrix} 0 \\ 0 \\ 0 \\ 0 \end{bmatrix}$$

Thus, $u_1 + u_2 = 0$, $u_2 + u_3 = 0$, $u_3 + u_4 = 0$. The last equation gives no new information. We set $u_4 = k_1$ and find $u_3 = -k_1$, $u_2 = k_1$, $u_1 = -k_1$. As usual we solve

$$\begin{bmatrix} 1 & 1 & 0 & 0 \\ 0 & 1 & 1 & 0 \\ 0 & 0 & 1 & 1 \\ -4 & -4 & 3 & 3 \end{bmatrix} \begin{bmatrix} v_1 \\ v_2 \\ v_3 \\ v_4 \end{bmatrix} = k_1 \begin{bmatrix} -1 \\ 1 \\ -1 \\ 1 \end{bmatrix}$$

to obtain a root vector. Here, $v_1 + v_2 = -k_1$, $v_2 + v_3 = k_1$, $v_3 + v_4 = -k_1$ and the last equation is redundant (why?). Now set $v_4 = k_2$. Then $v_3 = -k_1 - k_2$, $v_2 = 2k_1 + k_2$, and $v_1 = -3k_1 - k_2$. Hence

$$\mathbf{v} = \begin{bmatrix} -3k_1 - k_2 \\ 2k_1 + k_2 \\ -k_1 - k_2 \\ k_2 \end{bmatrix}$$

and

$$\mathbf{x}_1(t) = e^{-t}\left\{\begin{bmatrix} -3k_1 - k_2 \\ 2k_1 + k_2 \\ -k_1 - k_2 \\ k_2 \end{bmatrix} + \begin{bmatrix} -k_1 \\ k_1 \\ -k_1 \\ k_1 \end{bmatrix} t\right\}$$

Case 2

Select $\lambda = 2$. The details unravel without complications. To avoid fractions we set $\mathbf{u} = 2k_3[1, 2, 4, 8]^T$ and find, after setting $v_4 = 8k_4$:

$$\mathbf{v} = \begin{bmatrix} -3k_3 + k_4 \\ -4k_3 + 2k_4 \\ -4k_3 + 4k_4 \\ 8k_4 \end{bmatrix} \qquad \mathbf{u} = 2k_3\begin{bmatrix} 1 \\ 2 \\ 4 \\ 8 \end{bmatrix}$$

and thus

$$\mathbf{x}_2(t) = e^{2t}(\mathbf{v} + \mathbf{u}t)$$

Since sums of solutions are solutions, we have

$$\mathbf{x}(t) = \mathbf{x}_1(t) + \mathbf{x}_2(t)$$

At $t = 0$ we find

$$\mathbf{x}(0) = \begin{bmatrix} -3k_1 - k_2 \\ 2k_1 + k_2 \\ -k_1 - k_2 \\ k_2 \end{bmatrix} + \begin{bmatrix} -3k_3 + k_4 \\ -4k_3 + 2k_4 \\ -4k_3 + 4k_4 \\ 8k_4 \end{bmatrix}$$

$$= \begin{bmatrix} -3 & -1 & -3 & 1 \\ 2 & 1 & -4 & 2 \\ -1 & -1 & -4 & 4 \\ 0 & 1 & 0 & 8 \end{bmatrix}\begin{bmatrix} k_1 \\ k_2 \\ k_3 \\ k_4 \end{bmatrix}$$

Since the determinant of this latter matrix is not zero $\mathbf{x}(0) = \mathbf{x}_0 \in \mathscr{R}^4$ is always solvable for some choice of k_1, k_2, k_3, and k_4.

We make one further observation. Apparently, if an eigenvalue is repeated, various combinations of chain lengths are possible. We have illustrated in the examples of this section how the eigenvalue $\lambda = 1$ repeated four times can lead to:

(1) a chain of length 4
(2) a chain of length 3 and one of length 1
(3) two chains of length 2
(4) a chain of length 2 and two chains of length 1

The matrix $A = I$ illustrates the case of four chains of length 1. These five cases represent all ways of partitioning the number 4 into sums of the numbers 1, 2, 3, and 4. That is,

$$4 = 4 = 3+1 = 2+2 = 2+1+1 = 1+1+1+1$$

There are no other possibilities.

Exercise 3.3

Part 1

1. Verify that $\mathbf{u} = [0, 0, 0, 1]$ is a root vector of order 2 for the matrix

$$A = \begin{bmatrix} 1 & 1 & 0 & 0 \\ 0 & 1 & 0 & 0 \\ 0 & 0 & 1 & 1 \\ 0 & 0 & 0 & 1 \end{bmatrix}$$

2. In Example 3.3.1, verify that there is no chain of length 3.
3. In Example 3.3.2, verify that there is no chain of length 3.
4. In Example 3.3.3, verify that there is no chain of length 3.
5. Verify that there is no chain of length 5 for the matrix A of Example 3.3.4.
6. In Example 3.3.5, verify that there are no root vectors of order 3.
   In the next two problems find solutions and compare with the solution given for Example 3.3.2.

7. $\mathbf{x}' = \begin{bmatrix} 1 & 0 & 0 & 0 \\ 0 & 1 & 1 & 0 \\ 0 & 0 & 1 & 0 \\ 0 & 0 & 0 & 1 \end{bmatrix}\mathbf{x}$ 8. $\mathbf{x}' = \begin{bmatrix} 1 & 0 & 0 & 0 \\ 0 & 1 & 0 & 0 \\ 0 & 0 & 1 & 1 \\ 0 & 0 & 0 & 1 \end{bmatrix}$

9. Solve

$$\mathbf{x}' = \begin{bmatrix} 1 & 0 & 0 & 0 \\ 0 & 1 & 1 & 0 \\ 0 & 0 & 1 & 1 \\ 0 & 0 & 0 & 1 \end{bmatrix}\mathbf{x}$$

(a) $\mathbf{x}(0) = \mathbf{e}_1$
(b) $\mathbf{x}(0) = \mathbf{e}_2$
(c) $\mathbf{x}(0) = \mathbf{e}_3$
(d) $\mathbf{x}(0) = \mathbf{e}_4$

In the next eight problems solve $\mathbf{x}' = A\mathbf{x}$ given A as follows:

10. $\begin{bmatrix} 2 & 0 & 0 \\ 0 & 2 & 0 \\ 0 & 0 & 2 \end{bmatrix}$ 11. $\begin{bmatrix} 2 & 1 & 0 \\ 0 & 2 & 0 \\ 0 & 0 & 2 \end{bmatrix}$

12. $\begin{bmatrix} 2 & 0 & 0 \\ 0 & 2 & 1 \\ 0 & 0 & 2 \end{bmatrix}$ 13. $\begin{bmatrix} 2 & 1 & 0 \\ 0 & 2 & 1 \\ 0 & 0 & 2 \end{bmatrix}$

14. $\begin{bmatrix} -1 & 1 & 0 & 0 \\ 0 & -1 & 0 & 0 \\ 0 & 0 & 1 & 1 \\ 0 & 0 & 0 & 1 \end{bmatrix}$ 15. $\begin{bmatrix} -1 & 1 & 0 & 0 \\ 0 & -1 & 0 & 0 \\ 0 & 0 & 1 & 0 \\ 0 & 0 & 0 & 1 \end{bmatrix}$

16. $\begin{bmatrix} 0 & 1 & 0 & 0 \\ -4 & 4 & 0 & 0 \\ 0 & 0 & 2 & 1 \\ 0 & 0 & 0 & 2 \end{bmatrix}$ 17. $\begin{bmatrix} 0 & 1 & 0 & 0 \\ -4 & 4 & 0 & 0 \\ 0 & 0 & 2 & 0 \\ 0 & 0 & 0 & 2 \end{bmatrix}$

Part 2

1. Solve

$$\mathbf{x}' = \begin{bmatrix} 1 & 1 & 0 & 0 & 1 \\ 0 & 1 & 0 & 0 & 2 \\ 0 & 0 & 1 & 1 & 3 \\ 0 & 0 & 0 & 1 & 4 \\ 0 & 0 & 0 & 0 & 1 \end{bmatrix} \mathbf{x} \qquad \mathbf{x}(0) = \begin{bmatrix} 1 \\ 0 \\ 2 \\ 0 \\ 1 \end{bmatrix}$$

See, Exercise 3.2, Part 1, Problem 13.

2. Set $\mathbf{x}_0 = [2, -1, 1, 1, 0]^T$ and using the answer to Problem 1 above, solve $\mathbf{x}' = A\mathbf{x}, \mathbf{x}(0) = \mathbf{x}_0$.
3.* Find a simple square matrix of order 5 with $\lambda = 2$ as its sole eigenvalue such that:
   (a) There exists a chain of length 1 and two chains of length 2.
   (b) There is a chain of length 3 and two chains of length 1.
   (c) There is a chain of length 3 and a chain of length 2.
4.* Prove that a chain of root vectors is a sequence of linearly independent vectors.
5.* Prove: Under the hypothesis of Theorem 3.3.2,

$$\mathbf{x}_1(t) = e^{\hat{\lambda}t}\,\hat{\mathbf{u}}_1$$
$$\mathbf{x}_2(t) = e^{\hat{\lambda}t}\,(\hat{\mathbf{u}}_2 + t\hat{\mathbf{u}}_1)$$
$$\vdots$$
$$\mathbf{x}_{k-1}(t) = e^{\hat{\lambda}t}\left(\hat{\mathbf{u}}_{k-1} + t\hat{\mathbf{u}}_{k-2} + \cdots + \frac{t^{k-2}}{(k-2)!}\,\hat{\mathbf{u}}_1\right)$$

are each solutions of $\mathbf{x}' = A\mathbf{x}$.

3.4. A RECAPITULATION

Let A be a real, $n \times n$ matrix and suppose $\lambda = \hat{\lambda}$ is a k-fold root of the characteristic equation, $C(\lambda) = \det(A - \lambda I)$, $1 \leq k \leq n$. Then $\hat{\lambda}$ is an eigenvalue repeated k times. We wish to discover the root vectors associated with $\hat{\lambda}$ and from them construct solutions to $\mathbf{x}' = A\mathbf{x}$.

Let us begin by summarizing the technique described by the examples in Sections 3.2 and 3.3 for finding chains of root vectors.

STEP 1. *Compute the eigenvectors of* A *corresponding to* $\lambda = \hat{\lambda}$ *and denote this family by* $\hat{\mathbf{u}}$.

This "generic" vector $\hat{\mathbf{u}}$ contains at least one arbitrary constant since nonzero multiples of an eigenvector are also eigenvectors. As we have seen, $\hat{\mathbf{u}}$ may in fact contain as many as k parameters. All this may be put in vector space terminology: the set of all eigenvectors of A corresponding to a k-fold eigenvalue along with the zero vector comprise a subspace of $\mathscr{R}^n$ of dimension $\leq k$. (The proof of the dimension restriction follows as a corollary to the theorem stated in Exercise 3.6, Part 2, Problem 8.) To return to our main point, not all the vectors represented by $\hat{\mathbf{u}}$ are vectors which terminate nontrivial chains (i.e., chains of length 2 or more). In Example 3.3.3, for instance,

$\lambda = 1$ is repeated four times and $\hat{\mathbf{u}} = [k_1, 0, 0, k_2]^T$. But only the subset in which $k_2 = 6k_1$ are terminal vectors in chains of length $\geqslant 2$. To discover which eigenvectors belong to nontrivial chains, we attempt to find $\hat{\mathbf{v}}$ so that $\hat{\mathbf{v}}, \hat{\mathbf{u}}$ is a chain.

STEP 2. *Compute all vectors which satisfy*

$$(\mathrm{A} - \hat{\lambda}\mathrm{I})\mathbf{v} = \hat{\mathbf{u}} \neq \mathbf{0} \tag{3.4.1}$$

and denote this family by $\hat{\mathbf{v}}$.

Since the system's matrix, $\mathrm{A} - \hat{\lambda}\mathrm{I}$, is singular, (3.4.1) is either inconsistent (as is the case if all the root vectors are eigenvectors; for instance, if A is symmetric) or $\hat{\mathbf{v}}$ represents a family of vectors at least one of which is a root vector of order 2. In this latter case, constraints are often introduced on the parameters of $\hat{\mathbf{u}}$ (the family of eigenvectors is narrowed by this process to include only eigenvectors in nontrivial chains) as well as interconnections between the parameters of $\hat{\mathbf{v}}$ and $\hat{\mathbf{u}}$. Witness Example 3.3.2 wherein $\hat{\mathbf{u}} = [k_1, 0, k_3, k_4]^T$ represents the family of eigenvectors after Step 1, while $\hat{\mathbf{v}} = [c_1, c_2, c_3, c_4]^T$, $\hat{\mathbf{u}} = [c_2, 0, 0, 0]^T$ represents chains of length 2, if $c_2 \neq 0$. Step 3 and successive steps thereafter are analogous to Step 2. We search for vectors $\hat{\mathbf{w}}$ such that $(\mathrm{A} - \hat{\lambda}\mathrm{I})\hat{\mathbf{w}} = \hat{\mathbf{v}}$ and then vectors $\hat{\mathbf{q}}$ such that $(\mathrm{A} - \hat{\lambda}\mathrm{I})\hat{\mathbf{q}} = \hat{\mathbf{w}}$, and so on. After some point we find that $\hat{\mathbf{u}} = \mathbf{0}$ is required and the process is terminated. In Example 3.3.2 illustrated above, if $\hat{\mathbf{w}}$ were to exist, then

$$(\mathrm{A} - \hat{\lambda}\mathrm{I})\hat{\mathbf{w}} = \begin{bmatrix} 0 & 1 & 0 & 0 \\ 0 & 0 & 0 & 0 \\ 0 & 0 & 0 & 0 \\ 0 & 0 & 0 & 0 \end{bmatrix} \begin{bmatrix} w_1 \\ w_2 \\ w_3 \\ w_4 \end{bmatrix} = \begin{bmatrix} c_1 \\ c_2 \\ c_3 \\ c_4 \end{bmatrix} = \mathbf{v}$$

implies $c_2 = 0$, which in turn implies $\hat{\mathbf{u}} = \mathbf{0}$, a contradiction. Here the process terminates after two steps. It is instructive to note again that the chain given as $\hat{\mathbf{v}}, \hat{\mathbf{u}}$ in the above example degenerates if $c_2 = 0$. That is, $\hat{\mathbf{v}}, \hat{\mathbf{u}}$ is not a chain of length 2 for every choice of the parameters in $\hat{\mathbf{v}}$. In chains of length 3, some choice of parameters in $\hat{\mathbf{w}}$ will reduce $\hat{\mathbf{w}}, \hat{\mathbf{v}}, \hat{\mathbf{u}}$ to a chain of length 2 or 1. The chain of length 4 computed in Example 3.3.4 illustrates the case in which a reduction to a chain of length 3 occurs if $k_4 = 0$ and to a chain of length 2 if $k_4 = k_3 = 0$. When we have reached the point at which $\hat{\mathbf{u}} = \mathbf{0}$ is required we have ended the first stage in our attempt to solve $\mathbf{x}' = \mathrm{A}\mathbf{x}$.

The second stage is trivial. Supposing the existence of, say, $\hat{\mathbf{w}}, \hat{\mathbf{v}}, \hat{\mathbf{u}}$ such that $(\mathrm{A} - \hat{\lambda}\mathrm{I})\hat{\mathbf{w}} = \hat{\mathbf{v}}$, $(\mathrm{A} - \hat{\lambda}\mathrm{I})\hat{\mathbf{v}} = \hat{\mathbf{u}}$, Theorem 3.3.2 asserts:

$$\mathbf{x}(t) = \left(\hat{\mathbf{w}} + \hat{\mathbf{v}}t + \hat{\mathbf{u}}\frac{t^2}{2}\right)e^{\hat{\lambda}t} \tag{3.4.2}$$

is a solution of $\mathbf{x}' = \mathrm{A}\mathbf{x}$ for each choice the parameters of $\hat{\mathbf{w}}, \hat{\mathbf{v}}$, and $\hat{\mathbf{u}}$. By specifying, one at a time, one of the parameters equal to 1 and the remaining parameters zero, we obtain from (3.4.2) specific solutions of $\mathbf{x}' = \mathrm{A}\mathbf{x}$. After this stage, we turn to a new eigenvalue of A and repeat, if necessary, the various steps which led us to Equation (3.4.2). When all the eigenvalues of A are exhausted, the sum of the functions so obtained is, by superposition, a solution of $\mathbf{x}' = A\mathbf{x}$ containing some arbitrary parameters. Question: Can this solution be specified so that at t_0 it assumes the value $\mathbf{x}(t_0)$? In other words, are we able to solve $\mathbf{x}' = A\mathbf{x}$, $\mathbf{x}(t_0) = \mathbf{x}_0$ with the solution we have obtained by this method, for every $\mathbf{x}_0 \in \mathscr{R}^n$?

Although our experience in Section 3.3 would suggest an affirmative answer, a proof constructed along these lines is very awkward. This difficulty is typical of trial-and-error methods. They work well (perhaps!) in specific examples but tend to be unsuitable for the general case. For this reason, we choose not to answer the question we posed above and instead use an alternative approach to prove that $\mathbf{x}' = A\mathbf{x}$, $\mathbf{x}(t_0) = \mathbf{x}_0$ is always solvable. (This approach is developed in Sections 3.5 and 3.6.)

Exercise 3.4

1.* Prove: The set of all eigenvectors of $A_{n\times n}$ corresponding to $\lambda = \hat{\lambda}$, an eigenvalue of A, along with the zero vector, is a subspace of $\mathscr{C}^n$, called the eigenspace of A.
2.* Give an example of a real $n \times n$ matrix A where the eigenspace is of dimension k, $1 \leqslant k \leqslant n$. (Hint: Study the matrix with 1's on the main diagonal and some 1's on the superdiagonal.)
3. Prove: The set of all root vectors corresponding to a fixed eigenvalue of $A_{n\times n}$ with $\mathbf{0}$ is a subspace of $\mathscr{C}^n$.

3.5. ANOTHER VIEW OF THE INITIAL-VALUE PROBLEM

We have previously found a solution of $\mathbf{x}' = A\mathbf{x}$, $\mathbf{x}(0) = \mathbf{x}_0$ by finding the general solution of $\mathbf{x}' = A\mathbf{x}$ and then specifying the arbitrary constants so that $\mathbf{x}(0) = \mathbf{x}_0$. This point of view places the general solution in a fundamental position and regards the solution of the initial-value problem as a by-product. We take the opposite view in this section, holding that the initial-value problem is fundamental and that the general solution is its corollary.

Basically, the method we present is constructive. We start with $\mathbf{x}_0$ and construct from it the set $\{\mathbf{x}_0, A\mathbf{x}_0, \ldots, A^n\mathbf{x}_0\}$. These $n+1$ vectors are linearly dependent since they are each n-tuples. Hence some linear combination is $\mathbf{0}$ and from this fact we find just those eigenvectors and/or root vectors whose span contains $\mathbf{x}_0$. Finally, we express $\mathbf{x}_0$ as a linear combination of these root vectors and eigenvectors and write $\mathbf{x}(t)$ by means of Theorem 3.3.2.

This program is easily understood from some illustrative examples.

3.5.1. Example

Solve

$$\mathbf{x}' = \begin{bmatrix} 2 & 1 & 0 \\ 0 & 2 & 0 \\ 0 & 0 & 1 \end{bmatrix} \mathbf{x} \qquad \mathbf{x}_0 = \begin{bmatrix} 2 \\ 0 \\ 3 \end{bmatrix} = \mathbf{x}(0)$$

Solution

As an exercise (Exercise 3.5, Part 1, Problem 1) we ask the reader to solve this problem by the methods of the previous sections. The reader should compare this and the following procedures. Set and compute

$$\mathbf{x}_0 = \begin{bmatrix} 2 \\ 0 \\ 3 \end{bmatrix} \qquad A\mathbf{x}_0 = \mathbf{x}_1 = \begin{bmatrix} 4 \\ 0 \\ 3 \end{bmatrix} \qquad A^2\mathbf{x}_0 = A\mathbf{x}_1 = \mathbf{x}_2 = \begin{bmatrix} 8 \\ 0 \\ 3 \end{bmatrix}$$

and

$$A^3\mathbf{x}_0 = A\mathbf{x}_2 = \mathbf{x}_3 = \begin{bmatrix} 16 \\ 0 \\ 3 \end{bmatrix} \tag{3.5.1}$$

The dependency relation among $\{\mathbf{x}_0, \mathbf{x}_1, \mathbf{x}_2, \mathbf{x}_3\}$ can be found by many means. We choose for this problem the reduction

$$\begin{bmatrix} 2 & 0 & 3 & \mathbf{x}_0 \\ 4 & 0 & 3 & \mathbf{x}_1 \\ 8 & 0 & 3 & \mathbf{x}_2 \\ 16 & 0 & 3 & \mathbf{x}_3 \end{bmatrix} \to \begin{bmatrix} 2 & 0 & 3 & \mathbf{x}_0 \\ 0 & 0 & -3 & \mathbf{x}_1 - 2\mathbf{x}_0 \\ 0 & 0 & -9 & \mathbf{x}_2 - 4\mathbf{x}_0 \\ 0 & 0 & -21 & \mathbf{x}_3 - 8\mathbf{x}_0 \end{bmatrix} \to \begin{bmatrix} 2 & 0 & 3 & \mathbf{x}_0 \\ 0 & 0 & -3 & \mathbf{x}_1 - 2\mathbf{x}_0 \\ 0 & 0 & 0 & \mathbf{x}_2 - 4\mathbf{x}_0 - 3(\mathbf{x}_1 - 2\mathbf{x}_0) \\ 0 & 0 & -21 & \mathbf{x}_3 - 8\mathbf{x}_0 \end{bmatrix}$$

We have found that $\{\mathbf{x}_0, A\mathbf{x}_0 = \mathbf{x}_1\}$ is linearly independent but $\{\mathbf{x}_0, A\mathbf{x}_0 = \mathbf{x}_1, A^2\mathbf{x}_0 = \mathbf{x}_2\}$ is not. In fact, the third row states

$$\begin{aligned} \mathbf{x}_2 - 4\mathbf{x}_0 - 3(\mathbf{x}_1 - 2\mathbf{x}_0) &= \mathbf{x}_2 - 3\mathbf{x}_1 + 2\mathbf{x}_0 \\ &= (A^2 - 3A + 2I)\mathbf{x}_0 = \mathbf{0} \end{aligned} \tag{3.5.2}$$

Let us study (3.5.2) in detail. Note

$$(A^2 - 3A + 2I)\mathbf{x}_0 = (A - 2I)[(A - I)\mathbf{x}_0] = \mathbf{0} \tag{3.5.3}$$

and

$$(A^2 - 3A + 2I)\mathbf{x}_0 = (A - I)[(A - 2I)\mathbf{x}_0] = \mathbf{0} \tag{3.5.4}$$

Set $\mathbf{u}_1 = (A - 2I)\mathbf{x}_0$ and $\mathbf{u}_2 = (A - I)\mathbf{x}_0$. Now neither $\mathbf{u}_1$ nor $\mathbf{u}_2$ are $\mathbf{0}$. For if $\mathbf{u}_1 = \mathbf{0}$, then $\mathbf{u}_1 = A\mathbf{x}_0 - 2\mathbf{x}_0 = \mathbf{0}$ would imply $\{\mathbf{x}_0, A\mathbf{x}_0\}$ is linearly dependent. Likewise for $\mathbf{u}_2$. From (3.5.3) and (3.5.4) it follows that

$$(A - 2I)\mathbf{x}_0 = \mathbf{u}_1 \quad \text{is annihilated by} \quad A - I$$

and

$$(A - I)\mathbf{x}_0 = \mathbf{u}_2 \quad \text{is annihilated by} \quad A - 2I$$

Therefore $\lambda = 1$ is an eigenvalue of A. Also, $\mathbf{u}_1$, because it is annihilated by $A - I$, is a corresponding eigenvector; $\lambda = 2$ is an eigenvalue of A and $\mathbf{u}_2$ is an eigenvector corresponding to it. From (3.5.1) we find

$$\mathbf{u}_1 = A\mathbf{x}_0 - 2\mathbf{x}_0 = -\begin{bmatrix} 0 \\ 0 \\ 3 \end{bmatrix} \qquad \mathbf{u}_2 = A\mathbf{x}_0 - \mathbf{x}_0 = \begin{bmatrix} 2 \\ 0 \\ 0 \end{bmatrix}$$

so that, by inspection,

$$-\mathbf{u}_1+\mathbf{u}_2=\begin{bmatrix}2\\0\\3\end{bmatrix}=\mathbf{x}_0 \tag{3.5.5}$$

Corresponding to the eigenvector $\mathbf{u}_1$ is the solution $\mathbf{u}_1e^t$, and corresponding to $\mathbf{u}_2$ is the solution $\mathbf{u}_2e^{2t}$. Therefore, corresponding to (3.5.5) is the solution

$$\mathbf{x}_1(t)=-\mathbf{u}_1e^t+\mathbf{u}_2e^{2t} \tag{3.5.6}$$

which has $\mathbf{x}(0)=\mathbf{x}_0$ since $\mathbf{x}(0)=-\mathbf{u}_1+\mathbf{u}_2$.

3.5.2. Example

Same problem as Example 3.5.1, except

$$\mathbf{x}_0=\begin{bmatrix}0\\-1\\0\end{bmatrix}$$

Solution

We compute

$$\mathbf{x}_0=\begin{bmatrix}0\\-1\\0\end{bmatrix}\qquad \mathrm{A}\mathbf{x}_0=\begin{bmatrix}-1\\-2\\0\end{bmatrix}\qquad \mathrm{A}^2\mathbf{x}_0=\begin{bmatrix}-4\\-4\\0\end{bmatrix}$$

and hence by inspection, $\{\mathbf{x}_0, \mathrm{A}\mathbf{x}_0\}$ is linearly independent while

$$\mathrm{A}^2\mathbf{x}_0=4\mathrm{A}\mathbf{x}_0-4\mathbf{x}_0$$

Therefore,

$$(\mathrm{A}-2\mathrm{I})^2\mathbf{x}_0=(\mathrm{A}^2-4\mathrm{A}+4\mathrm{I})\mathbf{x}_0=\mathbf{0}$$

Now, $(\mathrm{A}-2\mathrm{I})\mathbf{x}_0\neq\mathbf{0}$ and $(\mathrm{A}-2\mathrm{I})^2\mathbf{x}_0=\mathbf{0}$. Therefore, $(\mathrm{A}-2\mathrm{I})\mathbf{x}_0=\mathbf{u}_{1,1}$ is an eigenvector corresponding to the eigenvalue $\lambda=2$ and $\mathbf{x}_0=\mathbf{u}_{1,2}$ is a root vector of order 2. We find

$$\mathbf{u}_{1,1}=\mathrm{A}\mathbf{x}_0-2\mathbf{x}_0=\begin{bmatrix}-1\\0\\0\end{bmatrix}\qquad \mathbf{u}_{1,2}=\mathbf{x}_0=\begin{bmatrix}0\\-1\\0\end{bmatrix}$$

and hence

$$\mathbf{x}_2(t)=(\mathbf{u}_{1,2}+t\mathbf{u}_{1,1})e^{2t} \tag{3.5.7}$$

with $\mathbf{x}(0)=\mathbf{u}_{1,2}$, solves the equation and meets the initial condition.

3.5.3. Example

Same as Example 3.5.1 except

$$\mathbf{x}_0=\begin{bmatrix}1\\0\\0\end{bmatrix}$$

Solution

As before

$$\mathbf{x}_0 = \begin{bmatrix} 1 \\ 0 \\ 0 \end{bmatrix} \qquad \mathrm{A}\mathbf{x}_0 = \begin{bmatrix} 2 \\ 0 \\ 0 \end{bmatrix}$$

leads to $\mathrm{A}\mathbf{x}_0 - 2\mathrm{I}\mathbf{x}_0 = (\mathrm{A} - 2\mathrm{I})\mathbf{x}_0 = \mathbf{0}$. Therefore $\mathbf{x}_3(t) = \mathbf{x}_0 e^{2t}$ is the solution of this initial-value problem.

We might point out that the function

$$\mathbf{x}(t) = k_1\mathbf{x}_1(t) + k_2\mathbf{x}_2(t) + k_3\mathbf{x}_3(t)$$

solves $\mathbf{x}' = \mathrm{A}\mathbf{x}$,

$$\mathrm{A} = \begin{bmatrix} 2 & 1 & 0 \\ 0 & 2 & 0 \\ 0 & 0 & 1 \end{bmatrix}$$

by the principle of superposition. Also,

$$\mathbf{x}(0) = k_1\begin{bmatrix} 2 \\ 0 \\ 3 \end{bmatrix} + k_2\begin{bmatrix} 0 \\ -1 \\ 0 \end{bmatrix} + k_3\begin{bmatrix} 1 \\ 0 \\ 0 \end{bmatrix}$$

Since

$$\begin{bmatrix} 2 \\ 0 \\ 3 \end{bmatrix} \quad \begin{bmatrix} 0 \\ -1 \\ 0 \end{bmatrix} \quad \begin{bmatrix} 1 \\ 0 \\ 0 \end{bmatrix}$$

spans $\mathscr{R}^2$, $\mathbf{x}(t)$ solves the general initial-value problem, $\mathbf{x}' = \mathrm{A}\mathbf{x}$, $\mathbf{x}(0) = \mathbf{x}_0$ ($\mathbf{x}_0$ arbitrary). For a solution $\mathbf{x}(t)$ of $\mathbf{x}' = \mathrm{A}\mathbf{x}$ to have this property it is apparent that $\mathbf{x}(0)$ must be a linear combination of vectors which span $\mathscr{R}^n$. This motivates the notion of a general solution of $\mathbf{x}' = \mathrm{A}\mathbf{x}$: The solution of the problem $\mathbf{x}' = \mathrm{A}\mathbf{x}$, $x(0) = \mathbf{x}_0$, where $\mathbf{x}_0$ is an arbitrary vector, is called, by some authors, *the general solution* of $\mathbf{x}' = \mathrm{A}\mathbf{x}$.† This language is sometimes convenient despite its vagueness. See the remark following Example 2.4.1.

It is tacit in this statement that $\mathbf{x}(0)$ is not a specific vector but one which may assume arbitrary values as in Example 3.5.3.

In the earlier sections, we obtained a general solution and specified the parameters to solve a given initial-value problem. *From the point of view of this section, the general solution is obtained from the solutions of initial-value problems.*

We now turn to a challenging example of an initial-value problem.

3.5.4. Example

Solve

$$\mathbf{x}' = \begin{bmatrix} 2 & -1 & 1 & 0 \\ 0 & 3 & -1 & 0 \\ -2 & -1 & 5 & 0 \\ 0 & 0 & 0 & 1 \end{bmatrix}\mathbf{x} \qquad \mathbf{x}_0 = \begin{bmatrix} 1 \\ 1 \\ 3 \\ 0 \end{bmatrix}$$

†In Section 4.1 we will formulate a precise definition of this concept.

Solution

We find

$$\mathbf{x}_0 = \begin{bmatrix} 1 \\ 1 \\ 3 \\ 0 \end{bmatrix} \qquad A\mathbf{x}_0 = \begin{bmatrix} 4 \\ 0 \\ 12 \\ 0 \end{bmatrix} \qquad A^2\mathbf{x}_0 = \begin{bmatrix} 20 \\ -12 \\ 52 \\ 0 \end{bmatrix} \qquad A^3\mathbf{x}_0 = \begin{bmatrix} 104 \\ -88 \\ 232 \\ 0 \end{bmatrix} \tag{3.5.8}$$

After the usual row reductions and some labor,

$$\begin{bmatrix} 1 & 1 & 3 & 0 & \mathbf{x}_0 \\ 4 & 0 & 12 & 0 & \mathbf{x}_1 \\ 20 & -12 & 52 & 0 & \mathbf{x}_2 \\ 104 & -88 & 232 & 0 & \mathbf{x}_3 \end{bmatrix} \to \begin{bmatrix} 1 & 1 & 3 & 0 & \mathbf{x}_0 \\ 0 & -4 & 0 & 0 & \mathbf{x}_1 - 4\mathbf{x}_0 \\ 0 & 0 & 8 & 0 & \mathbf{x}_2 - 8\mathbf{x}_1 + 12\mathbf{x}_0 \\ 0 & 0 & 0 & 0 & \mathbf{x}_3 - 10\mathbf{x}_2 + 32\mathbf{x}_1 - 32\mathbf{x}_0 \end{bmatrix}$$

Hence,

$$(A^3 - 10A^2 + 32A - 32I)\mathbf{x}_0 = (A - 2I)(A - 4I)^2\mathbf{x}_0 = \mathbf{0} \tag{3.5.9}$$

From this it follows that

(1) $\mathbf{u}_1 = (A - 4I)^2\mathbf{x}_0$ is an eigenvector corresponding to the eigenvalue $\lambda = 2$, since $(A - 2I)$ annihilates $\mathbf{u}_1 \neq \mathbf{0}$.

(2) $\mathbf{u}_2 = (A - 2I)(A - 4I)\mathbf{x}_0$ is an eigenvector corresponding to the eigenvalue $\lambda = 4$, since $(A - 4I)$ annihilates $\mathbf{u}_2 \neq \mathbf{0}$.

(3) $\mathbf{u}_{2,2} = (A - 2I)\mathbf{x}_0$ is a root vector of order 2 corresponding to the eigenvalue $\lambda = 4$, since $(A - 4I)\mathbf{u}_{2,2} = \mathbf{u}_2$ and $\mathbf{u}_{2,2} \neq \mathbf{0}$.

We use (3.5.8) to find

$$\mathbf{u}_1 = (A - 4I)^2\mathbf{x}_0 = A^2\mathbf{x}_0 - 8A\mathbf{x}_0 + 16\mathbf{x}_0 = \begin{bmatrix} 4 \\ 4 \\ 4 \\ 0 \end{bmatrix}$$

$$\mathbf{u}_2 = (A - 2I)(A - 4I) = A^2\mathbf{x}_0 - 6A\mathbf{x}_0 + 8\mathbf{x}_0 = \begin{bmatrix} 4 \\ -4 \\ 4 \\ 0 \end{bmatrix}$$

$$\mathbf{u}_{2,2} = (A - 2I)\mathbf{x}_0 = A\mathbf{x}_0 - 2\mathbf{x}_0 = \begin{bmatrix} 2 \\ -2 \\ 6 \\ 0 \end{bmatrix}$$

We solve

$$\mathbf{x}_0 = c_1\mathbf{u}_1 + c_2\mathbf{u}_2 + c_3\mathbf{u}_{2,2}$$

for c_1, c_2, c_3 and find $c_1 = \frac{1}{4}, c_2 = -\frac{1}{4}, c_3 = \frac{1}{2}$. Therefore

$$\mathbf{x}_0 = \tfrac{1}{4}\mathbf{u}_1 - \tfrac{1}{4}\mathbf{u}_2 + \tfrac{1}{2}\mathbf{u}_{2,2}$$

$$\mathbf{x}(t) = \tfrac{1}{4}\mathbf{u}_1 e^{2t} - \tfrac{1}{4}\mathbf{u}_2 e^{4t} + \tfrac{1}{2}[\mathbf{u}_{2,2} + t\mathbf{u}_2]e^{4t}$$

$$= e^{2t}\begin{bmatrix}1\\1\\1\\0\end{bmatrix} + e^{4t}\begin{bmatrix}2t\\-2t\\2t+2\\0\end{bmatrix}$$

Exercise 3.5

Part 1

1. Solve the initial-value problem of Example 3.5.1 by the technique of the previous section and compare with Equation (3.5.6).
2. Solve the problem of Example 3.5.2 by the technique of the previous section and compare with Equation (3.5.7).
3. Solve

$$\mathbf{x}' = \begin{bmatrix}1 & 1 & 0\\0 & 1 & 0\\0 & 1 & 1\end{bmatrix}\mathbf{x}$$

when

(a) $\mathbf{x}_0 = \mathbf{e}_1 = [1, 0, 0]^{\mathrm{T}}$
(b) $\mathbf{x}_0 = \mathbf{e}_2 = [0, 1, 0]^{\mathrm{T}}$
(c) $\mathbf{x}_0 = \mathbf{e}_3 = [0, 0, 1]^{\mathrm{T}}$

4. Write the general solution of the differential system given in Problem 3 above from the answers given to parts (a), (b), and (c) of this question.
5. Solve the initial-value problem of Example 3.5.4 with $\mathbf{x}_0 = [0, 0, 0, 0, 1]^{\mathrm{T}}$.

Use the method of this section to solve the initial-value problem, $\mathbf{x}' = \mathrm{A}\mathbf{x}$, $\mathbf{x}(0) = \mathbf{x}_0$, given A and $\mathbf{x}_0$ in Problems 6–15 below.

6. $\mathrm{A} = \begin{bmatrix}0 & 1\\-4 & 4\end{bmatrix} \quad \mathbf{x}_0 = \begin{bmatrix}1\\2\end{bmatrix}$

7. $\begin{bmatrix}0 & 1\\-4 & 4\end{bmatrix} \quad \mathbf{x}_0 = \begin{bmatrix}1\\1\end{bmatrix}$

8. $\mathrm{A} = \begin{bmatrix}1 & 1 & 0\\0 & 1 & 0\\0 & 0 & -2\end{bmatrix} \quad \mathbf{x}_0 = \begin{bmatrix}1\\0\\1\end{bmatrix}$

9. $\mathrm{A} = \begin{bmatrix}1 & 1 & 0\\0 & 1 & 0\\0 & 0 & -2\end{bmatrix} \quad \mathbf{x}_0 = \begin{bmatrix}1\\1\\1\end{bmatrix}$

10. $\mathrm{A} = \begin{bmatrix}0 & 1 & 0\\0 & 0 & 1\\4 & -8 & 5\end{bmatrix} \quad \mathbf{x}_0 = \begin{bmatrix}0\\1\\1\end{bmatrix}$

11. $\mathrm{A} = \begin{bmatrix}0 & 1 & 0\\0 & 0 & 1\\4 & -8 & 5\end{bmatrix} \quad \mathbf{x}_0 = \begin{bmatrix}0\\1\\0\end{bmatrix}$

12. $\mathrm{A} = \begin{bmatrix}-1 & 1 & 0\\0 & -1 & 0\\0 & 0 & -1\end{bmatrix} \quad \mathbf{x}_0 = \begin{bmatrix}-3\\1\\-2\end{bmatrix}$

13. $\mathrm{A} = \begin{bmatrix}0 & 1 & 0\\0 & 0 & 1\\1 & -3 & 3\end{bmatrix} \quad \mathbf{x}_0 = \begin{bmatrix}1\\1\\1\end{bmatrix}$

14. Same as Problem 13 except $\mathbf{x}_0 = [0, 1, 2]^{\mathrm{T}}$.
15. Same as Problem 13 except $\mathbf{x}_0 = [0, 0, 1]^{\mathrm{T}}$.

Part 2

1. If A is 2×2, $\mathbf{y}$ solves $\mathbf{x}' = A\mathbf{x}$, $\mathbf{x}(0) = \mathbf{y}_0$, $\mathbf{z}$ solves $\mathbf{x}' = A\mathbf{x}$, $\mathbf{x}(0) = \mathbf{z}_0$ then prove that constants a and b can be found such that $a\mathbf{y}+b\mathbf{z}$ solves $\mathbf{x}' = A\mathbf{x}$, $\mathbf{x}(0) = \mathbf{x}_0$ for any $\mathbf{x} \in \mathscr{R}^2$ if $\mathbf{y}_0$, $\mathbf{z}_0$ are linearly independent.
2. Generalize the theorem stated as Problem 1 above for A, arbitrary.

3.6.* THE GENERAL THEORY

Polynomials in the matrix A which, when operating on $\mathbf{x}_0$, yields $\mathbf{0}$, played a central role in the method presented in Section 3.5. We examine these polynomials more carefully in this section.

3.6.1. DEFINITION *If $f(z)$ is a polynomial of degree n, say, $f(z) = c_0z^n + c_1z^{n-1} + \cdots + c_n$, then $f(A)$ is defined by*

$$f(A) = c_0A^n + c_1A^{n-1} + \cdots + c_nI$$

In forming $f(A)$, then, we make the substitution $z^k \leftrightarrow A^k$ for $1 \leqslant k \leqslant n$ and $1 = z^0 \leftrightarrow I = A^0$. It is not difficult to show that polynomials in A behave like polynomials in z in the sense that

$$f(z) = f_1(z)f_2(z) \qquad \text{and} \qquad g(z) = c_1g_2(z) + c_2g_2(z)$$

imply

$$f(A) = f_1(A)f_2(A) \qquad \text{and} \qquad g(A) = c_1g_1(A) + c_2g_2(A)$$

for polynomials f_1, f_2, g_1, and g_2.

3.6.2. DEFINITION *Let A be $n\times n$, $\mathbf{x}_0 \in \mathscr{C}^n$ and $f(z)$ a polynomial. If $f(A)\mathbf{x}_0 = \mathbf{0}$ we say $f(z)$ is an annihilating polynomial for $\mathbf{x}_0$ with respect to A.*

That is to say, $f(z)$ is an annihilating polynomial for $\mathbf{x}_0$ if $f(A)$ annihilates $\mathbf{x}_0$. We sometimes say $f(z)$ annihilates $\mathbf{x}_0$ instead of the more precise statement in the definition because it is simpler to do so. No confusion should result because of this.

3.6.1. Example

(1) $\mathbf{0}$ is annihilated by any polynomial, A arbitrary.

(2) $\mathbf{x}_0$, an eigenvector of A corresponding to the eigenvalue λ_0, is annihilated by $(z-\lambda_0)$.

(3) $\mathbf{u}_k$ a root vector of A, order k, corresponding to the eigenvalue λ is annihilated by $(z-\lambda)^k$.

(4) $\begin{bmatrix} 1 \\ 0 \\ 3 \end{bmatrix}$ is annihilated by $z^2 - 3z + 2$ where $A = \begin{bmatrix} 2 & 1 & 0 \\ 0 & 2 & 0 \\ 0 & 0 & 1 \end{bmatrix}$.

(5) $\begin{bmatrix} 1 \\ 1 \\ 3 \\ 0 \end{bmatrix}$ is annihilated by $z^3 - 10z^2 + 32z - 32$ if

$$A = \begin{bmatrix} 2 & -1 & 1 & 0 \\ 0 & 3 & -1 & 0 \\ -2 & -1 & 5 & 0 \\ 0 & 0 & 0 & 1 \end{bmatrix}$$

(6) If $f(z)$ annihilates $\mathbf{x}_0$ then $g(z)f(z)$ annihilates $\mathbf{x}_0$ also.

Item (1) is trivial; items (2) and (3) follow from the definition of eigenvector and root vector, respectively; items (4) and (5) were proved in Section 3.5, Equations (3.5.2) and (3.5.9), respectively. Item (6) follows because

$$[g(A)f(A)]\mathbf{x}_0 = g(A)[f(A)\mathbf{x}_0] = g(A)\mathbf{0} = \mathbf{0}$$

independent of $g(A)$.

For us, it is particularly important that we have the lowest degree polynomial which annihilates $\mathbf{x}_0$. This motivates the following definition.

3.6.3. DEFINITION *If* $g(z) = z^m + a_1 z^{m-1} + \cdots + a_m$, $m \geqslant 0$, *annihilates* $\mathbf{x}_0$ *(with respect to* A*) and no polynomial of lesser degree does, then* $g(z)$ *is the minimal annihilating polynomial of* $\mathbf{x}_0$ *with respect to* A.

The choice of the word "the" in the definition signifies that the minimal polynomial is unique. This is easily proved.

3.6.4. THEOREM *If* $g(z)$ *and* $h(z)$ *are two minimal polynomials of degree* k *annihilating* $\mathbf{x}_0$ *then* $g(z) = h(z)$.

PROOF The difference, $g(z) - h(z)$ annihilates $\mathbf{x}_0$ because $[g(A) - h(A)]\mathbf{x}_0 = g(A)\mathbf{x}_0 - h(A)\mathbf{x}_0 = \mathbf{0}$. The degrees of g and h are the same, since we are assuming they are both minimal annihilators of $\mathbf{x}_0$. But, because $g(z)$ and $h(z)$ both start with z^k (i.e., they are monic), their difference is of degree $\leqslant k-1$. This is impossible, unless $g(z) - h(z) \equiv 0$, because otherwise $g(z) - h(z)$ would annihilate $\mathbf{x}_0$ and be of degree $< k$.

This establishes the fact that there is never more than one minimal annihilating polynomial for a given $\mathbf{x}_0$. Is there even one?

3.6.5. THEOREM *Every vector* $\mathbf{x}_0$ *has a minimal annihilating polynomial.*

PROOF Consider the set of $n+1$ vectors $\{\mathbf{x}_0, A\mathbf{x}_0, A^2\mathbf{x}_0, \ldots, A^n\mathbf{x}_0\}$. Since each vector is in $\mathscr{C}^n$, this set is linearly dependent. Consider the sequences

$$\begin{array}{l} \mathbf{x}_0 \\ \mathbf{x}_0, A\mathbf{x}_0 \\ \mathbf{x}_0, A\mathbf{x}_0, A^2\mathbf{x}_0 \\ \cdot \\ \cdot \\ \cdot \\ \mathbf{x}_0, A\mathbf{x}_0, A^2\mathbf{x}_0, \ldots, A^n\mathbf{x}_0 \end{array}$$

There must be a first sequence which is linearly dependent since the last one must be linearly dependent. Suppose, therefore, that $\{\mathbf{x}_0, A\mathbf{x}_0, \ldots, A^k\mathbf{x}_0\}$, $1 \leq k \leq n$, is linearly dependent but $\{\mathbf{x}_0, A\mathbf{x}_0, \ldots, A^{k-1}\mathbf{x}_0\}$ is not. Then there exists scalars $c_0, c_1, \ldots, c_k$ with $c_k \neq 0$ such that

$$c_0\mathbf{x}_0 + c_1 A\mathbf{x}_0 + \cdots + c_k A^k \mathbf{x}_0 = \mathbf{0} \tag{3.6.1}$$

(If $c_k = 0$, then some nontrivial linear combination of the vectors $\mathbf{x}_0, A\mathbf{x}_0, \ldots, A^{k-1}\mathbf{x}_0$ must be zero, contradicting our assumption that $\{\mathbf{x}_0, A\mathbf{x}_0, \ldots, A^k\mathbf{x}_0\}$ is the first dependent set.) From (3.6.1) it follows that

$$g(z) = z^k + \frac{c_{k-1}}{c_k} z^{k-1} + \frac{c_{k-2}}{c_k} z^{k-2} + \cdots + \frac{c_0}{c_k}$$

annihilates $\mathbf{x}_0$. Furthermore, if $f(z)$ has degree less than k, then $f(A)\,\mathbf{x}_0 \neq \mathbf{0}$ since $\mathbf{x}_0, \ldots, A^{k-1}\mathbf{x}_0$ is linearly independent. Thus $g(z)$ is minimal.

The proof of this theorem establishes not only the existence of an annihilating polynomial but indicates a constructive procedure for obtaining it. This was, in fact, precisely the method we have used and illustrated in Section 3.5.

3.6.2. Example

Find the minimal annihilating polynomial of

$$\mathbf{x}_0 = \begin{bmatrix} 1 \\ 0 \end{bmatrix} \quad \text{given} \quad A = \begin{bmatrix} -1 & -2 \\ 3 & 4 \end{bmatrix}$$

Solution

With these definitions,

$$\mathbf{x}_0 = \begin{bmatrix} 1 \\ 0 \end{bmatrix} \quad A\begin{bmatrix} 1 \\ 0 \end{bmatrix} = \begin{bmatrix} -1 \\ 3 \end{bmatrix} \quad \text{and} \quad A^2\begin{bmatrix} 1 \\ 0 \end{bmatrix} = A\begin{bmatrix} -1 \\ 3 \end{bmatrix} = \begin{bmatrix} -5 \\ 9 \end{bmatrix}$$

the sequences $\mathbf{x}_0$, and $\mathbf{x}_0$, $A\mathbf{x}_0$ are clearly independent, but $\mathbf{x}_0$, $A\mathbf{x}_0$, $A^2\mathbf{x}_0$ cannot be. To obtain the dependency relation, we compute

$$\begin{bmatrix} 1 & 0 & \mathbf{x}_0 \\ -1 & 3 & A\mathbf{x}_0 \\ -5 & 9 & A^2\mathbf{x}_0 \end{bmatrix} \to \begin{bmatrix} 1 & 0 & \mathbf{x}_0 \\ 0 & 3 & A\mathbf{x}_0 + \mathbf{x}_0 \\ 0 & 9 & A^2\mathbf{x}_0 + 5\mathbf{x}_0 \end{bmatrix} \to \begin{bmatrix} 1 & 0 & \mathbf{x} \\ 0 & 3 & A\mathbf{x}_0 + \mathbf{x}_0 \\ 0 & 0 & A^2\mathbf{x}_0 + 5\mathbf{x}_0 - 3(A\mathbf{x}_0 + \mathbf{x}_0) \end{bmatrix}$$

Thus, $(A^2 - 3A + 2I)\mathbf{x}_0 = \mathbf{0}$ and $z^2 - 3z + 2$ is the minimal annihilating polynomial.

The connection between the roots of the minimal polynomial and the eigenvalues, eigenvectors, and root vectors of A has been amply illustrated in Section 3.5. We state these results in a general context in the next two lemmas.

First recall that λ is called an m-fold root of $g(z)$ if $g(z) = (z - \lambda_1)^m g_1(z)$ where $g_1(\lambda_1) \neq 0$.

3.6.6. LEMMA *If $\mathbf{x}_0$ has the minimal annihilating polynomial $g(z)$ and λ_1 is an m-fold root of $g(z)$, then λ_1 is an eigenvalue of A and $g_1(A)\mathbf{x}_0 = \mathbf{u}_{1,m}$ is a root vector of order m corresponding to λ_1.*

PROOF The proof is a consequence of two facts:

(1) $(\mathbf{A}-\lambda_1\mathbf{I})^m \mathbf{u}_{1,m} = (\mathbf{A}-\lambda_1\mathbf{I})^m [g_1(\mathbf{A})\mathbf{x}_0]$
$= g(\mathbf{A})\mathbf{x}_0 = \mathbf{0}$

(2) $(\mathbf{A}-\lambda_1\mathbf{I})^{m-1} \mathbf{u}_{1,m} \neq \mathbf{0}$ for otherwise, $(\mathbf{A}-\lambda_1\mathbf{I})^{m-1} \mathbf{u}_{1,m} = \mathbf{0}$ would imply that $h(z) = (z-\lambda_1)^{m-1}g_1(z)$ annihilates $\mathbf{x}_0$, contradicting the fact that $g(z)$ is minimal.

Extending this result to all the factors of $g(z)$ we have

3.6.7. LEMMA *If* $g(z) = (z-\lambda_1)^{m_1}(z-\lambda_2)^{m_2}\cdots(z-\lambda_r)^{m_r}$, $m_1+m_2+\cdots+m_r = k$, *is the minimal annihilating polynomial of* $\mathbf{x}_0$ *then*

$$
\begin{aligned}
\mathbf{u}_{1,m_1} &= g_1(\mathbf{A})\mathbf{x}_0 \quad \textit{where} \quad g_1(z) = g(z)/(z-\lambda_1)^{m_1}\\
\mathbf{u}_{2,m_2} &= g_2(\mathbf{A})\mathbf{x}_0 \quad \textit{where} \quad g_2(z) = g(z)/(z-\lambda_2)^{m_2}\\
&\;\;\vdots\\
\mathbf{u}_{r,m_r} &= g_r(\mathbf{A})\mathbf{x}_0 \quad \textit{where} \quad g_r(z) = g(z)/(z-\lambda_r)^{m_r}
\end{aligned}
\tag{3.6.2}
$$

are root vectors of order $m_1, m_2, \ldots, m_r$ *corresponding to the eigenvalues* $\lambda_1, \lambda_2, \ldots, \lambda_r$, *respectively.*

Let us label the vectors in the chains determined by each of the root vectors in (3.6.2). We use Equation (3.6.2).

$$
\begin{aligned}
&(1)\ \text{Chain for } \lambda = \lambda_1:\\
&\mathbf{u}_{1,m_1} = g_1(\mathbf{A})\mathbf{x}_0\\
&\mathbf{u}_{1,m_1-1} = (\mathbf{A}-\lambda_1\mathbf{I})\,\mathbf{u}_{1,m_1} = (\mathbf{A}-\lambda_1\mathbf{I})\,g_1(\mathbf{A})\mathbf{x}_0\\
&\;\;\vdots\\
&\mathbf{u}_{1,1} = (\mathbf{A}-\lambda_1\mathbf{I})\mathbf{u}_{1,2} = (\mathbf{A}-\lambda_1\mathbf{I})^{m_1-1}\,g_1(\mathbf{A})\mathbf{x}_0\\
&(2)\ \text{Chain for } \lambda = \lambda_2:\\
&\mathbf{u}_{2,m_2} = g_2(\mathbf{A})\mathbf{x}_0\\
&\mathbf{u}_{2,m_2-1} = (\mathbf{A}-\lambda_2\mathbf{I})\mathbf{u}_{2,m_2} = (\mathbf{A}-\lambda_2\mathbf{I})g_2(\mathbf{A})\mathbf{x}_0\\
&\;\;\vdots\\
&\mathbf{u}_{2,1} = (\mathbf{A}-\lambda_2\mathbf{I})\mathbf{u}_{2,2} = (\mathbf{A}-\lambda_2\mathbf{I})^{m_2-1}g_2(\mathbf{A})\mathbf{x}_0\\
&(r)\ \text{Chain for } \lambda = \lambda_r:\\
&\mathbf{u}_{r,m_r} = g_r(\mathbf{A})\mathbf{x}_0\\
&\mathbf{u}_{r,m_r-1} = (\mathbf{A}-\lambda_r\mathbf{I})\mathbf{u}_{r,m_r} = (\mathbf{A}-\lambda_r\mathbf{I})g_r(\mathbf{A})\mathbf{x}_0\\
&\;\;\vdots\\
&\mathbf{u}_{r,1} = (\mathbf{A}-\lambda_r\mathbf{I})\mathbf{u}_{r,2} = (\mathbf{A}-\lambda_r\mathbf{I})^{m_r-1}g_r(\mathbf{A})\mathbf{x}_0
\end{aligned}
\tag{3.6.3}
$$

This defines a set of $m_1+m_2+\cdots+m_r = k$ vectors where k = degree of the minimal polynomial $g(z)$.

3.6.8. LEMMA *The k vectors in (3.6.3) are linearly independent.*

PROOF Let $\alpha_{i,j}$ be scalars such that

$$\begin{aligned} &\alpha_{1,1}\mathbf{u}_{1,m_1} + \alpha_{1,2}\mathbf{u}_{1,m_1-1} + \cdots + \alpha_{1,m_1}\mathbf{u}_{1,1} \\ &+ \alpha_{2,1}\mathbf{u}_{2,m_2} + \alpha_{2,2}\mathbf{u}_{2,m_2-1} + \cdots + \alpha_{2,m_2}\mathbf{u}_{2,1} + \cdots \\ &+ \alpha_{r,1}\mathbf{u}_{r,m_r} + \alpha_{r,2}\mathbf{u}_{r,m_r-1} + \cdots + \alpha_{r,m_r}\mathbf{u}_{r,1} = \mathbf{0} \end{aligned} \tag{3.6.4}$$

In terms of A, we find that (3.6.4) may be written via (3.6.3) as

$$\begin{aligned} &\{\alpha_{1,1}g_1(\mathrm{A}) + \alpha_{1,2}(\mathrm{A}-\lambda_1\mathrm{I})g_1(\mathrm{A}) + \cdots + \alpha_{1,m_1}(\mathrm{A}-\lambda_1\mathrm{I})^{m_1-1}g_1(\mathrm{A}) \\ &+ \alpha_{2,1}g_2(\mathrm{A}) + \cdots + \alpha_{2,m_2}(\mathrm{A}-\lambda_2\mathrm{I})^{m_2-1}g_2(\mathrm{A}) \\ &+ \cdots \\ &+ \alpha_{r,1}g_r(\mathrm{A}) + \cdots + \alpha_{r,m_r}(\mathrm{A}-\lambda_r\mathrm{I})^{m_r-1}g_r(\mathrm{A})\}\mathbf{x}_0 = \mathbf{0} \end{aligned} \tag{3.6.5}$$

The polynomial in the braces is a sum of polynomials each of whose degree is less than k, in view of the definition of $g(z), g_1(z), \ldots, g_r(z)$, and is therefore itself of degree $< k$. Call this polynomial $G(\mathrm{A})$. Then $G(z)$ annihilates $\mathbf{x}_0$ and is of lesser degree than the minimal polynomial annihilating $\mathbf{x}_0$, namely, $g(z)$, and is therefore identically zero. We now show that $\alpha_{ij} = 0$ for all i and j which would then complete the proof. Refer to the definition of $g_i(z)$. This shows that $G(z) \equiv 0$ may be written

$$\begin{aligned} g(z)&\left\{\frac{\alpha_{1,1}}{(z-\lambda_1)^{m_1}} + \frac{\alpha_{1,2}}{(z-\lambda_1)^{m_1-1}} + \cdots + \frac{\alpha_{1,m_1}}{z-\lambda_1}\right. \\ &\left. + \frac{\alpha_{r,1}}{(z-\lambda_r)^{m_r}} + \frac{\alpha_{r,2}}{(z-\lambda_r)^{m_r-1}} + \cdots + \frac{\alpha_{r,m_r}}{z-\lambda_r}\right\} \equiv 0 \end{aligned}$$

Since $g(z) \not\equiv 0$, the expression in the braces is identically zero. If the expression in braces is multiplied through by $(z-\lambda_1)^{m_1}$ and then z set equal to λ_1, we find $\alpha_{11} = 0$. Using this, multiplying the brackets by $(z-\lambda_1)^{m_1-1}$, setting $z = \lambda_1$ again leads to $\alpha_{1,2} = 0$. We repeat this until $0 = \alpha_{1,1} = \alpha_{1,2} = \cdots = \alpha_{1,m_1}$ is concluded. But the same argument with $\alpha_{1,k}$ replaced by $\alpha_{j,k}$ shows that $\alpha_{j,1} = \alpha_{j,2} = \cdots = \alpha_{j,m_j} = 0$ for all j.

3.6.9. THEOREM *The k vectors enumerated in (3.6.3) form a basis for* $\langle \mathbf{x}_0, \mathrm{A}\mathbf{x}_0, \ldots, \mathrm{A}^{k-1}\mathbf{x}_0\rangle$ *and hence some linear combination of them represents* $\mathbf{x}_0$.

PROOF The k vectors described in (3.6.3) are constructed from $\mathbf{x}_0$ by operating on $\mathbf{x}_0$ by some polynomial in A of degree $< k$. Therefore each of these vectors is a linear combination of the vectors $\mathbf{x}_0, \mathrm{A}\mathbf{x}_0, \ldots, \mathrm{A}^{k-1}\mathbf{x}_0$. Lemma 3.6.8 establishes their independence. Since there are k of them they form a basis for $\langle \mathbf{x}_0, \mathrm{A}\mathbf{x}_0, \ldots, \mathrm{A}^{k-1}\mathbf{x}_0\rangle$. Finally, $\mathbf{x}_0$ is in $\langle \mathbf{x}_0, \mathrm{A}\mathbf{x}_0, \ldots, \mathrm{A}^{k-1}\mathbf{x}_0\rangle$ and this proves the theorem.

Now corresponding to each root vector and eigenvector of A there is a solution of $\mathbf{x}' = \mathrm{A}\mathbf{x}$ which takes on this eigenvector or root vector at $t_0 = 0$.

If, in the linear combination of root vectors and eigenvectors used to express $\mathbf{x}_0$, we replace the root vectors and eigenvectors by the corresponding solutions of $\mathbf{x}' = \mathrm{A}\mathbf{x}$, the resulting sum will be a solution of $\mathbf{x}' = \mathrm{A}\mathbf{x}$ and at $t_0 = 0$ reduce to $\mathbf{x}_0$. Hence we can always find a solution of $\mathbf{x}' = \mathrm{A}\mathbf{x}, \mathbf{x}(0) = \mathbf{x}_0$. We express this by stating the fundamental existence theorem for linear systems with constant coefficients:

3.6.10. THEOREM *For any choice of* $\mathbf{x}_0 \in \mathscr{R}^n$ *and matrix* A, *there exists a vector function* $\mathbf{x}(t)$ *such that*

$$\mathbf{x}'(t) = \mathrm{A}\mathbf{x}(t) \qquad \mathbf{x}(0) = \mathbf{x}_0$$

In fact, the solution whose existence is assured by this theorem is unique. The proof of this is given in Chapter 4. We shall also see in Chapter 5 that every nth-order linear equation with constant coefficients can be studied by examining a closely related system. Hence Theorem 3.6.10 will prove the existence of solutions in this case also. Moreover, the restriction $t_0 = 0$ is not essential.

3.6.11. COROLLARY *For any choice of* $\mathbf{x}_0 \in \mathscr{R}^n$ *and matrix* A, *there exists a vector function* $\mathbf{x}(t)$ *such that*

$$\mathbf{x}'(t) = \mathrm{A}\mathbf{x}(t), \mathbf{x}(t_0) = \mathbf{x}_0$$

for any real number t_0.

PROOF The proof is accomplished by a simple change of variables. We first let $\mathbf{y}(t)$ be a solution of

$$\mathbf{y}' = \mathrm{A}\mathbf{y} \qquad \mathbf{y}(0) = \mathbf{x}_0 \tag{3.6.6}$$

Now set $\mathbf{x}(t) = \mathbf{y}(t - t_0)$. We claim $\mathbf{x}(t)$ is the required solution. To see this compute

$$\begin{aligned} \mathbf{x}'(t) &= \frac{d}{dt}\mathbf{y}(t - t_0) = \mathbf{y}'(t - t_0) \\ &= \mathrm{A}\mathbf{y}(t - t_0) = \mathrm{A}\mathbf{x}(t) \end{aligned} \tag{3.6.7}$$

and

$$\mathbf{x}(t_0) = \mathbf{y}(0) = \mathbf{x}_0$$

We also have the following corollary.

3.6.12. COROLLARY *If* A *is a real* $n \times n$ *matrix and* $\mathbf{x}_0$ *is in* $\mathscr{R}^n$ *then the initial-value problem*

$$\mathbf{x}' = \mathrm{A}\mathbf{x} \qquad \mathbf{x}(t_0) = \mathbf{x}_0$$

has a real solution $\mathbf{x}(t)$.

PROOF The preceding corollary guarantees a solution, $\mathbf{x}(t)$, but the real part $\operatorname{Re}\mathbf{x}(t)$ is also a solution. (We will later see that $\mathbf{x}(t)$ must be real to begin with.) We conclude this chapter with the long-delayed proof of Theorem 2.5.4. First, a preliminary result.

3.6.13. THEOREM *If* A *is a real* $n \times n$ *symmetric matrix, then* A *has no root vectors of order* $\geqslant 2$.

PROOF Suppose the contrary, that is, suppose $\mathbf{b} \neq \mathbf{0}$ exists such that

$$(\mathrm{A}-\lambda \mathrm{I})\mathbf{b} = \mathbf{a} \neq \mathbf{0}$$
$$(\mathrm{A}-\lambda \mathrm{I})\mathbf{a} \neq \mathbf{0}$$

Then $(\mathrm{A}-\lambda\mathrm{I})^2\mathbf{b} = \mathbf{0}$ and $(\mathrm{A}-\lambda\mathrm{I})\mathbf{b} \neq \mathbf{0}$. Consider the 1×1 matrix

$$[\alpha] = \mathbf{a}^{\mathrm{T}}\bar{\mathbf{a}} = \left[\sum_{i=1}^{n} a_i \bar{a}_i\right] \neq [0] \tag{3.6.8}$$

since $\mathbf{a} \neq \mathbf{0}$. Now according to Lemma 2.5.3, λ is real and thus, since A and λI are real, $\overline{(\mathrm{A}-\lambda\mathrm{I})\mathbf{b}} = (\mathrm{A}-\lambda\mathrm{I})\bar{\mathbf{b}}$. We have, from the symmetry of $\mathrm{A}-\lambda\mathrm{I}$ and Lemma 2.5.1,

$$\begin{aligned}[\alpha] &= [\{(\mathrm{A}-\lambda\mathrm{I})\mathbf{b}\}^{\mathrm{T}}\{\overline{(A-\lambda\mathrm{I})\mathbf{b}}\}] \\ &= [\{(\mathrm{A}-\lambda\mathrm{I})\mathbf{b}\}^{\mathrm{T}}(\mathrm{A}-\lambda\mathrm{I})\bar{\mathbf{b}}] \\ &= [\bar{\mathbf{b}}^{\mathrm{T}}(\mathrm{A}-\lambda\mathrm{I})(\mathrm{A}-\lambda\mathrm{I})\mathbf{b}]\end{aligned}$$

Thus, because $(\mathrm{A}-\lambda\mathrm{I})^2\mathbf{b} = \mathbf{0}$, the contradiction

$$\begin{aligned}[0] \neq [\alpha] &= [\bar{\mathbf{b}}^{\mathrm{T}}(\mathrm{A}-\lambda\mathrm{I})^2\mathbf{b}] \\ &= [\mathbf{b}^{\mathrm{T}}\mathbf{0}] = [0]\end{aligned}$$

proves the theorem.

2.5.4. THEOREM *If* A *is a real* $n \times n$ *symmetric matrix, there exists* n *linearly independent eigenvectors of* A *each belonging to* $\mathscr{R}^n$.

PROOF In the proof of Theorem 3.6.9 let $\mathbf{x}_0 = \mathbf{e}_i$, $i = 1, 2, \ldots, n$. We conclude that each vector $\mathbf{e}_1, \mathbf{e}_2, \ldots, \mathbf{e}_n$ is a linear combination of some of the chains constructed according to Equation (3.6.3). But since there are only eigenvectors in these chains, by Theorem 3.6.13, we deduce that each of the unit vectors $\mathbf{e}_i$, $i = 1, 2, \ldots, n$ is a linear combination of (at most)n eigenvectors of A. Since $\langle \mathbf{e}_1, \mathbf{e}_2, \ldots, \mathbf{e}_n \rangle = \mathscr{R}^n$, the $\mathrm{N} \leqslant n^2$ eigenvectors computed in Equation (3.6.3) (at most n for each $\mathbf{e}_i$) also span $\mathscr{R}^n$. From this family of eigenvectors we may extract a linearly independent family spanning $\mathscr{R}^n$. The dimension of $\mathscr{R}^n$ is n and hence the number of linearly independent eigenvectors spanning $\mathscr{R}^n$ must also be n. We conclude with the observation that this basis of eigenvectors can always be selected with real entries. This follows from the fact that an eigenvector is a solution of $(A-\lambda I)\mathbf{u}=\mathbf{0}$ and, as is the case when A is real and symmetric, $A-\lambda I$ is a real matrix. Thus **u** may always be found with real entries. A perusal of the proof of Theorem 3.6.9 shows that all the chains in Equation (3.6.3) can therefore be constructed from real eigenvectors.

Exercise 3.6

1. Let A be a given $n\times n$ matrix and $g(z)=(z-\mu_1)\,(z-\mu_2)\cdots(z-\mu_k)$ be a minimal annihilating polynomial of $\mathbf{x}_0\neq\mathbf{0}\in\mathscr{R}^n$. Show $\mu_1,\mu_2\ldots,\mu_k$ are eigenvalues of A.
2. Same as Problem 1 above except that $g(z)=(z-\mu)^k g_1(z)$, $g_1(\mu)\neq 0$. Prove that there exists a root vector of order k of A.
3. If the eigenvalues of A are all equal, show that every vector belonging to $\mathscr{R}^n$ is a root vector of A.
4. If $C(\lambda)$ is the characteristic polynomial of A and $g(z)=(z-\mu_1)(z-\mu_2)\cdots(z-\mu_k)$ is a minimal annihilating polynomial of $\mathbf{x}\neq\mathbf{0}$ ($\mathbf{x}\in\mathscr{R}^n$, μ_i, distinct) then $g(z)$ divides $C(z)$.
5. Prove that the minimal annihilating polynomial of $\mathbf{x}\neq\mathbf{0}$ divides every annihilating polynomial of **x**.
6. Let $\mathbf{v}_k$ be a root vector of order k corresponding to the eigenvalue μ of A. Show that $g(z)=(z-\mu)^k$ is the minimum, monic annihilating polynomial of $\mathbf{v}_k$.
7.* If $C(\lambda)$ is the characteristic polynomial of A and the eigenvalues of A are distinct then $C(A)$ is the zero matrix. (This is a special case of the Cayley-Hamilton Theorem which asserts the same conclusion without the hypothesis of distinct eigenvalues.)
8.* Prove: If $\lambda=\lambda$ is an eigenvalue of the $n\times n$ matrix A repeated exactly k times, $1\leq k\leq n$, then there exists k linearly independent root vectors corresponding to λ. (An eigenvector is a root vector of order 1.)

3.7. ELECTRICAL NETWORKS

The theory of electrical circuits is the result of a century of investigation by many scientists including Volta, Galvani, Franklin, Cavendish, Faraday, and Maxwell. In very few circumstances does the mathematics so faithfully mirror physics as in this beautiful theory. Of necessity our treatment is brief.

We shall treat a circuit consisting of three idealized elements: An *inductor*, represented by the symbol in Figure 3.7.1(a); a *resistor*, represented by the symbol in Figure 3.7.1(b); a *capacitor*, represented by the sumbol in Figure 3.7.1(c). If we consider any one of these three elements, denoted by the noncommital symbol in Figure 3.7.1(d), then we may speak of a current flowing from T_1 to T_2 and denote this current by i_{12}. This current is proportional to the number of electrons moving from T_2 to T_1 per second. Current is measured in *amperes*. Historical accident accounts for the peculiarity of current flow taken opposite to the movement of electrons and

Figure 3.7.1

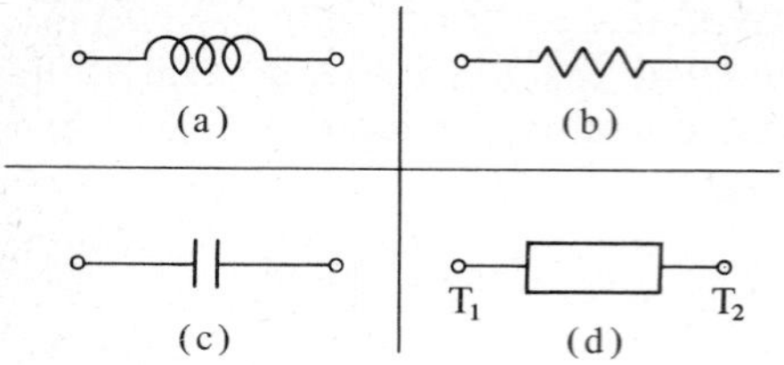

long usage perpetuates it. Across each circuit element is a difference of potential, or *voltage drop* as it is commonly known, written V_{12}. The voltage drop, measured in *volts*, is proportional to the amount of work done by the electric force field on an electron in moving the electron at constant velocity from T_2 to T_1. By agreement, $V_{12} = -V_{21}$ and $i_{12} = -i_{21}$.

The current through an element is related to the voltage drop across it. This relationship depends upon the nature of the element in the following way.

RESISTORS As the name suggests, this element tends to resist the flow of current through it. According to *Ohm's Law*, the mathematical interrelationship between current and voltage drop is given by

$$Ri_{12} = V_{12} \tag{3.7.1}$$

The constant R, measured in *ohms*, is called the resistance of the resistor; it is a physical constant dependent upon the composition and dimensions of the resistor.

INDUCTORS An inductor tends to impede a change in current magnitude. The slower the change, the less the obstruction given to the current by the inductor. Indeed, a constant current flows unimpeded through an inductor. This phenomenon is a consequence of the *Law of Induction* and is expressed mathematically by

$$L\frac{di_{12}}{dt} = V_{12} \tag{3.7.2}$$

Here L is measured in *henries* and is called the inductance. It is a parameter depending on the physical makeup of the inductor.

CAPACITORS A capacitor stores electrons. The effect of this storage is to block slowly varying currents and pass currents whose magnitudes are rapidly changing. The precise relationship follows from *Coulomb's Law* and is given by

$$\frac{1}{C}i_{12} = \frac{d}{dt}V_{12} \tag{3.7.3}$$

C is called the capacitance of the capacitor and is measured in *farads*.

When these components are interconnected in some circuit, the magnitudes of the various voltage drops and the currents are also interconnected and these relationships are known as *Kirchhoff's Laws*.

CURRENT LAW When several elements have a common terminal T_4 as in Figure 3.7.2(a), the sum of the currents in each element flowing toward T_4 is zero. That is, $i_{14}+i_{24}+i_{34}=0$. This law prohibits the accumulation of charge at a terminal.

Figure 3.7.2

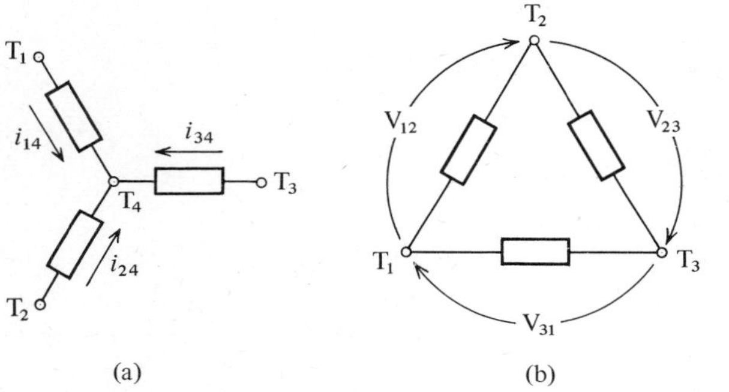

POTENTIAL LAW When elements are connected in a loop such as in Figure 3.7.2(b) the sum of the voltage drops measured across each element is zero. Specifically, $V_{12}+V_{23}+V_{31}=0$.

From these five laws, systems of differential equations arise which must be satisfied at each instant by the current flowing through each element and the voltage drop across each element.

3.7.1. Example

Analyze the circuit described by Figure 3.7.3.

Figure 3.7.3

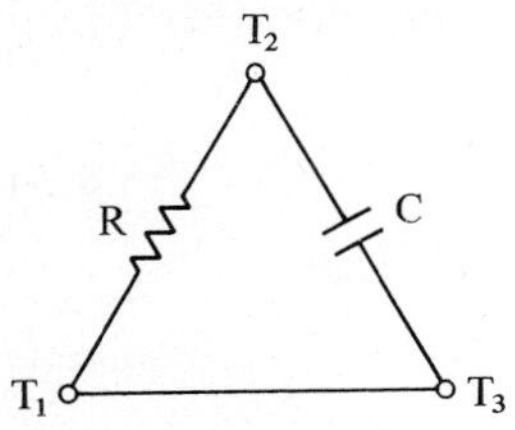

Solution

There are two currents, one flowing through the resistor and one flowing through the capacitor. Corresponding to these currents, we have two voltage drops, one across each of the elements. (The element T_3T_1 is an ideal conductor which acts as a resistor with zero resistance and hence we take $V_{13} = 0$.) The two Kirchhoff Laws yield $i_{12} + i_{32} = 0$ and $V_{12} + V_{23} = 0$. The relevant relationships between currents and voltages are given by (3.7.1) and (3.7.3) and leads to $V_{12} = Ri_{12}$, $dV_{23}/dt = i_{23}/C$. Differentiating the Kirchhoff Voltage Law and substituting, we get

$$\begin{aligned} V'_{12} + V'_{23} &= Ri'_{12} + \frac{1}{C} i_{23} \\ &= Ri'_{23} + \frac{1}{C} i_{23} \\ &= 0 \end{aligned}$$

Thus, $i'_{23} = -(1/RC)i_{23}$ so that $i_{23}(t) = i_{23}(0)e^{-(1/RC)t}$.

3.7.2. Example

Analyze the circuit described by Figure 3.7.4.

Figure 3.7.4

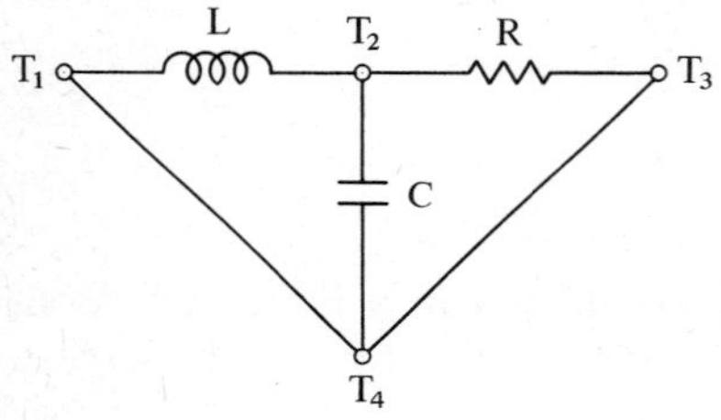

Solution

At the terminal T_2, we have

$$i_{12} + i_{32} + i_{42} = 0 \tag{3.7.4}$$

Consider the "exterior" loop $T_1T_2T_3T_4T_1$. The voltage drops satisfy the potential law, $V_{12} + V_{23} + V_{34} + V_{41} = 0$. The voltage drops from T_3 to T_4 and from T_4 to T_1 are both zero assuming the corresponding elements to be perfect conductors. Hence,

$$V_{12} + V_{23} = 0 \tag{3.7.5}$$

The voltage drops around $T_2T_3T_4T_2$ yield, from Kirchhoff's Law,

$$V_{23} + V_{42} = 0 \tag{3.7.6}$$

We now use (3.7.1), (3.7.2), and (3.7.3) to write two differential equations in two unknowns, i_{12} and i_{23}. From (3.7.5) we get

$$L\frac{di_{12}}{dt} + Ri_{23} = 0 \tag{3.7.7}$$

From the derivative of (3.7.6), we obtain

$$R\frac{di_{23}}{dt}+\frac{1}{C}i_{42}=0$$

which becomes from (3.7.4)

$$R\frac{di_{23}}{dt}+\frac{1}{C}(-i_{12}+i_{23})=0 \tag{3.7.8}$$

Let $\mathbf{i}=\begin{bmatrix} i_{12} \\ i_{23} \end{bmatrix}$. Then

$$\mathbf{i}'=\begin{bmatrix} 0 & -\dfrac{R}{L} \\ \dfrac{1}{CR} & -\dfrac{1}{CR} \end{bmatrix}\mathbf{i} \tag{3.7.9}$$

If we solve (3.7.9) we obtain i_{12} and i_{23} and via (3.7.4), i_{42}. With these known, V_{12}, V_{23}, and V_{42} follow directly from (3.7.1), (3.7.2), and (3.7.3).

We will examine three different cases which will exemplify the three basic situations that can occur in networks of the present type.

3.7.3. Example

In Example 3.7.2 choose (i) $R=\frac{1}{4}$, $L=1$, $C=3$; (ii) $R=\frac{1}{2}$, $C=L=1$; and (iii) $R=C=1$, $L=2$. In each case find the general solution and discuss the behavior of the voltages and currents as $t\to\infty$.

Solution

The characteristic equation of the matrix in (3.6.9) is $\lambda^2+\lambda/RC+1/LC=0$, hence the eigenvalues are given by

$$\lambda_1=-\frac{1}{2RC}-\frac{1}{2RC}\left(1-\frac{4R^2C}{L}\right)^{1/2}$$

$$\lambda_2=-\frac{1}{2RC}+\frac{1}{2RC}\left(1-\frac{4R^2C}{L}\right)^{1/2}$$

(i) If $1>4R^2C/L$, there are two negative unequal roots, which is the case if $R=\frac{1}{4}$, $L=1$, $C=3$. Indeed, $\lambda_1=-1$, $\lambda_2=-\frac{1}{3}$. The eigenvectors are found to be $\begin{bmatrix} 1 \\ 4 \end{bmatrix}$ corresponding to $\lambda_1=-1$ and $\begin{bmatrix} 3 \\ 4 \end{bmatrix}$ corresponding to $\lambda_2=-\frac{1}{3}$. The general solution is the general linear combination of the solutions obtained from these eigenvectors. Since λ_1 and λ_2 are negative, the currents (and hence the voltages) through each circuit element decay exponentially to zero.

(ii) Here $1=4R^2C/L$ and $\lambda_1=\lambda_2=-1$. The general solution is

$$\mathbf{i}(t)=k_1\begin{bmatrix} 1 \\ 2 \end{bmatrix}e^{-t}+k_2\left\{\begin{bmatrix} 1 \\ 0 \end{bmatrix}+t\begin{bmatrix} 1 \\ 2 \end{bmatrix}\right\}e^{-t}$$

and again all the currents and voltages decay to zero.

(iii) In this case $1 < 4R^2C/L = 2$ and we find $\lambda = -\frac{1}{2} \pm \frac{1}{2}i$. We obtain the eigenvector $\begin{bmatrix} \frac{1}{2} - \frac{i}{2} \\ 1 \end{bmatrix}$ corresponding $\lambda = -\frac{1}{2} - \frac{1}{2}i$. Therefore,

$$\mathbf{i}(t) = k_1 \operatorname{Re}\left\{\begin{bmatrix} \frac{1}{2} - \frac{i}{2} \\ 1 \end{bmatrix} \exp\left[\left(-\frac{1}{2} - \frac{i}{2}\right)t\right]\right\} + k_2 \operatorname{Im}\left\{\begin{bmatrix} \frac{1}{2} - \frac{i}{2} \\ 1 \end{bmatrix} \exp\left[\left(-\frac{1}{2} - \frac{i}{2}\right)t\right]\right\}$$

$$= k_1 \exp\left[-\frac{1}{2}t\right] \begin{bmatrix} \frac{1}{2}\cos\frac{t}{2} - \frac{1}{2}\sin\frac{t}{2} \\ \cos\frac{t}{2} \end{bmatrix} + k_2 \exp\left[-\frac{1}{2}t\right] \begin{bmatrix} -\frac{1}{2}\cos\frac{t}{2} - \frac{1}{2}\sin\frac{t}{2} \\ -\sin\frac{t}{2} \end{bmatrix}$$

and the currents die off, oscillating with exponentially decreasing amplitude.

Exercise 3.7

1. Solve the differential equations for the cases (i), (ii), and (iii) of Example 3.7.3.
2. Express the solution of case (iii) in the form

$$\mathbf{i}(t) = k_1 e^{-(1/2)t} \begin{bmatrix} A \sin\left(\frac{t+\theta_1}{2}\right) \\ B \sin\left(\frac{t+\theta_2}{2}\right) \end{bmatrix} + k_1 e^{-(1/2)t} \begin{bmatrix} C \sin\left(\frac{t+\phi_1}{2}\right) \\ D \sin\left(\frac{t+\phi_2}{2}\right) \end{bmatrix}$$

Chapter 4

APPLICATIONS OF THE THEORY OF THE HOMOGENEOUS EQUATION

4.1. FUNDAMENTAL MATRICES

A particularly convenient way of discussing the behavior of $\mathbf{x}' = \mathrm{A}\mathbf{x}$ is to group its solutions together as columns of a matrix. This is not simply a notational nicety; by this formulation we will be able to organize the previous theorems, discover new results, and prepare the way for the study of differential systems more complicated than those with constant coefficients. Two major goals of this chapter are the proof of the uniqueness theorem for the initial-value problem and the complete solution of $\mathbf{x}' = \mathrm{A}\mathbf{x} + \mathbf{f}(t)$. We begin with a notion central to our entire theory, that of the fundamental matrix.

4.1.1. DEFINITION *An $n \times n$ matrix, Φ, whose columns are solutions of the system $\mathbf{x}' = \mathrm{A}\mathbf{x}$ and are linearly independent at t_0, is called a fundamental matrix of $\mathbf{x}' = \mathrm{A}\mathbf{x}$ at t_0.*

4.1.1. Example

Find a fundamental matrix for the system $\mathbf{x}' = \mathrm{A}\mathbf{x}$ at $t_0 = 0$, where

$$\mathrm{A} = \begin{bmatrix} 1 & 0 \\ -1 & 3 \end{bmatrix}$$

Solution

The vector functions

$$\mathbf{x}_1(t) = \begin{bmatrix} 2 \\ 1 \end{bmatrix} e^{t} \qquad \text{and} \qquad \mathbf{x}_2(t) = \begin{bmatrix} 0 \\ 1 \end{bmatrix} e^{3t}$$

are solutions of the given system which reduce to the linearly independent vectors $[2, 1]^T$ and $[0, 1]^T$ when $t = 0$. Thus

$$\begin{bmatrix} 2e^t & 0 \\ e^t & e^{3t} \end{bmatrix}$$

is a fundamental matrix. The vector functions

$$\mathbf{x}_3(t) = \begin{bmatrix} 2 \\ 1 \end{bmatrix} e^t + \begin{bmatrix} 0 \\ 1 \end{bmatrix} e^{3t} \quad \text{and} \quad \mathbf{x}_4(t) = \begin{bmatrix} 0 \\ 4 \end{bmatrix} e^{3t}$$

are also solutions of the given system and linearly independent at $t = 0$. Hence

$$\begin{bmatrix} 2e^t & 0 \\ e^t + e^{3t} & 4e^{3t} \end{bmatrix}$$

is another fundamental matrix at $t_0 = 0$. Clearly fundamental matrices are not unique.

To construct fundamental matrices we must find n solutions of $\mathbf{x}' = \mathrm{A}\mathbf{x}$ which are linearly independent at t_0. We know enough about the behavior of systems with constant coefficients that we can always succeed in doing so.

4.1.2. THEOREM *Every system* $\mathbf{x}' = \mathrm{A}\mathbf{x}$ *has a fundamental matrix at* t_0.

PROOF The following n initial-value problems have a real solution by Corollary 3.6.12, "the existence theorem":

$$\begin{array}{lll} \mathbf{x}' = \mathrm{A}\mathbf{x}, \mathbf{x}(t_0) = \mathbf{e}_1 & \text{has a solution} & \mathbf{x}_1(t) \\ \mathbf{x}' = \mathrm{A}\mathbf{x}, \mathbf{x}(t_0) = \mathbf{e}_2 & \text{has a solution} & \mathbf{x}_2(t) \\ \vdots & \vdots & \vdots \\ \mathbf{x}' = \mathrm{A}\mathbf{x}, \mathbf{x}(t_0) = \mathbf{e}_n & \text{has a solution} & \mathbf{x}_n(t) \end{array}$$

where, as usual, $\mathbf{e}_i = [0, \ldots, 0, 1, 0, \ldots, 0]^T$ is a vector with zero entries except for a "1" in the ith position. The matrix whose columns are the functions $\mathbf{x}_1, \mathbf{x}_2, \ldots, \mathbf{x}_n$ is a fundamental matrix because at t_0 the columns reduce to $\mathbf{e}_1, \mathbf{e}_2, \ldots, \mathbf{e}_n$. This completes the proof.

For matrices, a differential equation analogous to the equation $\mathbf{x}' = \mathrm{A}\mathbf{x}$ is the equation $\mathrm{X}' = \mathrm{AX}$ where X is a matrix function of t. The alternative characterization of a fundamental matrix can be formulated in terms of $\mathrm{X}' = \mathrm{AX}$.

4.1.3. THEOREM Φ *is a fundamental matrix of* $\mathbf{x}' = \mathrm{A}\mathbf{x}$ *at* t_0 *if and only if*

(a) Φ *is a solution of* $\mathrm{X}' = \mathrm{AX}$

and

(b) $\det \Phi(t_0) \neq 0$

PROOF To say that Φ is a solution of $X' = AX$ is to say that $\Phi'(t) = A\Phi(t)$. Suppose we write

$$\Phi(t) = [\mathbf{x}_1(t), \ldots, \mathbf{x}_n(t)]$$

then

$$\Phi'(t) = [\mathbf{x}_1'(t), \ldots, \mathbf{x}_n'(t)] \qquad \text{and} \qquad A\Phi(t) = [A\mathbf{x}_1(t), \ldots, A\mathbf{x}_n(t)]$$

But two matrices are equal if and only if corresponding columns are equal. Thus

$$\Phi'(t) = A\Phi(t)$$

if and only if

$$\mathbf{x}_j'(t) = A\mathbf{x}_j(t)$$

that is, if and only if each column of Φ is a solution.

The condition (b) is equivalent to $\mathbf{x}_1(t_0), \mathbf{x}_2(t_0), \ldots, \mathbf{x}_n(t_0)$ being a linearly independent sequence of vectors. This completes the proof.

The condition (b) is also equivalent to "$\Phi^{-1}(t_0)$ exists," a criterion we sometimes find convenient to use in place of either "$\det \Phi(t_0) \neq 0$" or "the columns of Φ are linearly independent at t_0."

4.1.2. Example

Verify Theorem 4.1.3 for the system with

$$A = \begin{bmatrix} 0 & 1 \\ -1 & -2 \end{bmatrix}$$

Solution

After some labor we find the solutions

$$\mathbf{x}_1(t) = \begin{bmatrix} 1 \\ -1 \end{bmatrix} e^{-t} \qquad \text{and} \qquad \mathbf{x}_2(t) = \begin{bmatrix} t \\ 1-t \end{bmatrix} e^{-t}$$

which are linearly independent at each fixed t. Therefore

$$\Phi(t) = \begin{bmatrix} e^{-t} & te^{-t} \\ -e^{-t} & (1-t)e^{-t} \end{bmatrix}$$

We verify

$$\Phi'(t) = \begin{bmatrix} -e^{-t} & (1-t)e^{-t} \\ e^{-t} & (-2+t)e^{-t} \end{bmatrix}$$

and

$$A\Phi(t) = \begin{bmatrix} 0 & 1 \\ -1 & -2 \end{bmatrix} \begin{bmatrix} e^{-t} & te^{-t} \\ -e^{-t} & (1-t)e^{-t} \end{bmatrix}$$

$$= \begin{bmatrix} -e^{-t} & (1-t)e^{-t} \\ e^{-t} & (-2+t)e^{-t} \end{bmatrix}$$

so that $\phi'(t) \equiv A\Phi(t)$. Also, $\det \Phi(t) = e^{-2t} > 0$ for all t.

Because of Theorem 4.1.3, we may now formulate an elegant representation for the solutions of $\mathbf{x}' = A\mathbf{x}$ in terms of the fundamental matrix.

4.1.4. THEOREM *If Φ is a fundamental matrix of $\mathbf{x}' = A\mathbf{x}$ at t_0, then $\mathbf{x}(t) = \Phi(t)\mathbf{k}$ is a solution of $\mathbf{x}' = A\mathbf{x}$ for every constant vector $\mathbf{k}$.*

PROOF We easily compute $\mathbf{x}'(t) = \Phi'(t)\mathbf{k} = A\Phi(t)\mathbf{k}$ by Theorem 4.1.3. But $\Phi(t)\mathbf{k} = \mathbf{x}(t)$, hence $\mathbf{x}'(t) \equiv A\mathbf{x}(t)$, which proves the theorem.

4.1.5. COROLLARY *Under the hypothesis of Theorem 4.1.3, $\mathbf{x}(t) = \Phi(t)\mathbf{k}$, $\mathbf{k} = \Phi^{-1}(t_0)\mathbf{x}_0$ is a solution of the initial-value problem $\mathbf{x}' = A\mathbf{x}$, $\mathbf{x}(t_0) = \mathbf{x}_0$.*

PROOF The matrix $\Phi^{-1}(t_0)$ exists by definition of a fundamental matrix at t_0. Hence the constant vector $\Phi^{-1}(t_0)\mathbf{x}_0$ is defined and by Theorem 4.1.4, $\mathbf{x}(t) = \Phi(t)[\Phi^{-1}(t_0)\mathbf{x}_0]$ is a solution of $\mathbf{x}' = A\mathbf{x}$. Since

$$\mathbf{x}(t_0) = \Phi(t_0)\Phi^{-1}(t_0)\mathbf{x}_0 = \mathbf{x}_0$$

the proof is completed.

If the constant matrix B is nonsingular, then ΦB is a fundamental matrix of $\mathbf{x}' = A\mathbf{x}$ whenever Φ is. This follows directly from Theorem 4.1.3 because of

$$[\Phi(t)B]' = \Phi'(t)B = A\Phi(t)B = A[\Phi(t)B]$$

and

$$\det\,[\Phi(t_0)B] = \det\Phi(t_0)\det B \neq 0$$

We state this as follows.

4.1.6. COROLLARY *If B is a nonsingular constant matrix and Φ is a fundamental matrix of $\mathbf{x}' = A\mathbf{x}$ at t_0, then so is ΦB.*

By choosing $B = \Phi^{-1}(t_0)$ we may always normalize a fundamental matrix so that its value at t_0 is the identity. The matrix constructed in the proof of Theorem 4.1.2 is just such a matrix. We call such a fundamental matrix a *normalized fundamental matrix* of $\mathbf{x}' = A\mathbf{x}$ at t_0. Normalized fundamental matrices make the statements of many theorems simpler to phrase. For instance, the conclusion of Corollary 4.1.5 would read "$\cdots \mathbf{x}(t) = \Phi(t)\mathbf{x}_0$ is a solution of the initial-value problem $\mathbf{x}' = A\mathbf{x}$, $\mathbf{x}(t_0) = \mathbf{x}_0$," since $\mathbf{k} = \Phi^{-1}(t_0)\mathbf{x}_0 = \mathbf{x}_0$, if $\Phi(t_0) = I$.

4.1.3. Example

Solve $\mathbf{x}' = A\mathbf{x}$, $\mathbf{x}(0) = [2, 2]^T$ where

$$A = \begin{bmatrix} 1 & 0 \\ -1 & 3 \end{bmatrix}$$

Solution

In Example 4.1.1, we have found a fundamental matrix

$$\Phi(t) = \begin{bmatrix} 2e^t & 0 \\ e^t & e^{3t} \end{bmatrix}$$

and

$$\Phi^{-1}(0) = \begin{bmatrix} \frac{1}{2} & 0 \\ -\frac{1}{2} & 1 \end{bmatrix}$$

Thus

$$\Phi(t)\Phi^{-1}(0) = \begin{bmatrix} e^t & 0 \\ \dfrac{e^t - e^{3t}}{2} & e^{3t} \end{bmatrix}$$

is a normalized fundamental matrix at $t_0 = 0$. Hence

$$\mathbf{x}(t) = \begin{bmatrix} e^t & 0 \\ \dfrac{e^t - e^{3t}}{2} & e^{3t} \end{bmatrix} \begin{bmatrix} 2 \\ 2 \end{bmatrix} = \begin{bmatrix} 2e^t \\ e^t + e^{3t} \end{bmatrix}$$

is a solution of the given initial-value problem.

4.1.4. Example

Solve $\mathbf{x}' = \mathrm{A}\mathbf{x}$, $\mathbf{x}(0) = [2, -1]^{\mathrm{T}}$

$$\mathrm{A} = \begin{bmatrix} 0 & 1 \\ -1 & -2 \end{bmatrix}$$

Solution

From the work done in Example 4.1.2,

$$\Phi(t) = \begin{bmatrix} e^{-t} & te^{-t} \\ -e^{-t} & (1-t)e^{-t} \end{bmatrix}$$

is a fundamental matrix for the given system at $t_0 = 0$ and

$$\Phi^{-1}(0) = \begin{bmatrix} 1 & 0 \\ 1 & 1 \end{bmatrix}$$

Thus,

$$\mathbf{x}(t) = \Phi(t)\Phi^{-1}(0) \begin{bmatrix} 2 \\ -1 \end{bmatrix} = e^{-t} \begin{bmatrix} 1+t & t \\ -t & 1-t \end{bmatrix} \begin{bmatrix} 2 \\ -1 \end{bmatrix}$$

$$= e^{-t} \begin{bmatrix} 2+t \\ -1-t \end{bmatrix}$$

is a solution of

$$\mathbf{x}' = \begin{bmatrix} 0 & 1 \\ -1 & -2 \end{bmatrix} \mathbf{x} \qquad \mathbf{x}(0) = \begin{bmatrix} 2 \\ -1 \end{bmatrix}$$

4.1.5. Example

Normalize the fundamental matrix

$$\Phi(t) = \begin{bmatrix} 2e^t & 0 \\ e^t & e^{3t} \end{bmatrix}$$

at t_0.

Solution

The inverse of Φ at t_0 is easily computed by row reductions to be

$$\Phi^{-1}(t_0) = \begin{bmatrix} \frac{1}{2}e^{-t_0} & 0 \\ -\frac{1}{2}e^{-3t_0} & e^{-3t_0} \end{bmatrix}$$

Therefore

$$\Phi(t)\Phi^{-1}(t_0) = \begin{bmatrix} e^{t-t_0} & 0 \\ \dfrac{e^{t-t_0} - e^{3(t-t_0)}}{2} & e^{3(t-t_0)} \end{bmatrix}$$

is the required matrix.

Exercise 4.1

Part 1

In Problems 1–22 you are given the system matrix A. In each case, find a normalized fundamental matrix of the system $\mathbf{x}' = A\mathbf{x}$ at $t_0 = 0$.

Problems 1–9 are taken from Exercise 2.4.

1. $A = \begin{bmatrix} 1 & 3 \\ 1 & -1 \end{bmatrix}$
2. $A = \begin{bmatrix} 1 & 1 \\ 3 & -1 \end{bmatrix}$
3. $A = \begin{bmatrix} 2 & 1 \\ 2 & 3 \end{bmatrix}$
4. $A = \begin{bmatrix} 1 & \alpha^2 - 1 \\ 1 & -1 \end{bmatrix}$
5. $A = \begin{bmatrix} 4 & -3 & -2 \\ 2 & -1 & -2 \\ 3 & -3 & -1 \end{bmatrix}$
6. $A = \begin{bmatrix} \cos\theta & +\sin\theta \\ \sin\theta & -\cos\theta \end{bmatrix}$, $0 < \theta < \frac{\pi}{2}$
7. $A = \begin{bmatrix} 1 & 1 & 1 \\ 0 & 2 & 1 \\ 0 & 0 & 3 \end{bmatrix}$
8. $A = \begin{bmatrix} 1 & 1 & 1 \\ 1 & -1 & 1 \\ 0 & 0 & 0 \end{bmatrix}$
9. $A = \begin{bmatrix} 1 & 1 & -1 \\ 0 & 0 & 1 \\ 0 & -2 & -3 \end{bmatrix}$

Problems 10–12 are taken from Exercise 2.5.

10. $A = \begin{bmatrix} 2 & 0 & 0 \\ 0 & 1 & 0 \\ 0 & 0 & 1 \end{bmatrix}$
11. $A = \begin{bmatrix} 1 & 0 & 0 \\ 0 & 3 & \sqrt{2} \\ 0 & \sqrt{2} & 2 \end{bmatrix}$
12. $A = \begin{bmatrix} 1 & 1 & 1 \\ 1 & 1 & 1 \\ 1 & 1 & 1 \end{bmatrix}$

13. Example 3.1.2

$$A = \begin{bmatrix} 1 & 1 & 0 \\ 0 & 1 & 0 \\ 0 & 1 & 1 \end{bmatrix}$$

14. $A = \begin{bmatrix} 0 & 1 & 0 \\ 0 & 0 & 1 \\ 0 & 0 & 0 \end{bmatrix}$

15. Example 3.3.1

$$A = \begin{bmatrix} 1 & 1 & 0 & 0 \\ 0 & 1 & 0 & 0 \\ 0 & 0 & 1 & 1 \\ 0 & 0 & 0 & 1 \end{bmatrix}$$

Problems 16–22 are taken from Exercise 3.3.

16. $A = \begin{bmatrix} 2 & 0 & 0 \\ 0 & 2 & 0 \\ 0 & 0 & 2 \end{bmatrix}$

17. $A = \begin{bmatrix} 2 & 1 & 0 \\ 0 & 2 & 0 \\ 0 & 0 & 2 \end{bmatrix}$

18. $A = \begin{bmatrix} 2 & 0 & 0 \\ 0 & 2 & 1 \\ 0 & 0 & 2 \end{bmatrix}$

19. $A = \begin{bmatrix} 2 & 1 & 0 \\ 0 & 2 & 1 \\ 0 & 0 & 2 \end{bmatrix}$

20. $A = \begin{bmatrix} -1 & 1 & 0 & 0 \\ 0 & -1 & 0 & 0 \\ 0 & 0 & 1 & 1 \\ 0 & 0 & 0 & 1 \end{bmatrix}$

21. $A = \begin{bmatrix} -1 & 1 & 0 & 0 \\ 0 & -1 & 0 & 0 \\ 0 & 0 & 1 & 0 \\ 0 & 0 & 0 & 1 \end{bmatrix}$

22. $A = \begin{bmatrix} 0 & 1 & 0 & 0 \\ -4 & 4 & 0 & 0 \\ 0 & 0 & 2 & 1 \\ 0 & 0 & 0 & 2 \end{bmatrix}$

Part 2

1. If Φ is a fundamental matrix of $\mathbf{x}' = A\mathbf{x}$ and B is nonsingular, Corollary 4.1.6 asserts that ΦB is a fundamental matrix for the same system. Why isn't $B\Phi$ a fundamental matrix for this system also?
2. Referring to Problem 1, above, suppose $AB = BA$. Prove that $B\Phi$ is now a fundamental matrix of $\mathbf{x}' = A\mathbf{x}$.
3. State the converse to the theorem suggested in Problem 2 above. Is the converse true? Explain.
4. Let the eigenvalues of A be $\lambda_1, \bar{\lambda}_1, \lambda_3, \ldots, \lambda_n$ where λ_k is real, $k > 3$ and λ_1 is complex. Let $\mathbf{u}_1, \bar{\mathbf{u}}_1, \mathbf{u}_3, \ldots, \mathbf{u}_n$ be corresponding eigenvectors.
   (a) Show that

$$\Phi(t) = [\mathbf{u}_1 e^{\lambda_1 t}, \bar{\mathbf{u}}_1 e^{\bar{\lambda}_1 t}, \mathbf{u}_3 e^{\lambda_3 t}, \ldots, \mathbf{u}_n e^{\lambda_n t}]$$

is a fundamental matrix of $\mathbf{x}' = A\mathbf{x}$.

(b) Define B as

$$\left[\begin{array}{cc|c} \frac{1}{2} & \frac{1}{2i} & 0 \\ \frac{1}{2} & \frac{-1}{2i} & \\ \hline 0 & & I \end{array}\right]$$

where I is the $(n-2)\times(n-2)$ identity. Prove ΦB is a *real* fundamental matrix of $\mathbf{x}' = A\mathbf{x}$.

(c) Generalize to the case where the eigenvalues of A are $\lambda_1, \bar{\lambda}_1, \lambda_2, \bar{\lambda}_2, \ldots, \lambda_r, \bar{\lambda}_r$.

4.2. THE UNIQUENESS THEOREM

The main point of this section is to prove that there is only one solution to the initial-value problem, $\mathbf{x}' = A\mathbf{x}$, $\mathbf{x}(t_0) = \mathbf{x}_0$. The proof requires a preliminary result to the effect that a fundamental matrix is nonsingular at each t. To motivate the proof of this theorem suppose Φ is a normalized fundamental matrix for $\mathbf{x}' = A\mathbf{x}$ at t_0. Suppose, also, that Φ^{-1} exists for all t. Then by differentiating $\Phi\Phi^{-1} = I$ we obtain $\Phi'\Phi^{-1} + \Phi(\Phi^{-1})' = O$. We solve for $(\Phi^{-1})'$ to obtain

$$(\Phi^{-1})' = -\Phi^{-1}\Phi'\Phi^{-1}$$

Since $\Phi' = A\Phi$, this latter equation results in

$$(\Phi^{-1})' = -\Phi^{-1}A\Phi\Phi^{-1} = -\Phi^{-1}A$$

This latter equation is more readily interpretable if we take transposes. For typographical reasons, set $\Phi^{-1} = \Psi^{T}$ so that the latter equation may be written

$$(\Psi^{T})' = -\Psi^{T}A$$

Then transposing and noting $(\Psi^{T})' = (\Psi')^{T}$, we obtain

$$\Psi' = -A^{T}\Psi$$

The expression, $\Psi' = -A^{T}\Psi$ leads us to believe that Ψ (i.e., $[\Phi^{-1}]^{T}$) may be a fundamental matrix for $\mathbf{x}' = -A^{T}\mathbf{x}$. Of course the above argument does not prove this for it begins with the assumption that Φ^{-1} exists—the very result we wish to prove! But with this hint we see how the following direct argument proves the existence of Φ^{-1}.

4.2.1. THEOREM *Let Φ and Ψ be normalized fundamental matrices for $\mathbf{x}' = A\mathbf{x}$ and $\mathbf{x}' = -A^{T}\mathbf{x}$ at t_0, respectively. Then*

$$\Psi^{T}(t)\Phi(t) = I \quad \textit{for all} \quad t \tag{4.2.1}$$

PROOF Since Ψ is differentiable, so is its transpose and therefore so is $\Psi^T\Phi$. Indeed,

$$\begin{aligned}\frac{d}{dt}[\Psi^T\Phi] &= \frac{d}{dt}[\Psi^T]\Phi + \Psi^T\frac{d}{dt}[\Phi]\\ &= \left\{\frac{d}{dt}[\Psi]\right\}^T\Phi + \Psi^T A\Phi\\ &= -\{A^T\Psi\}^T\Phi + \Psi^T A\Phi\\ &= -\Psi^T A\Phi + \Psi^T A\Phi\\ &= O\end{aligned}$$

Hence $\Psi^T\Phi = C$, a constant matrix. To evaluate C choose $t = t_0$. Because Φ and Ψ are normalized fundamental matrices, $\Psi^T(t_0)\Phi(t_0) = I = C$ which proves the theorem. Note the important fact stated in Equation (4.2.1), namely, the normalized fundamental matrix has an inverse for each t given by the transpose of the normalized fundamental matrix of $\mathbf{x}' = -A^T\mathbf{x}$. The equation $\mathbf{x}' = -A^T\mathbf{x}$ is called the *adjoint* of $\mathbf{x}' = A\mathbf{x}$.

4.2.2. COROLLARY *Fundamental matrices are nonsingular for all t.*

PROOF If Φ is a fundamental matrix at t_0 then $\Phi(t)\Phi^{-1}(t_0)$ is a normalized fundamental matrix at t_0 and is therefore invertible for every t. But then so is Φ.

4.2.3. COROLLARY *Solutions of* $\mathbf{x}' = A\mathbf{x}$ *which are linearly independent at* t_0 *are linearly independent for all t.*

PROOF This is just an interpretation of Corollary 4.2.2. Put in still another form is the following.

4.2.4. COROLLARY *Fundamental matrices at* t_0 *are fundamental matrices for all t.*

These corollaries point out an interesting and important fact about solutions of $\mathbf{x}' = A\mathbf{x}$: *the vector functions that solve such systems cannot at one point be linearly independent vectors and then at some other point be dependent*. This is certainly false for randomly picked vectors. For instance, $\mathbf{x}_1(t) = [1, t, 0]^T$, $\mathbf{x}_2(t) = [0, 1, 0]^T$, $\mathbf{x}_3(t) = [0, 0, t]^T$ are linearly dependent at $t = 0$ and linearly independent at $t = 1$. This means that these vectors do not simultaneously solve any system of the form $\mathbf{x}' = A\mathbf{x}$. We know this from Corollary 4.2.3. Let us verify this fact directly by attempting to find a matrix A such that these vectors are solutions of $\mathbf{x}' = A\mathbf{x}$. Let

$$A = \begin{bmatrix} a_{11} & a_{12} & a_{13}\\ a_{21} & a_{22} & a_{23}\\ a_{31} & a_{32} & a_{33}\end{bmatrix}$$

then by Theorem 4.1.3,

$$\Phi'(t) = \begin{bmatrix} 0 & 0 & 0 \\ 1 & 0 & 0 \\ 0 & 0 & 1 \end{bmatrix} = \begin{bmatrix} a_{11} & a_{12} & a_{13} \\ a_{21} & a_{22} & a_{23} \\ a_{31} & a_{32} & a_{33} \end{bmatrix} \begin{bmatrix} 1 & 0 & 0 \\ t & 1 & 0 \\ 0 & 0 & t \end{bmatrix} \tag{4.2.2}$$

If the right-hand side is multiplied out and compared with the left-hand side we reach a contradiction. Of course we should not jump to the conclusion that any set of vector functions which are linearly independent for all t are automatically solutions of some system $\mathbf{x}' = \mathrm{A}\mathbf{x}$. Refer to Exercise 4.2, Part 2, Problem 2, for an example illustrating otherwise.

4.2.1. Example

Illustrate Theorem 4.2.1 by means of the system whose matrix is

$$\mathrm{A} = \begin{bmatrix} 1 & 0 \\ -1 & 3 \end{bmatrix}$$

Solution

In Example 4.1.3 we have computed a normalized fundamental matrix at $t_0 = 0$, namely,

$$\Phi(t) = \begin{bmatrix} e^t & 0 \\ \dfrac{e^t - e^{3t}}{2} & e^{3t} \end{bmatrix}$$

The system $\mathbf{x}' = -\mathrm{A}^{\mathrm{T}}\mathbf{x}$ has solutions

$$\mathbf{x}_1(t) = \begin{bmatrix} 1 \\ 0 \end{bmatrix} e^{-t} \quad \text{and} \quad \mathbf{x}_2(t) = \begin{bmatrix} 1 \\ -2 \end{bmatrix} e^{-3t}$$

so that a fundamental matrix for this system is

$$\Psi_1(t) = \begin{bmatrix} e^{-t} & e^{-3t} \\ 0 & -2e^{-3t} \end{bmatrix}$$

Hence

$$\Psi(t) = \begin{bmatrix} e^{-t} & \dfrac{e^{-t} - e^{-3t}}{2} \\ 0 & e^{-3t} \end{bmatrix} = \Psi_1(t)\Psi_1^{-1}(0)$$

is a normalized fundamental matrix at $t_0 = 0$. We may now verify readily that

$$\Psi^{\mathrm{T}}(t)\Phi(t) = \mathrm{I}$$

4.2.5. THEOREM *If* $\mathbf{x}(t)$ *is any solution of* $\mathbf{x}' = \mathrm{A}\mathbf{x}$ *and* Φ *is a normalized fundamental matrix of* $\mathbf{x}' = \mathrm{A}\mathbf{x}$ *at* t_0 *then*

$$\mathbf{x}(t) = \Phi(t)\mathbf{x}(t_0) \tag{4.2.3}$$

PROOF The motivation for the proof stems from the observation that Equation (4.2.3) may be written as $\Phi^{-1}(t)\mathbf{x}(t) = \mathbf{x}(t_0)$, a constant vector. The proof, then, consists of showing $\mathbf{z}(t) \equiv \Phi^{-1}(t)\mathbf{x}(t)$ is a constant vector, by demonstrating that $\mathbf{z}'(t) = \mathbf{0}$. According to Theorem 4.2.1, $\Phi^{-1}(t) = \Psi^{T}(t)$, where $\Psi' = -A^{T}\Psi$. Differentiating $\mathbf{z}(t) = \Phi^{-1}(t)\mathbf{x}(t) = \Psi^{T}(t)\mathbf{x}(t)$ we have,

$$\begin{aligned}\mathbf{z}' &= (\Psi')^{T}\mathbf{x} + \Psi^{T}\mathbf{x}' \\ &= (-A^{T}\Psi)^{T}\mathbf{x} + \Psi^{T}A\mathbf{x} \\ &= -\Psi^{T}A\mathbf{x} + \Psi^{T}A\mathbf{x} = \mathbf{0}\end{aligned}$$

This implies $\mathbf{z}(t)$ is constant. But $\mathbf{z}(t) = \Phi^{-1}(t)\mathbf{x}(t)$; hence $\Phi^{-1}(t)\mathbf{x}(t) = \mathbf{c}$, $\mathbf{c}$ a constant vector. To evaluate $\mathbf{c}$ set $t = t_0$ and recall that $\Phi(t_0) = I$. Thus, $\mathbf{c} = \mathbf{x}(t_0)$ and finally, $\Phi^{-1}(t)\mathbf{x}(t) = \mathbf{x}(t_0)$ yields $\mathbf{x}(t) = \Phi(t)\mathbf{x}(t_0)$ as required.

We knew before this theorem that $\Phi(t)\mathbf{k}$ was a solution of $\mathbf{x}' = A\mathbf{x}$ for every choice of the constant $\mathbf{k}$ (Theorem 4.1.4). The force of Theorem 4.2.5 is that these are *all* the solutions. Put somewhat differently, every solution of $\mathbf{x}' = A\mathbf{x}$ is obtainable from $\Phi(t)\mathbf{k}$ by proper selection of $\mathbf{k}$. If we choose Φ to be a normalized fundamental matrix then Theorem 4.2.5 may be interpreted to read that every solution of $\mathbf{x}' = A\mathbf{x}$ is some linear combination of a basic solution set. This suggests the following definition.

4.2.6. DEFINITION *The vector function $\Phi(t)\mathbf{k}$ is called a general solution of $\mathbf{x}' = A\mathbf{x}$ if Φ is any fundamental matrix of the system $\mathbf{x}' = A\mathbf{x}$.*

An important corollary follows immediately from Theorem 4.2.5. If $\mathbf{x}(t)$ is a solution of $\mathbf{x}' = A\mathbf{x}$ which vanishes at some point, say $t = t_0$ then from Equation (4.2.3), $\mathbf{x}(t) = \Phi(t)\mathbf{x}(t_0) = \Phi(t)\mathbf{0} = \mathbf{0}$, for all t.

4.2.7. COROLLARY *Any solution of $\mathbf{x}' = A\mathbf{x}$ which vanishes at t_0 is identically zero. In particular, then, the initial-value problem $\mathbf{x}' = A\mathbf{x}$, $\mathbf{x}(0) = \mathbf{0}$ has the unique solution $\mathbf{x}(t) \equiv \mathbf{0}$.*

It is a short step from this corollary to the important "uniqueness theorem" for linear systems with constant coefficients. Compare the following theorem with the special case stated as Theorem 2.1.1.

4.2.8. THEOREM (The uniqueness theorem.) *If $\mathbf{x}(t)$ and $\mathbf{y}(t)$ are both solutions of the initial-value problem $\mathbf{x}' = A\mathbf{x}$, $\mathbf{x}(t_0) = \mathbf{x}_0$ then $\mathbf{x}(t) \equiv \mathbf{y}(t)$.*

PROOF Define $\mathbf{z}(t) = \mathbf{x}(t) - \mathbf{y}(t)$. Then

$$\begin{aligned}\mathbf{z}'(t) &= \mathbf{x}'(t) - \mathbf{y}'(t) \\ A\mathbf{z}(t) &= A\mathbf{x}(t) - A\mathbf{y}(t) \\ &= \mathbf{x}'(t) - \mathbf{y}'(t)\end{aligned}$$

Therefore $\mathbf{z}(t)$ satisfies $\mathbf{x}' = A\mathbf{x}$. Also, $\mathbf{z}(t_0) = \mathbf{x}(t_0) - \mathbf{y}(t_0) = \mathbf{0}$. Hence, by Corollary 4.2.7, $\mathbf{z}(t) \equiv \mathbf{0}$.

In terms of fundamental matrices, the uniqueness theorem takes the following form.

4.2.9. COROLLARY *Let Φ be a normalized fundamental matrix for the system* $\mathbf{x}' = \mathrm{A}\mathbf{x}$ *at* t_0. *Then*

$$\mathrm{X}(t) = \Phi(t)\mathrm{X}_0$$

is the unique solution to the matrix initial-value problem

$$\mathrm{X}' = \mathrm{AX} \qquad \mathrm{X}(t_0) = \mathrm{X}_0 \tag{4.2.4}$$

PROOF By direct substitution we see immediately that $\Phi(t)\mathrm{X}_0$ is a solution of the given initial-value problem. That it is the only solution follows from the uniqueness theorem applied to each column of $\Phi(t)\mathrm{X}_0$.

A normalized fundamental matrix at t_0 satisfies

$$\mathrm{X}' = \mathrm{AX} \qquad \mathrm{X}(t_0) = \mathrm{I}$$

by definition. From Corollary 4.2.9 the solution of this latter matrix initial-value problem is unique. This proves the following corollary.

4.2.10. COROLLARY *The normalized fundamental matrix* Φ *of* $\mathbf{x}' = \mathrm{A}\mathbf{x}$ *at* t_0 *is unique.*

Exercise 4.2

Part 1

Problems 1–9 are taken from Exercise 4.1, Part 1. Find Φ^{-1} and then verify Equation (4.2.1).

1. Problem 1	2. Problem 2	3. Problem 3
4. Problem 4	5. Problem 10	6. Problem 11
7. Problem 12	8. Problem 16	9. Problem 19

Part 2

1. Multiply out the right-hand side of Equation (4.2.2) and verify the inconsistency stated in the text.
2. Show that

$$\Phi(t) = \begin{bmatrix} e^{t^2} & 0 \\ 0 & 1 \end{bmatrix}$$

is nonsingular for all t but is not a fundamental matrix for any system $\mathbf{x}' = \mathrm{A}\mathbf{x}$, A a matrix of constants.
3. Complete the details of the proof of Corollary 4.2.7.
4. If $\mathbf{x}(t)$ is a solution of $\mathbf{x}' = \mathrm{A}\mathbf{x}$ where A is a real $n \times n$ matrix prove that $\mathrm{Re}\,\mathbf{x}(t)$ and $\mathrm{Im}\,\mathbf{x}(t)$ are also solutions.
5. Suppose $\mathbf{x}(t)$ is a solution of $\mathbf{x}' = \mathrm{A}\mathbf{x}$, A real, $\mathbf{x}(t_0) = \mathbf{x}_0 \in \mathscr{R}^n$. Prove that $\mathrm{Re}\,\mathbf{x}(t)$ is a solution of this same initial-value problem. Hence by the uniqueness theorem $\mathrm{Re}\,\mathbf{x}(t) = \mathbf{x}(t)$. What does this imply about the range of $\mathbf{x}(t)$?
6. Use the result of Problem 5 above to prove: If $\mathbf{x}(t)$ is a solution of $\mathbf{x}' = \mathrm{A}\mathbf{x}$, which is real for a single value of t, it is real for all t.
7. In Problems 4, 5, and 6 above, A is a real matrix. If A were not real, would the same conclusions follow? Explain.

8. Suppose Φ is a fundamental matrix of $\mathbf{x}' = \mathrm{A}\mathbf{x}$. Show that Φ is a constant matrix if and only if $\mathrm{A} = 0$.
9. Let Φ and Λ be fundamental matrices of $\mathbf{x}' = \mathrm{A}\mathbf{x}$. Find a constant matrix B such that $\Phi(t)\mathrm{B} \equiv \Lambda(t)$ and prove that B is unique and nonsingular. Hint: Normalize Φ and Λ at any fixed t.

4.3. SOME CONSEQUENCES OF THE UNIQUENESS THEOREM

We establish a number of interesting theorems in this section. These theorems play three roles: they offer practice in the use of the uniqueness theorem, they have importance in their own right, and they lead to a complete solution of the inhomogeneous problem, $\mathbf{x}' = \mathrm{A}\mathbf{x} + \mathbf{f}$, a topic discussed at great length in Section 5.

4.3.1. THEOREM *Let Φ be a normalized fundamental matrix of $\mathbf{x}' = \mathrm{A}\mathbf{x}$ at $t_0 = 0$. Then for all real numbers s and t,*

$$\Phi(s+t) = \Phi(t)\Phi(s) \tag{4.3.1}$$

PROOF Let s be fixed, say $s = s_0$. We claim that both sides of Equation (4.3.1) are solutions of the same matrix initial-value problem

$$\mathrm{X}' = \mathrm{AX} \qquad \mathrm{X}(0) = \Phi(s_0)$$

For,

$$\frac{d}{dt}[\Phi(s_0+t)] = \mathrm{A}\Phi(s_0+t)$$

and

$$\frac{d}{dt}[\Phi(t)\Phi(s_0)] = \frac{d}{dt}[\Phi(t)]\Phi(s_0)$$
$$= \mathrm{A}[\Phi(t)\Phi(s_0)]$$

by Theorem 4.1.2. Also, $\Phi(s_0+t)$ at $t=0$ is $\Phi(s_0)$ and $\Phi(t)\Phi(s_0)$ at $t=0$ is $\Phi(0)\Phi(s_0) = \Phi(s_0)$ since $\Phi(0) = \mathrm{I}$. By Corollary 4.2.9 these two matrices are identical, which is what we set out to prove.

4.3.1. Example

Verify Theorem 4.3.1 for the following fundamental matrix normalized at $t_0 = 0$:

$$\Phi(t) = \begin{bmatrix} e^t & 0 \\ \dfrac{e^t - e^{3t}}{2} & e^{3t} \end{bmatrix}$$

Solution

We have by direct multiplication,

$$\Phi(t)\Phi(s) = \begin{bmatrix} e^t & 0 \\ \dfrac{e^t - e^{3t}}{2} & e^{3t} \end{bmatrix} \begin{bmatrix} e^s & 0 \\ \dfrac{e^s - e^{3s}}{2} & e^{3s} \end{bmatrix}$$

$$= \begin{bmatrix} e^{t+s} & 0 \\ \dfrac{e^{t+s} - e^{3(t+s)}}{2} & e^{3(t+s)} \end{bmatrix}$$

$$= \Phi(t+s)$$

Note that this theorem is false if Φ is not normalized at $t_0 = 0$. Consider, for instance, the fundamental matrix

$$\Phi(t) = \begin{bmatrix} e^{t-t_0} & 0 \\ \dfrac{e^{t-t_0} - e^{3(t-t_0)}}{2} & e^{3(t-t_0)} \end{bmatrix}$$

normalized at $t_0 \neq 0$. We find by direct multiplication that $\phi(t+s) \neq \Phi(t)\Phi(s)$.

4.3.2. COROLLARY *If Φ is a normalized fundamental matrix at $t_0 = 0$, then*

$$\Phi^{-1}(t) = \Phi(-t) \tag{4.3.2}$$

PROOF We simply note that s is arbitrary in Equation (4.3.1) so that the choice $s = -t$ reduces Equation (4.3.1) to

$$I = \Phi(0) = \Phi(t)\Phi(-t)$$

4.3.3. COROLLARY *Let Φ be any fundamental matrix of* $\mathbf{x}' = A\mathbf{x}$. *Then*

$$\Phi^{-1}(t) = \Phi^{-1}(0)\Phi(-t)\Phi^{-1}(0)$$

PROOF The matrix $\Lambda(t) = \Phi(t)\Phi^{-1}(0)$ is a normalized fundamental matrix of $\mathbf{x}' = A\mathbf{x}$ at $t_0 = 0$. Hence $\Lambda^{-1}(t) = \Lambda(-t)$ from the previous corollary. But $\Lambda(-t) = \Phi(-t)\Phi^{-1}(0)$. Therefore,

$$\Lambda^{-1}(t) = [\Phi(t)\Phi^{-1}(0)]^{-1} = \Phi(0)\Phi^{-1}(t)$$

and

$$\Lambda^{-1}(t) = \Lambda(-t) = \Phi(-t)\Phi^{-1}(0)$$

Equating these expressions for $\Lambda^{-1}(t)$ completes the proof.

4.3.2. Example

Verify Equation (4.3.2) for the normalized fundamental matrix of Example 4.1.3, namely,

$$\Phi(t)\Phi^{-1}(0) = \Lambda(t) = \begin{bmatrix} e^t & 0 \\ \dfrac{e^t - e^{3t}}{2} & e^{3t} \end{bmatrix}$$

Solution

We find by direct computation

$$\Lambda(t)\Lambda(-t) = \begin{bmatrix} e^t & 0 \\ \dfrac{e^t - e^{3t}}{2} & e^{3t} \end{bmatrix} \begin{bmatrix} e^{-t} & 0 \\ \dfrac{e^{-t} - e^{-3t}}{2} & e^{-3t} \end{bmatrix} = \begin{bmatrix} 1 & 0 \\ 0 & 1 \end{bmatrix}$$

4.3.3. Example

Verify Corollary 4.3.3 for the fundamental matrix

$$\Phi(t) = \begin{bmatrix} 2e^t & 0 \\ e^t & e^{3t} \end{bmatrix}$$

Solution

We have

$$\Phi^{-1}(0) = \begin{bmatrix} \frac{1}{2} & 0 \\ -\frac{1}{2} & 1 \end{bmatrix}$$

Therefore,

$$\Phi^{-1}(t) = \begin{bmatrix} \frac{1}{2} & 0 \\ -\frac{1}{2} & 1 \end{bmatrix} \begin{bmatrix} 2e^{-t} & 0 \\ e^{-t} & e^{-3t} \end{bmatrix} \begin{bmatrix} \frac{1}{2} & 0 \\ -\frac{1}{2} & 1 \end{bmatrix}$$

$$= \begin{bmatrix} \dfrac{e^{-t}}{2} & 0 \\ \dfrac{-e^{-3t}}{2} & e^{-3t} \end{bmatrix}$$

We check by noting,

$$\Phi(t)\Phi^{-1}(t) = \begin{bmatrix} 2e^t & 0 \\ e^t & e^{3t} \end{bmatrix} \begin{bmatrix} \dfrac{e^{-t}}{2} & 0 \\ \dfrac{-e^{-3t}}{2} & e^{-3t} \end{bmatrix}$$

$$= \begin{bmatrix} 1 & 0 \\ 0 & 1 \end{bmatrix}$$

So far we have considered the relationships between the values taken on at various points by a normalized fundamental matrix. We now turn to the relationships connecting fundamental matrices normalized at different points.

4.3.4. THEOREM (Transition property) *Suppose the system* $\mathbf{x}' = \mathrm{A}\mathbf{x}$ *has the fundamental matrices*

$$\begin{array}{ll} \Phi_0: & \textit{normalized at } \; t_0 \\ \Phi_1: & \textit{normalized at } \; t_1 \end{array}$$

Then for each value of t

$$\Phi_1(t)\Phi_0(t_1) = \Phi_0(t) \tag{4.3.3}$$

PROOF It is instructive to construct an argument analogous to the proof of Theorem 4.3.1. To do this show that both sides of Equation (4.3.3) solve the same matrix initial-value problem

$$\mathrm{X}' = \mathrm{AX} \qquad \mathrm{X}(t_1) = \Phi_0(t_1)$$

We leave the details to the reader. See Exercise 4.3, Problem 7. For an even simpler proof see Exercise 4.3, Problem 8.

4.3.5. COROLLARY *With the same notation as in Theorem* 4.3.4, *we have*

$$\Phi_1(t_0)\Phi_0(t_1) = \mathrm{I} \tag{4.3.4}$$

PROOF Set $t = t_0$ in Equation (4.3.3).

As a last application of the uniqueness theorem consider the following result.

4.3.6. THEOREM *Let* Φ *be a normalized fundamental matrix of* $\mathbf{x}' = \mathrm{A}\mathbf{x}$ *at* t_0. *Then*

$$\mathrm{A}\Phi(t) = \Phi(t)\mathrm{A} \tag{4.3.5}$$

PROOF Verify that both sides of Equation (4.3.5) are solutions of the same matrix initial-value problem

$$\mathrm{X}' = \mathrm{AX} \qquad \mathrm{X}(t_0) = \mathrm{A}$$

Now quote Corollary 4.2.9.

4.3.4. Example

Verify Theorem 4.3.6 for the system matrix A of Example 4.1.3.

Solution

In the example referred to,

$$\mathrm{A} = \begin{bmatrix} 1 & 0 \\ -1 & 3 \end{bmatrix} \qquad \text{and} \qquad \Phi(t) = \begin{bmatrix} e^t & 0 \\ \dfrac{e^t - e^{3t}}{2} & e^{3t} \end{bmatrix}$$

Then

$$A\Phi = \begin{bmatrix} e^t & 0 \\ \frac{1}{2}e^t - \frac{3}{2}e^{3t} & 3e^{3t} \end{bmatrix} = \Phi A$$

The hypothesis that Φ be a normalized fundamental matrix cannot be dropped. To see this, it is sufficient to find an example in which Φ is not normalized at any t_0 and in which Equation (4.3.5) is violated. Consider Example 4.1.3 again. This time set

$$\Phi(t) = \begin{bmatrix} 2e^t & 0 \\ e^t & e^{3t} \end{bmatrix}$$

a fundamental matrix which is not a normalized matrix for any t. (Why?) Then

$$A\Phi(t) = \begin{bmatrix} 1 & 0 \\ -1 & 3 \end{bmatrix}\begin{bmatrix} 2e^t & 0 \\ e^t & e^{3t} \end{bmatrix}$$

$$= \begin{bmatrix} 2e^t & 0 \\ e^t & 3e^{3t} \end{bmatrix}$$

But

$$\Phi(t)A = \begin{bmatrix} 2e^t & 0 \\ e^t & e^{3t} \end{bmatrix}\begin{bmatrix} 1 & 0 \\ -1 & 3 \end{bmatrix}$$

$$= \begin{bmatrix} 2e^t & 0 \\ e^t - e^{3t} & 3e^{3t} \end{bmatrix}$$

so that $A\Phi(t) \neq \Phi(t)A$. In fact, $A\Phi(t) \neq \Phi(t)A$ for all t.

Exercise 4.3

1. Let Φ be a normalized fundamental matrix of $\mathbf{x}' = A\mathbf{x}$ at $t_0 = 0$. Prove that $\Phi(t)\Phi(s) = \Phi(s)\Phi(t)$.
2. Is there a t_0 such that

$$\Phi(t_0) = \begin{bmatrix} 2e^{t_0} & 0 \\ e^{t_0} & e^{3t_0} \end{bmatrix} = I?$$

3. Let Φ be a normalized fundamental matrix of $\mathbf{x}' = A\mathbf{x}$ at $t_0 = 0$. Prove that $\Phi(t_1)\Phi(t)$ is another fundamental matrix of this system. Hint: Exercise 4.1, Part 2, Problem 2 and Theorem 4.3.6.
4. Does Equation (4.3.2) hold if Φ is a normalized matrix at $t_0 \neq 0$?
5. Let $\mathbf{y}(t) = \mathbf{x}(-t)$. Show that $\mathbf{y}'(t) = A\mathbf{y}(t)$ if $\mathbf{x}'(t) = -A\mathbf{x}(t)$.
6. Suppose A is symmetric and Φ is a normalized fundamental matrix of $\mathbf{x}' = A\mathbf{x}$ at $t_0 = 0$. Prove that $\Phi^T(t) = \Phi(t)$. Hint: Consider the system $\mathbf{x}' = -A\mathbf{x}$ and use Problem 5 above and $\Psi^T\Phi = I$ and $\Phi^{-1}(t) = \Phi(-t)$.
7. Prove Theorem 4.3.4 by constructing a proof as outlined in the text.
8. Prove Theorem 4.3.4 by observing that $\Phi_0(t)\Phi_0^{-1}(t_1)$ is a normalized fundamental matrix at t_1. What last step is required to complete the proof that $\Phi_0(t)\Phi_0^{-1}(t_1) = \Phi_1(t)$?

4.4. THE INHOMOGENEOUS PROBLEM: VARIATION OF PARAMETERS

We devote this section to finding the general solution of the inhomogeneous system

$$\mathbf{x}' = \mathbf{A}\mathbf{x} + \mathbf{f} \tag{4.4.1}$$

where $\mathbf{f}$ is any continuous vector function of t. Associated with this inhomogeneous system is the *complementary* homogeneous system obtained by setting $\mathbf{f} = \mathbf{0}$ in (4.4.1),

$$\mathbf{x}' = \mathbf{A}\mathbf{x} \tag{4.4.2}$$

Remarkably, the general solution of the complementary system is always sufficient to solve the inhomogeneous system. The method for doing so is known as *variation of parameters* and is of great theoretical as well as practical importance. We illustrate this technique for the case $n = 1$ for which the system (4.4.1) reduces to the single equation

$$x' = ax + f \tag{4.4.3}$$

A solution of the complementary equation $x' = ax$ is $x_c(t) = e^{at}$. We now search for a function u such that $x_p(t) = u(t)e^{at}$ is a solution of Equation (4.4.3). This is the heart of the method. We compute

$$\begin{aligned} x_p'(t) &= au(t)e^{at} + e^{at}u'(t) \\ ax_p(t) + f(t) &= au(t)e^{at} + f(t) \end{aligned}$$

But $x_p'(t) = ax_p(t) + f(t)$ implies from the above calculation,

$$e^{at}u'(t) = f(t) \tag{4.4.4}$$

Hence we select any function u such that $u'(t) = e^{-at}f(t)$. Having found u, we obtain $x_p(t) = e^{at}u(t)$ as a solution.

4.4.1. Example

Solve $x' = 2x + e^{2t}$.

Solution

We know $x_c(t) = e^{2t}$ solves the complementary system $x' = 2x$. Next, $u'(t) = e^{-2t} \cdot e^{2t} = 1$. Pick $u(t) = t$ and then $x_p(t) = te^{2t}$ and a quick check verifies that this function is indeed a solution. Incidentally, the reader should verify that

$$\begin{aligned} x(t) &= kx_c(t) + x_p(t) = ke^{2t} + te^{2t} \\ &= (k+t)e^{2t} \end{aligned}$$

is a solution for any constant k.

Now we return to the n-dimensional problem and follow an analogous procedure called the variation of parameters. The role of e^{at} is played by a fundamental matrix Φ of the complementary system. We *assume* a vector function $\mathbf{u}$ such that $\mathbf{x}_p$ given by

$$\mathbf{x}_p(t) = \Phi(t)\mathbf{u}(t) \tag{4.4.5}$$

is a solution of $\mathbf{x}' = A\mathbf{x} + \mathbf{f}$. This means

$$\mathbf{x}'(t) = \Phi'(t)\mathbf{u}(t) + \Phi(t)\mathbf{u}'(t)$$
$$= A\Phi(t)\mathbf{u}(t) + \Phi(t)\mathbf{u}'(t)$$

and

$$\mathbf{x}_p'(t) = A[\Phi(t)\mathbf{u}(t)] + \mathbf{f}(t)$$

Therefore,

$$\Phi(t)\mathbf{u}'(t) = \mathbf{f}(t) \tag{4.4.6}$$

the direct analog of Equation (4.4.4). Since Φ^{-1} exists, Equation (4.4.6) implies

$$\mathbf{u}'(t) = \Phi^{-1}(t)\mathbf{f}(t) \tag{4.4.7}$$

and

$$\mathbf{x}_p(t) = \Phi(t)\mathbf{u}(t) = \Phi(t)\left\{\int_{t_0}^{t} \Phi^{-1}(s)\mathbf{f}(s)\, ds\right\}$$

which suggests

4.4.1. THEOREM *If* $\mathbf{f}$ *is continuous and* Φ *is a fundamental matrix of* $\mathbf{x}' = A\mathbf{x}$ *then*

$$\mathbf{x}' = A\mathbf{x} + \mathbf{f} \tag{4.4.8}$$

has a solution

$$\mathbf{x}_p(t) = \Phi(t) \int_{t_0}^{t} \Phi^{-1}(s)\mathbf{f}(s)\, ds \tag{4.4.9}$$

PROOF The derivative of $\mathbf{x}_p(t)$ is, according to the fundamental theorem of calculus,

$$\mathbf{x}_p'(t) = \Phi'(t) \int_{t_0}^{t} \Phi^{-1}(s)\mathbf{f}(s)\, ds + \Phi(t)\Phi^{-1}(t)\mathbf{f}(t)$$
$$= A\left[\Phi(t) \int_{t_0}^{t} \Phi^{-1}(s)\mathbf{f}(s)\, ds\right] + \mathbf{f}(t)$$
$$= A\mathbf{x}_p(t) + \mathbf{f}(t)$$

which proves the theorem.

Now $\mathbf{x}_p(t)$ as given by Equation (4.4.9) is just one of infinitely many particular solutions of the system (4.4.8). It has the useful property that it vanishes at $t = t_0$, because the integral is zero at t_0. We may obtain other particular solutions of (4.4.8) by adding to any particular solution of the inhomogeneous system any solution of the complementary system. That is,

$$\mathbf{x}(t) = \Phi(t)\mathbf{k} + \mathbf{x}_p(t)$$

is a solution of (4.4.8) for any choice of $\mathbf{k}$.

The form of $\mathbf{x}_p(t)$ as given by Equation (4.4.9) can be altered somewhat if Φ is a normalized fundamental matrix of $\mathbf{x}' = A\mathbf{x}$ at $t_0 = 0$. We state this as a corollary to Theorem 4.4.1.

4.4.2. COROLLARY *If* Φ *is a normalized fundamental matrix of* $\mathbf{x}' = \mathbf{A}\mathbf{x}$ *at* $t_0 = 0$ *then*

$$\mathbf{x}_p(t) = \int_0^t \Phi(t-s)\mathbf{f}(s)\, ds \tag{4.4.10}$$

is a solution of

$$\mathbf{x}' = \mathbf{A}\mathbf{x} + \mathbf{f}$$

which vanishes at $t = 0$.

PROOF We start from Equation (4.4.9) in which we replace $\Phi^{-1}(s)$ by $\Phi(-s)$, from Corollary 4.3.2, obtaining

$$\mathbf{x}_p(t) = \int_0^t \Phi(t)\Phi(-s)\mathbf{f}(s)\, ds$$

which in turn becomes

$$\mathbf{x}_p(t) = \int_0^t \Phi(t-s)\mathbf{f}(s)\, ds$$

by Theorem 4.3.1, Equation (4.3.1).

4.4.2. Example

Find a solution of

$$\mathbf{x}' = \begin{bmatrix} 1 & 0 \\ -1 & 3 \end{bmatrix}\mathbf{x} + \begin{bmatrix} e^t \\ 1 \end{bmatrix}$$

Solution

Rather than use either Equation (4.4.9) or (4.4.10) we obtain a particular solution directly from Equation (4.4.6). In Example 4.1.1 we have computed Φ, hence

$$\Phi(t)\mathbf{u}' = \begin{bmatrix} 2e^t & 0 \\ e^t & e^{3t} \end{bmatrix}\mathbf{u}' = \begin{bmatrix} e^t \\ 1 \end{bmatrix}$$

which by row operations is equivalent to

$$\begin{bmatrix} 2e^t & 0 \\ 0 & e^{3t} \end{bmatrix}\mathbf{u}' = \begin{bmatrix} e^t \\ 1 - \dfrac{e^t}{2} \end{bmatrix}$$

If the entries of $\mathbf{u}'$ are written as u_1' and u_2', we have $u_1' = \frac{1}{2}$ and $u_2' = e^{-3t} - (e^{-2t}/2)$ so that, upon integration, $u_1 = (t/2)$ and $u_2 = -\frac{1}{3}e^{-3t} + \frac{1}{4}e^{-2t}$, making $\mathbf{u} = [t/2, -\frac{1}{3}e^{-3t} + \frac{1}{4}e^{-2t}]^T$. We thus have a function $\mathbf{u}$ whose derivative satisfies Equation (4.4.6). From Equation (4.4.5), then,

$$\begin{aligned} \mathbf{x}_p(t) &= \Phi(t)\mathbf{u}(t) \\ &= \begin{bmatrix} 2e^t & 0 \\ e^t & e^{3t} \end{bmatrix}\begin{bmatrix} \dfrac{t}{2} \\ -\dfrac{1}{3}e^{-3t} + \dfrac{1}{4}e^{-2t} \end{bmatrix} = \begin{bmatrix} te^t \\ \left(\dfrac{t}{2} + \dfrac{1}{4}\right)e^t - \dfrac{1}{3} \end{bmatrix} \end{aligned} \tag{4.4.11}$$

is a solution to the given inhomogeneous system. One may verify this by substitution directly into the system. For contrast, suppose we use Corollary 4.4.2 to find a particular solution. Let Λ be a normalized fundamental matrix for the given system at $t_0 = 0$. Then

$$\Lambda(t) = \begin{bmatrix} e^t & 0 \\ \frac{1}{2}e^t - \frac{1}{2}e^{3t} & e^{3t} \end{bmatrix}$$

Hence,

$$\begin{aligned}
\mathbf{x}_p(t) &= \int_0^t \Lambda(t-s)\mathbf{f}(s)\,ds \\
&= \int_0^t \begin{bmatrix} e^{t-s} & 0 \\ \frac{1}{2}e^{t-s} - \frac{1}{2}e^{3(t-s)} & e^{3(t-s)} \end{bmatrix} \begin{bmatrix} e^s \\ 1 \end{bmatrix} ds \\
&= \int_0^t \begin{bmatrix} e^t \\ \frac{1}{2}e^t - \frac{1}{2}e^{3t-2s} + e^{3(t-s)} \end{bmatrix} ds \\
&= \begin{bmatrix} te^t \\ \left(\frac{t}{2}+\frac{1}{4}\right)e^t + \left(\frac{1}{12}e^{3t} - \frac{1}{3}\right) \end{bmatrix} \qquad (4.4.12)
\end{aligned}$$

Compare the two solutions obtained. They are not the same. Their difference is the function $[0, (1/12)e^{3t}]^{\mathrm{T}}$, a solution of the complementary equation. Why? Indeed it is easy to prove that the difference of any two solutions of an inhomogeneous system is a solution of the complementary homogeneous system. It is this very fact which enables us to conclude directly from the uniqueness theorem for homogeneous initial-value problems, a uniqueness theorem for the inhomogeneous problem.

4.4.3. THEOREM *The initial-value problem*

$$\mathbf{x}' = \mathbf{A}\mathbf{x} + \mathbf{f} \qquad \mathbf{x}(t_0) = \mathbf{x}_0 \qquad (4.4.13)$$

has the unique solution

$$\mathbf{x}(t) = \mathbf{x}_p(t) + \Phi(t)\mathbf{x}_0 \qquad (4.4.14)$$

where Φ *is a normalized fundamental matrix of the complementary system,* $\mathbf{x}' = \mathbf{A}\mathbf{x}$ *at* t_0 *and*

$$\mathbf{x}_p(t) = \Phi(t) \int_{t_0}^t \Phi^{-1}(s)\mathbf{f}(s)\,ds \qquad (4.4.15)$$

PROOF The function given by Equation (4.4.14) is surely a solution of the given inhomogeneous system and meets the initial condition since the integral vanishes at t_0 and $\Phi(t_0) = \mathbf{I}$. To prove uniqueness, suppose $\mathbf{y}(t)$ is any solution of (4.4.13). Define $\mathbf{z}(t) = \mathbf{x}(t) - \mathbf{y}(t)$. Then,

$$\begin{aligned}
\mathbf{z}'(t) &= \mathbf{x}'(t) - \mathbf{y}'(t) \\
&= \mathbf{A}\mathbf{x}(t) + \mathbf{f}(t) - \mathbf{A}\mathbf{y}(t) - \mathbf{f}(t) \\
&= \mathbf{A}(\mathbf{x}(t) - \mathbf{y}(t)) = \mathbf{A}\mathbf{z}(t)
\end{aligned}$$

and

$$\mathbf{z}(t_0) = \mathbf{x}(t_0) - \mathbf{y}(t_0) = \mathbf{0}$$

Thus $\mathbf{z}(t)$ is a solution of a homogeneous system and vanishes at t_0. By Corollary 4.2.7 $\mathbf{z}(t)$ is identically zero which means that $\mathbf{x}(t) \equiv \mathbf{y}(t)$, as required for the proof.

We conclude this section with a definition of the general solution of an inhomogeneous system.

4.4.4. DEFINITION *Let $\mathbf{x}_p(t)$ be any solution of an inhomogeneous system and Φ be any fundamental matrix of the complementary system. The function*

$$\mathbf{x}(t) = \mathbf{x}_p(t) + \Phi(t)\mathbf{k} \tag{4.4.16}$$

is called the general solution of the inhomogeneous system.

The motivation for the name "general solution" lies in the fact that every solution of $\mathbf{x}' = A\mathbf{x} + \mathbf{f}$ is obtainable from Equation (4.4.16) by proper selection of $\mathbf{k}$. How would you demonstrate the proof of this assertion? Notice that the general solution is the sum of the general solution of the complementary system with a particular solution of the inhomogeneous equation. If we denote the general solution of the complementary system by $\mathbf{x}_c(t)$, Equation (4.4.16) takes the suggestive form $\mathbf{x}_c(t) + \mathbf{x}_p(t) = \mathbf{x}(t)$.

Exercise 4.4

Part 1

1. Solve the following inhomogeneous differential equations and initial-value problems. Use the method illustrated in Example 4.4.1.

   (a) $x' - x = t \qquad x(1) = 0$
   (b) $x' - x = \sin t \qquad x(\pi) = 1$

2. Find a solution of

$$\mathbf{x}' = \begin{bmatrix} 1 & 0 \\ -1 & 3 \end{bmatrix}\mathbf{x} + \begin{bmatrix} 1 \\ 0 \end{bmatrix}$$

3. Same as Problem 2 above, except use Equation (4.4.10).
4. Find a solution of

$$\mathbf{x}' = \begin{bmatrix} 1 & 0 \\ -1 & 3 \end{bmatrix}\mathbf{x} + \begin{bmatrix} 0 \\ 1 \end{bmatrix}$$

   by any method.

5. Show that the sum of the solutions of Problems 2 and 4 above is a solution of

$$\mathbf{x}' = \begin{bmatrix} 1 & 0 \\ -1 & 3 \end{bmatrix}\mathbf{x} + \begin{bmatrix} 1 \\ 1 \end{bmatrix}$$

6. Find the general solution of

$$\mathbf{x}' = \begin{bmatrix} 0 & 1 \\ 1 & 0 \end{bmatrix}\mathbf{x} + \mathbf{f}(t)$$

where

(a) $\mathbf{f}(t) = [1, -1]^{\mathrm{T}}$ (b) $\mathbf{f}(t) = [0, e^t]^{\mathrm{T}}$
(c) $\mathbf{f}(t) = [1, 1]^{\mathrm{T}}$ (d) $\mathbf{f}(t) = [1, 1-e^t]^{\mathrm{T}}$
(e) $\mathbf{f}(t) = [t, t]^{\mathrm{T}}$

7. Given the fundamental matrix

$$\Phi(t) = e^t\begin{bmatrix} 1 & -1+t \\ 1 & t \end{bmatrix}$$

of the system $\mathbf{x}' = \mathrm{A}\mathbf{x}$, where

$$\mathrm{A} = \begin{bmatrix} 0 & 1 \\ -1 & 2 \end{bmatrix}$$

find the solution of $\mathbf{x}' = \mathrm{A}\mathbf{x} + \mathbf{f}$, $\mathbf{x}(0) = \mathbf{x}_0$, where
(a) $\mathbf{f}(t) = [0, 1]^{\mathrm{T}}$ $\mathbf{x}(0) = [0, 1]^{\mathrm{T}}$
(b) $\mathbf{f}(t) = [1, 1]^{\mathrm{T}}$ $\mathbf{x}(0) = [0, 1]^{\mathrm{T}}$
(c) $\mathbf{f}(t) = [1, -1]^{\mathrm{T}} e^{-t}$ $\mathbf{x}(0) = [0, 0]^{\mathrm{T}}$

8. Substitute $\mathbf{x}_p(t)$ as given by Equation (4.4.11) into the system of Example 4.4.2 and verify that it is a solution.
9. Same as Problem 8, above, except use $\mathbf{x}_p(t)$ of Equation (4.4.12).

Part 2

1. Given

$$\mathbf{x}' = \begin{bmatrix} 0 & -4 \\ -4 & 0 \end{bmatrix}\mathbf{x} + \mathbf{f}(t)\begin{bmatrix} 1 \\ 1 \end{bmatrix}$$

write the general solution by means of Theorem 4.4.3.

2.* Suppose $\mathbf{x}(0) = [x_0, y_0]^{\mathrm{T}}$. If for some t_0, $\mathbf{x}(t_0) = \mathbf{0}$, prove that the solution of the system given in Problem 1 above implies $x_0 = y_0$.
3. A *principle of superposition* for inhomogeneous equations: If $\mathbf{f} = \mathbf{f}_1 + \mathbf{f}_2$ and $\mathbf{x}_1$ solves $\mathbf{x}' = \mathrm{A}\mathbf{x} + \mathbf{f}_1$ and $\mathbf{x}_2$ solves $\mathbf{x}' = \mathrm{A}\mathbf{x} + \mathbf{f}_2$ then $\mathbf{x}_1 + \mathbf{x}_2$ solves $\mathbf{x}' = \mathrm{A}\mathbf{x} + \mathbf{f}$.
4. Differentiate $\mathbf{x}_p(t) = \int_0^t \Phi(t-s)\mathbf{f}(s)\,ds$ under the assumption that Φ is a normalized fundamental matrix at $t_0 = 0$.
5. Use the result of Problem 4, above, to prove Corollary 4.4.2 directly.
6. If $\mathbf{y}$ is a particular solution of $\mathbf{x}' = \mathrm{A}\mathbf{x} + \mathbf{f}$ show that $\mathbf{k}$ in Equation (4.4.16) can be selected so that $\mathbf{x} \equiv \mathbf{y}$.

4.5. THE INHOMOGENEOUS PROBLEM: SOME SPECIAL METHODS

In solving $\mathbf{x}' = \mathrm{A}\mathbf{x} + \mathbf{f}$ we look for any particular solution of $\mathbf{x}' = \mathrm{A}\mathbf{x} + \mathbf{f}$ to which we then add the general solution of the complementary system, $\mathbf{x}' = \mathrm{A}\mathbf{x}$. The method of variation of parameters always yields a particular solution once a knowledge of the general solution of the complementary equation is available. It sometimes happens that we may bypass entirely the general solution of the complementary system and discover particular solutions of the inhomogeneous equation by inspection. This happens frequently enough to warrant the inclusion of the technique as a special section.

The method is historically known as the method of *undetermined coefficients*. It takes full advantage of the special form of the "forcing" function **f**. For instance, in the equation $\mathbf{x}' = A\mathbf{x}+\mathbf{b}$, **b** a constant, we might reasonably expect a constant solution, say $\mathbf{x}(t) = \mathbf{k}$. Since $\mathbf{x}'(t) = \mathbf{0}$ we have $A\mathbf{k}+\mathbf{b} = \mathbf{0}$ as the requirement that $\mathbf{x}(t) = \mathbf{k}$ be a solution. If A^{-1} exists we can always determine **k** as $\mathbf{k} = -A^{-1}\mathbf{b}$. The following is a particular case of a more general result.

4.5.1. THEOREM *The inhomogeneous system* $\mathbf{x}' = A\mathbf{x}-\mathbf{b}e^{wt}$ *has a solution of the form* $\mathbf{x}(t) = \mathbf{k}e^{wt}$ *if w is not an eigenvalue of* A.

PROOF The proof is accomplished by substitution of $\mathbf{k}e^{wt}$ into the system with the resulting conclusion,

$$\mathbf{k}we^{wt} = Ae^{wt}\mathbf{k}-\mathbf{b}e^{wt}$$

Hence, after dividing by e^{wt} and rearranging the terms,

$$(A-wI)\mathbf{k} = \mathbf{b} \tag{4.5.1}$$

If w is not an eigenvalue, then $A-wI$ is invertible and $\mathbf{k} = (A-wI)^{-1}\mathbf{b}$.

In the event that $\mathbf{f} = \mathbf{b}$, we have $w = 0$ and thus $(A-wI) = A$ is invertible if A is nonsingular. In practice it is often simpler to solve Equation (4.5.1) by row reductions than to compute $(A-wI)^{-1}$. In fact, we need not even commit Equation (4.5.1) to memory if we recall the idea of trying $\mathbf{k}e^{wt}$ as a solution when $\mathbf{f}(t) = \mathbf{b}e^{wt}$.

4.5.1. Example

Find a solution of

$$\mathbf{x}' = \begin{bmatrix} 0 & -1 \\ -1 & 0 \end{bmatrix}\mathbf{x}+e^{2t}\begin{bmatrix} 1 \\ 1 \end{bmatrix}$$

Solution

We try $\mathbf{x}(t) = \mathbf{k}e^{2t}$. Then, $\mathbf{x}'(t) = 2\mathbf{k}e^{2t}$ and hence **k** is determined from the equation

$$\begin{bmatrix} -2 & -1 \\ -1 & -2 \end{bmatrix}\mathbf{k} = -\begin{bmatrix} 1 \\ 1 \end{bmatrix}$$

That is, $\mathbf{k} = \frac{1}{3}[1, 1]^T$ and $\mathbf{x}(t) = (e^{2t}/3)[1, 1]^T$.

The uses of Theorem 4.5.1 can be extended in a number of ways. Consider the systems

$$\begin{aligned} &\text{(a)}\ \mathbf{x}' = A\mathbf{x}-\mathbf{b}e^{iwt} \\ &\text{(b)}\ \mathbf{x}' = A\mathbf{x}-\mathbf{b}\cos wt \\ &\text{(c)}\ \mathbf{x}' = A\mathbf{x}-\mathbf{b}\sin wt \end{aligned} \tag{4.5.2}$$

where A is a real matrix, **b** is a real constant vector and w is a real scalar. If we assume that $\mathbf{x}(t)$ solves Equation (4.5.2)(a) then

$$\mathbf{x}'(t) = A\mathbf{x}(t)-\mathbf{b}e^{iwt} \tag{4.5.3}$$

and by taking real and imaginary parts of Equation (4.5.3) we obtain

$$[\text{Re}\,\mathbf{x}(t)]' = A[\text{Re}\,\mathbf{x}(t)] - \mathbf{b}\cos wt$$
$$[\text{Im}\,\mathbf{x}(t)]' = A[\text{Im}\,\mathbf{x}(t)] - \mathbf{b}\sin wt$$

That is, the real part of a solution of Equation (4.5.2) (a) solves Equation (4.5.2) (b), while the imaginary part solves Equation (4.5.2) (c). But the system (4.5.2) (a) is in the form in which Theorem 4.5.1 is applicable.

4.5.2. Example

Find a particular solution of

$$\mathbf{x}' = \begin{bmatrix} 0 & -2 \\ 1 & -1 \end{bmatrix}\mathbf{x} + \sin t\begin{bmatrix} 0 \\ 1 \end{bmatrix}$$

Solution

We replace the given system by

$$\mathbf{x}' = \begin{bmatrix} 0 & -2 \\ 1 & -1 \end{bmatrix}\mathbf{x} + e^{it}\begin{bmatrix} 0 \\ 1 \end{bmatrix}$$

We try a solution of the form $\mathbf{x}_p(t) = \mathbf{k}e^{it}$. For this function,

$$ie^{it}\mathbf{k} = e^{it}\begin{bmatrix} 0 & -2 \\ 1 & -1 \end{bmatrix}\mathbf{k} + e^{it}\begin{bmatrix} 0 \\ 1 \end{bmatrix}$$

and hence

$$\begin{bmatrix} -i & -2 \\ 1 & -1-i \end{bmatrix}\mathbf{k} = \begin{bmatrix} 0 \\ -1 \end{bmatrix}$$

From this $\mathbf{k} = [-1+i, (1+i)/2]$ and therefore

$$\mathbf{x}_p(t) = e^{it}\begin{bmatrix} -1+i \\ \dfrac{1+i}{2} \end{bmatrix} = (\cos t + i\sin t)\begin{bmatrix} -1+i \\ \dfrac{1+i}{2} \end{bmatrix}$$

$$= \begin{bmatrix} -\cos t - \sin t + i(-\sin t + \cos t) \\ \dfrac{1}{2}(\cos t - \sin t) + \dfrac{i}{2}(\cos t + \sin t) \end{bmatrix}$$

Hence

$$\text{Im}\,\mathbf{x}_p(t) = \begin{bmatrix} -\sin t + \cos t \\ \dfrac{\cos t}{2} + \dfrac{\sin t}{2} \end{bmatrix}$$

is a solution of the given inhomogeneous system. If the function $\mathbf{f}$ is a combination of functions, say $\mathbf{f} = \mathbf{b}_1 e^{w_1 t} + \mathbf{b}_2 e^{w_2 t}$ then by superposition (see Exercise 4.4, Part 2, Problem 3) we may solve $\mathbf{x}' = A\mathbf{x} + \mathbf{f}$. We add together the solutions of $\mathbf{x}' = A\mathbf{x} + \mathbf{b}_1 e^{w_1 t}$ and $\mathbf{x}' = A\mathbf{x} + \mathbf{b}_2 e^{w_2 t}$; both systems are of the form in which Theorem 4.5.1 is applicable.

4.5.3. Example

Input/Response Analysis

If a time-dependent voltage source, $V(t)$, is inserted in the circuit of Section 3.7, as shown in Figure 4.5.1,

Figure 4.5.1

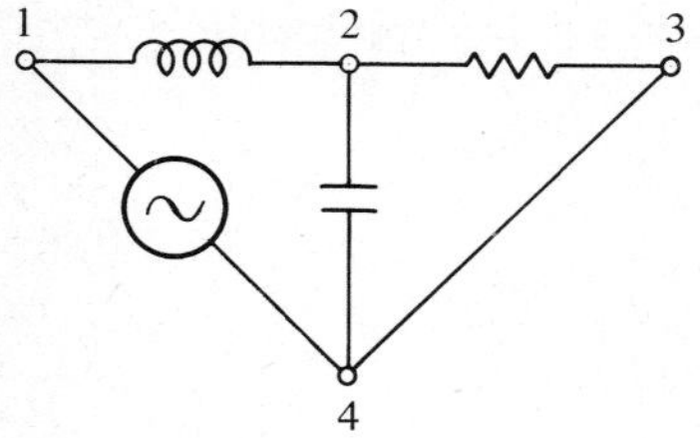

the circuit equations become

$$V_{12} + V_{23} = L\frac{di_{12}}{dt} + Ri_{23} = V(t)$$

and

$$\frac{d}{dt}V_{42} + \frac{d}{dt}V_{23} = \frac{1}{C}i_{42} + R\frac{di_{23}}{dt} = 0$$

since $i_{42} = i_{23} - i_{12}$, we get

$$i'_{12} = -\frac{R}{L}i_{23} + \frac{1}{L}V(t)$$

$$i'_{23} = \frac{-1}{RC}(i_{23} - i_{12})$$

or in vector matrix form

$$\mathbf{i}' = \begin{bmatrix} 0 & -\dfrac{R}{L} \\ \dfrac{1}{RC} & -\dfrac{1}{RC} \end{bmatrix}\mathbf{i} + \begin{bmatrix} \dfrac{1}{L}V(t) \\ 0 \end{bmatrix}$$

Let us consider the circuit of Figure 4.5.1, taking $V(t) = \cos \omega t$. For this oscillating input voltage with frequency proportional to ω we seek the output voltage $V_{23} = Ri_{23}$ across the resistor.

We have the equation

$$\mathbf{i}' = \begin{bmatrix} 0 & -\dfrac{R}{L} \\ \dfrac{1}{RC} & -\dfrac{1}{RC} \end{bmatrix}\mathbf{i} + \begin{bmatrix} \dfrac{1}{L}\cos \omega t \\ 0 \end{bmatrix}$$

which corresponds to the complex equation

$$\mathbf{z}' = \begin{bmatrix} 0 & -\dfrac{R}{L} \\ \dfrac{1}{RC} & -\dfrac{1}{RC} \end{bmatrix} \mathbf{z} + e^{i\omega t} \begin{bmatrix} \dfrac{1}{L} \\ 0 \end{bmatrix}$$

For convenience we set $\mu = i\omega$ and we seek a solution of the form $\mathbf{z}(t) = e^{\mu t}\mathbf{b}$. This leads to the equation

$$\begin{bmatrix} \mu & \dfrac{R}{L} \\ -\dfrac{1}{RC} & \mu + \dfrac{1}{RC} \end{bmatrix} \begin{bmatrix} b_1 \\ b_2 \end{bmatrix} = \begin{bmatrix} \dfrac{1}{L} \\ 0 \end{bmatrix}$$

We solve for b_2 as follows:

$$\left[\begin{array}{cc|c} \mu & \dfrac{R}{L} & \dfrac{1}{L} \\ -\dfrac{1}{RC} & \mu + \dfrac{1}{RC} & 0 \end{array}\right] \to \left[\begin{array}{cc|c} 1 & \dfrac{R}{L\mu} & \dfrac{1}{L\mu} \\ -1 & RC\mu + 1 & 0 \end{array}\right]$$

$$\to \left[\begin{array}{cc|c} 1 & \dfrac{R}{L\mu} & \dfrac{1}{L\mu} \\ 0 & RC\mu + 1 + \dfrac{R}{L\mu} & \dfrac{1}{L\mu} \end{array}\right] \to \left[\begin{array}{cc|c} 1 & \dfrac{R}{L\mu} & \dfrac{1}{L\mu} \\ 0 & RC\mu^2 + \mu + \dfrac{R}{L} & \dfrac{1}{L} \end{array}\right]$$

so that

$$b_2 = \frac{1}{LRC\mu^2 + \mu L + R} = \frac{1}{R - RLC\omega^2 + i\omega L}$$

Thus the second component of $\mathbf{z}(t)$ is

$$z_2(t) = \frac{1}{R - RLC\omega^2 + i\omega L} e^{i\omega t} = b_2 e^{i\omega t}$$

Let us compute the real part of $z_2(t)$ as follows: First we note that

$$z_2(t) = |b_2| \left(\frac{b_2}{|b_2|} \right) e^{i\omega t}$$

$$= |b_2| e^{i\phi} e^{i\omega t} = |b_2| e^{i(\omega t + \phi)}$$

where the phase angle ϕ satisfies

$$\cos \phi = \operatorname{Re}\left(\frac{b_2}{|b|} \right) \qquad \sin \phi = \operatorname{Im}\left(\frac{b_2}{|b|} \right)$$

Now, since $\mathbf{i}(t) = \operatorname{Re} \mathbf{z}(t)$ will be a solution of the original equation,

$$i_{23}(t) = \operatorname{Re} \mathbf{z}_2(t) = |b_2| \cos(\omega t + \phi)$$

and, therefore,

$$V_{23}(t) = R\,|b_2| \cos(\omega t + \phi)$$
$$= \frac{R}{[(R^2 - RLC\omega^2) + \omega^2 L^2]^{1/2}} \cos(\omega t + \phi)$$

The maximum value assumed by V_{23} is $R|b_2|$. It is called the *amplitude* of V_{23}. Since the amplitude of the input is 1, $R|b_2|$ represents the ratio of output to input amplitudes.

Exercise 4.5

Use Theorem 4.5.1 to find a particular solution of the following inhomogeneous equations.

1. $\mathbf{x}' = \begin{bmatrix} 1 & 0 \\ -1 & 3 \end{bmatrix} \mathbf{x} + \begin{bmatrix} 1 \\ 1 \end{bmatrix}$

2. $\mathbf{x}' = \begin{bmatrix} 1 & 0 \\ -1 & 3 \end{bmatrix} \mathbf{x} + \begin{bmatrix} 1 \\ 1 - e^{-t} \end{bmatrix}$

3. $\mathbf{x}' = \begin{bmatrix} 1 & 0 \\ -1 & 3 \end{bmatrix} \mathbf{x} + \begin{bmatrix} \sin t \\ \cos t \end{bmatrix}$

4. $\mathbf{x}' = \begin{bmatrix} 0 & -2 \\ 1 & -1 \end{bmatrix} \mathbf{x} + \cos t \begin{bmatrix} 1 \\ 0 \end{bmatrix}$

5. $\mathbf{x}' = \begin{bmatrix} 0 & -2 \\ 1 & -1 \end{bmatrix} \mathbf{x} + \begin{bmatrix} \cos t \\ \sin t \end{bmatrix}$

6. $x' = \alpha x + Ke^{at} \qquad a \neq \alpha$
7. $x' = \alpha x + K_1 + K_2 e^{at} + K_3 \sin \omega t \qquad \alpha \neq 0$
8. $x' = -x + 6 + 4e^{-t}$
9. $x' = 2x + e^t + \cos 2t$

10. $\mathbf{x}' = \begin{bmatrix} 2 & 1 & 0 \\ 0 & 2 & 0 \\ 0 & 0 & -1 \end{bmatrix} \mathbf{x} + \begin{bmatrix} e^t \\ 1 \\ 0 \end{bmatrix}$

11. $\mathbf{x}' = \begin{bmatrix} 0 & 1 \\ -1 & 2 \end{bmatrix} \mathbf{x} + e^t \cos t \begin{bmatrix} 1 \\ 1 \end{bmatrix}$

    Hint: Solve the same system with $\mathbf{f} = e^{(1+i)t}[1, 1]^T$ and then take the real part of the solution. Note

$$e^{a+ib} = e^a \cdot e^{ib} = e^a(\cos b + i \sin b) = e^a \cos b + ie^a \sin b$$

4.6.* GRONWALL'S INEQUALITY

The "size" of a solution is measured by the value of its norm. (In the case of a scalar function the norm is simply the absolute value.) In a number of applications of differential equations it is important to have, in lieu of an exact solution, an upper bound to the solution's norm. Such estimates can be obtained by an inequality whose statement and proof are the contents of this section.

4.6.1. LEMMA *Suppose f and g are continuous real-valued functions in $a \leqslant t \leqslant b$ and f' exists there. Suppose also that*

$$f'(t) \leqslant f(t)g(t) \qquad a \leqslant t \leqslant b \tag{4.6.1}$$

Then for all t in $[a, b]$,

$$f(t) \leqslant f(a) \exp\left[\int_a^t g(s)\, ds\right] \tag{4.6.2}$$

PROOF For typographical purposes, set

$$p(t) = \exp\left(-\int_a^t g(s)\, ds\right)$$

Then, $p'(t) = -g(t)p(t)$ and $p(a) = 1$. The function $\mathrm{F}(t) = f(t)p(t)$ is nonincreasing as we can see by computing F'. For,

$$\begin{aligned} \mathrm{F}' &= f'p + fp' \\ &= f'p - fpg \\ &= p(f' - gf) \leqslant 0 \end{aligned}$$

since $p > 0$ and $f' - fg \leqslant 0$ by hypothesis. Because F is nonincreasing for $a \leqslant t \leqslant b$, $\mathrm{F}(a) \geqslant \mathrm{F}(t)$. But $\mathrm{F}(a) = f(a)$ since $p(a) = 1$. From $\mathrm{F}(a) \geqslant \mathrm{F}(t)$ we have

$$\mathrm{F}(a) = f(a) \geqslant f(t)p(t) = \mathrm{F}(t) = f(t) \exp\left[-\int_a^t g(s)\, ds\right]$$

from which we conclude inequality (4.6.2).

4.6.2. THEOREM (Gronwall's Inequality) *If f and g* ($g \geqslant 0$) *are continuous real-valued functions in $a \leqslant t \leqslant b$ and K is some real constant for which*

$$f(t) \leqslant \mathrm{K} + \int_a^t f(s)g(s)\, ds \qquad a \leqslant t \leqslant b \tag{4.6.3}$$

then

$$f(t) \leqslant \mathrm{K} \exp\left[\int_a^t g(s)\, ds\right] \qquad a \leqslant t \leqslant b \tag{4.6.4}$$

PROOF Set $\mathrm{G}(t) = \mathrm{K} + \int_a^t f(s)g(s)\, ds$. Then G is differentiable in $[a, b]$; indeed, $\mathrm{G}' = fg$. Since $g \geqslant 0$ and $f \leqslant \mathrm{G}$ (by hypothesis), $fg \leqslant \mathrm{G}g$ and therefore $\mathrm{G}' \leqslant \mathrm{G}g$.

All the hypotheses of Lemma 4.6.1 are satisfied (with G playing the role of f) and hence

$$\mathrm{G}(t) \leqslant \mathrm{G}(a) \exp\left[\int_a^t g(s)\, ds\right] \tag{4.6.5}$$

By definition of G and hypothesis, $f(t) \leqslant \mathrm{G}(t)$ and $\mathrm{G}(a) = \mathrm{K}$. With these substitutions, the latter inequality becomes the conclusion of the theorem. (Note that the theorem does not require the existence of f'.)

4.6.1. Example

Show that the only continuous function, f, satisfying

$$0 \leqslant f(t) \leqslant \int_a^t f(s)\, ds \qquad t_0 \leqslant t$$

is the identically zero function.

Solution

In equality (4.6.3) set $g(s) = 1$ and $K = 0$. All the hypotheses of Theorem 4.6.2 are satisfied and hence inequality (4.6.4) yields the desired conclusion.

4.6.2. Example

Use the Gronwall Inequality to prove that only the identically zero function is a solution of the initial-value problem

$$x' = g(t)x \qquad x(t_0) = 0$$

with g continuous in $t \geqslant t_0$.

Solution

Integration of the differential equation gives

$$x(t) = \int_{t_0}^t g(s)x(s)\, ds \tag{4.6.6}$$

for any function, $x(t)$, which satisfies the differential equation and vanishes at $t = t_0$. We obtain from Equation (4.6.6)

$$0 \leqslant |x(t)| = \left| \int_{t_0}^t g(s)x(s)\, ds \right| \leqslant \int_{t_0}^t |g(s)|\, |x(s)|\, ds \tag{4.6.7}$$

In Theorem 4.6.2, set $f(t) = |x(t)|$ and $g(t) = |g(t)|$ and $K = 0$. Then inequality (4.6.7) implies $x(t) = 0$.

In this example, the important point is not the content of the conclusion, which could have been reached in a more direct way, but the method of using the Gronwall Inequality. This technique generalizes readily to systems with nonconstant coefficients. The essential features are the replacement of the differential equation by an equation involving an integral and then taking absolute values (norms in the case of systems). We illustrate some of these ideas in the next section.

Exercise 4.6

1. In the proof of the lemma where did we make use of the hypothesis that g was continuous?
2. In the proof of the lemma we used $p > 0$. Why is $p > 0$? Can $p(t) = 0$ for any t? Why?

3. In the proof of Theorem 4.6.2, where did we use the fact that f is continuous? How do we know that G is continuous?
4. If x is continuous, prove that $|x|$ is also. If x' exists does $|x|'$? If so, prove it. If not, offer an example for which it is false.

4.7.* THE GROWTH OF SOLUTIONS

We apply the Gronwall Inequality to the theory of differential equations. The central result is an upper estimate to the rate of growth of the solutions of the system $\mathbf{x}' = \mathrm{A}\mathbf{x}$.

4.7.1. THEOREM *If* $\mathbf{x}(t)$ *is any solution of* $\mathbf{x}' = \mathrm{A}\mathbf{x}$ *then for all* t *and* t_0,

$$\|\mathbf{x}(t)\| \leqslant \|\mathbf{x}(t_0)\| \exp(\|\mathrm{A}\|\,|t-t_0|) \tag{4.7.1}$$

PROOF Assume that $t > t_0$. Then, since $\mathbf{x}'(t) = \mathrm{A}\mathbf{x}(t)$, we have, by integration,

$$\mathbf{x}(t) - \mathbf{x}(t_0) = \int_{t_0}^{t} \mathrm{A}\mathbf{x}(s)\, ds$$

Therefore, using a number of properties of $\|\ \|$,

$$\|\mathbf{x}(t)\| \leqslant \|\mathbf{x}(t_0)\| + \int_{t_0}^{t} \|\mathrm{A}\|\,\|\mathbf{x}(s)\|\, ds$$

Set $f(t) = \|\mathbf{x}(t)\|$, $g(t) = \|\mathrm{A}\|$ and $\mathrm{K} = \|\mathbf{x}(t_0)\|$ in Theorem 4.6.2 to deduce

$$\|\mathbf{x}(t)\| \leqslant \|\mathbf{x}(t_0)\| \exp\left(\int_{t_0}^{t} \|\mathrm{A}\|\, ds\right) = \|\mathbf{x}(t_0)\| \exp\,[\|\mathrm{A}\|(t-t_0)] \tag{4.7.2}$$

which completes the proof in event $t \geqslant t_0$. For $t < t_0$, we set $\mathbf{y}(t) = \mathbf{x}(-t)$. Assuming $\mathbf{x}(t)$ solves $\mathbf{x}' = \mathrm{A}\mathbf{x}$, the chain rules implies that $\mathbf{y}(t)$ is a solution of $\mathbf{y}' = -\mathrm{A}\mathbf{y}$ since

$$\mathbf{y}'(t) = -\mathrm{A}\mathbf{x}(-t) = -\mathrm{A}\mathbf{y}(t)$$

Because $t < t_0$ we have $-t > -t_0$ and from inequality (4.7.2) applied to the equation $\mathbf{y}' = -\mathrm{A}\mathbf{y}$ we conclude

$$\begin{aligned}\|\mathbf{x}(t)\| = \|\mathbf{y}(-t)\| &\leqslant \|\mathbf{y}(-t_0)\| \exp\,[\|-\mathrm{A}\|(-t+t_0)] \\ &= \|\mathbf{x}(t_0)\| \exp\,[\|\mathrm{A}\|(-t+t_0)] \\ &= \|\mathbf{x}(t_0)\| \exp\,[\|\mathrm{A}\|\,|t-t_0|]\end{aligned} \tag{4.7.3}$$

Combining the inequalities (4.7.2) and (4.7.3) leads to the desired inequality (4.7.1).

Remarks

When $t = t_0$ inequality (4.7.1) becomes a trivial assertion. As $|t-t_0|$ grows large the inequality asserts that $|\mathbf{x}(t)|$ cannot grow large faster than a certain exponential function. (Functions so restricted in growth are called functions of exponential type.) Thus, for instance, $\|\mathbf{x}(t)\| = e^{t^2}$ is an impossibility for any solution of $\mathbf{x}' = \mathrm{A}\mathbf{x}$.

For some systems, inequality (4.7.1) is very much an overestimate and for some it is quite precise. The next two examples present such cases.

4.7.1. Example

Solve

$$\mathbf{x}' = \begin{bmatrix} 0 & 1 \\ -1 & 0 \end{bmatrix}\mathbf{x} \qquad \mathbf{x}(0) = \begin{bmatrix} 0 \\ 1 \end{bmatrix}$$

Solution

We compute

$$\mathbf{x}(t) = \begin{bmatrix} \sin t \\ \cos t \end{bmatrix} \qquad \|\mathbf{x}(t)\| = 1$$

and

$$\|\mathrm{A}\| = [0^2 + 1^2 + (-1)^2 + 0^2]^{1/2} = \sqrt{2}$$

so that

$$\|\mathbf{x}(t)\| = 1 \leqslant \|\mathbf{x}(0)\| e^{\sqrt{2}|t|} = e^{\sqrt{2}|t|}$$

an obviously imprecise bound when t is large and positive.

4.7.2. Example

Same as the previous example except

$$\mathbf{x}' = \begin{bmatrix} 1 & 0 \\ 0 & 0 \end{bmatrix}\mathbf{x} \qquad \mathbf{x}(0) = \begin{bmatrix} 1 \\ 0 \end{bmatrix}$$

Solution

Here

$$\mathbf{x}(t) = \begin{bmatrix} 1 \\ 0 \end{bmatrix} e^t \qquad \|\mathbf{x}(0)\| = e^t \qquad \|\mathrm{A}\| = 1$$

Therefore,

$$\|\mathbf{x}(t)\| = e^t \leqslant \|\mathbf{x}(0)\| e^{\|\mathrm{A}\|\,|t|} = e^{|t|}$$

and Gronwall's Inequality is exact for positive t.

The price we pay for an inequality valid for a wide class of functions is the possible imprecision of the estimate. The major uses of this theorem fortunately do not depend critically on the precision of the upper bound.

Two significant applications illustrating this fact are presented below as Problems 5 and 6 of this exercise set.

Exercise 4.7

1. Suppose $\mathbf{x}(t)$ is a solution of $\mathbf{x}' = A\mathbf{x}$ where $A = 0$. Verify inequality (4.7.1) in this case.
2. Suppose $\mathbf{x}' = I\mathbf{x}$. Repeat the proof of Theorem 4.7.1 but use the specific nature of A to deduce: If $\mathbf{x}(t)$ is a solution of $\mathbf{x}' = I\mathbf{x}$, then

$$\|\mathbf{x}(t)\| \leq \|\mathbf{x}(t_0)\| \exp(|t-t_0|)$$

3. Let $\mathbf{x}' = B\mathbf{x}$, where B is any matrix obtainable from I by a rearrangement of its rows. Deduce the same conclusion as in Problem 2. (Hint: $\|B\mathbf{x}\| = \|x\|$.)
4.* Let D be a diagonal matrix with diagonal entries, $d_1, d_2, \ldots, d_n$ and set $d = \max\{|d_1|, |d_2|, \ldots, |d_n|\}$. If $\mathbf{x}(t)$ is a solution of $\mathbf{x}' = D\mathbf{x}$, show that

$$\|\mathbf{x}(t)\| \leq \|\mathbf{x}(t_0)\| \exp(d|t-t_0|)$$

Note that Theorem 4.7.2 yields the less precise inequality

$$\|\mathbf{x}(t)\| \leq \|\mathbf{x}(t_0)\| \exp(\|D\|\,|t-t_0|)$$

where $\|D\|^2 = d_1^2 + d_2^2 + \cdots + d_n^2 \leq nd^2$.

Chapter 5
HIGHER-ORDER EQUATIONS

5.1. THE *n*th-ORDER LINEAR EQUATIONS IN ONE DIMENSION

The study of the nth-order equation with constant coefficients,

$$y^{(n)} + a_{n-1}y^{(n-1)} + \cdots + a_1 y^{(1)} + a_0 y = f(t) \tag{5.1.1}$$

is the main goal of this chapter. As usual, $f(t)$ is assumed to be continuous and by definition, a *solution* is any function with continuous nth derivative which upon substitution into (5.1.1) reduces it to an identity. For example, the function $y(t) = -\frac{1}{6}t^3 - \frac{1}{4}t^2 - \frac{1}{4}t$ is a solution of

$$y^{(2)} - 2y^{(1)} = t^2$$

The theory of (5.1.1) has a remarkable connection with the theory of the system $\mathbf{x}' = A\mathbf{x} + \mathbf{f}$. That such a connection should exist may seem surprising. It in fact arises quite naturally in the description of many physical phenomena. Visualize, for instance, a particle constrained to move on a line and subject to various forces. Suppose its distance from a fixed point on the line is measured by $y(t)$ and its movement is governed by the "law"

$$y^{(3)} - 6y^{(2)} + 5y^{(1)} = \sin t \tag{5.1.2}$$

At time $t = 0$, we further assume that $y(0) = y'(0) = y''(0) = 0$; in non-technical terms the experiment starts with the clock set to zero and the particle begins its motion from rest at the origin. Since y measures distance, $y^{(1)}$ measures velocity and $y^{(2)}$ measures acceleration. If we emphasize the physical variables by assigning $y^{(1)} = v$ and $y^{(2)} = a$, we may write the "law," Equation (5.1.2), as

$$\begin{aligned} y^{(1)} &= v \\ y^{(2)} &= a \\ y^{(3)} &= -5v + 6a + \sin t \end{aligned} \tag{5.1.3}$$

If we now denote by **x** the vector whose entries are y, v, and a, then Equations (5.1.3) take a familiar form

$$\begin{bmatrix} y' \\ v' \\ a' \end{bmatrix} = \begin{bmatrix} 0 & 1 & 0 \\ 0 & 0 & 1 \\ 0 & -5 & 6 \end{bmatrix} \begin{bmatrix} y \\ v \\ a \end{bmatrix} + \begin{bmatrix} 0 \\ 0 \\ \sin t \end{bmatrix} \tag{5.1.4}$$

a special case of $\mathbf{x}' = \mathbf{A}\mathbf{x} + \mathbf{f}$. Note also that

$$\mathbf{x}(0) = \begin{bmatrix} y(0) \\ v(0) \\ a(0) \end{bmatrix} = \begin{bmatrix} y(0) \\ y^{(1)}(0) \\ y^{(2)}(0) \end{bmatrix} = \mathbf{0}$$

Thus the simple act of renaming the derivatives of y according to their physical meanings results in the conversion of a third-order equation into a system of three first-order equations. This is precisely the connection between nth-order equations and systems we have previously alluded to. We spend the next few sections extracting the consequences of this seemingly trivial observation.

Exercise 5.1

1. Find the eigenvalues and eigenvectors of the systems matrix in Equation (5.1.4).
2. Find a fundamental matrix for the system given by Equation (5.1.4) and thus write the general solution of the corresponding homogeneous system.
3. Use variation of parameters to find the general solution of the inhomogeneous system given by Equation (5.1.4).
4. Verify that the first component of the solution given in answer to Problem 3 above is a solution of Equation (5.1.2).
5. Select the arbitrary constants in your answer to Problem 3 above so that $\mathbf{x}(0) = \mathbf{0}$. Is the first component of this solution a solution of Equation (5.1.2)? Does it place the particle at the origin at $t = 0$ and at rest then?

5.2. THE COMPANION SYSTEM

Let us apply the ideas of Section 5.1 to a more general situation. Suppose

$$y^{(3)} + a_2 y^{(2)} + a_1 y^{(1)} + a_0 y = f(t) \tag{5.2.1}$$

Set $\mathbf{x} = [x_1, x_2, x_3]^{\mathrm{T}} = [y, y^{(1)}, y^{(2)}]^{\mathrm{T}}$. Then,

$$\mathbf{x}' = \begin{bmatrix} y^{(1)} \\ y^{(2)} \\ y^{(3)} \end{bmatrix} = \begin{bmatrix} x_2 \\ x_3 \\ -a_0 x_1 - a_1 x_2 - a_2 x_3 \end{bmatrix} + \begin{bmatrix} 0 \\ 0 \\ f(t) \end{bmatrix}$$

$$= \begin{bmatrix} 0 & 1 & 0 \\ 0 & 0 & 1 \\ -a_0 & -a_1 & -a_2 \end{bmatrix} \mathbf{x} + f(t) \begin{bmatrix} 0 \\ 0 \\ 1 \end{bmatrix} \tag{5.2.2}$$

In the second-order equation

$$y^{(2)} + a_1 y^{(1)} + a_0 y = f(t) \tag{5.2.3}$$

the definition $\mathbf{x} = [x_1, x_2]^T = [y, y^{(1)}]^T$ leads to the system

$$\begin{bmatrix} 0 & 1 \\ -a_0 & -a_1 \end{bmatrix} \mathbf{x} + f(t) \begin{bmatrix} 0 \\ 1 \end{bmatrix} \tag{5.2.4}$$

Systems derived from nth-order equations by the definition $\mathbf{x} = [y, y^{(1)}, \ldots, y^{(n-1)}]^T$ are called their *companion systems*. Thus system (5.2.2) is the companion system for the third-order equation (5.2.1) while the system (5.2.4) is the companion system for Equation (5.2.3). These notions generalize in the obvious manner to the nth-order equation.

5.2.1. DEFINITION *A matrix of the form*

$$A_c = \begin{bmatrix} 0 & 1 & 0 & \cdots & 0 \\ 0 & 0 & 1 & \cdots & 0 \\ \vdots & \vdots & \vdots & \ddots & \vdots \\ 0 & 0 & 0 & \cdots & 1 \\ -a_0 & -a_1 & -a_2 & \cdots & -a_{n-1} \end{bmatrix} \tag{5.2.5}$$

is called a companion matrix

5.2.2. DEFINITION *The system*

$$\mathbf{x}' = \begin{bmatrix} 0 & 1 & 0 & \cdots & 0 \\ 0 & 0 & 1 & \cdots & 0 \\ \vdots & \vdots & \vdots & \ddots & \vdots \\ 0 & 0 & 0 & \cdots & 1 \\ -a_0 & -a_1 & -a_2 & \cdots & -a_{n-1} \end{bmatrix} \mathbf{x} = f(t) \begin{bmatrix} 0 \\ 0 \\ \vdots \\ 0 \\ 1 \end{bmatrix} \tag{5.2.6}$$

is the companion system for the nth-order *equation*

$$y^{(n)} + a_{n-1} y^{(n-1)} + \cdots + a_1 y^{(1)} + a_0 y = f(t) \tag{5.2.7}$$

The importance of the companion system is due to the next theorem, which has been anticipated implicitly, at least, in all we have done.

5.2.3. THEOREM *If* y *is any solution of Equation* (5.2.7), *then the vector* $\mathbf{x} = [y, y^{(1)}, \ldots, y^{(n-1)}]^T$ *is a solution of the companion system* (5.2.6). *Conversely,* if $\mathbf{x} = [x_1, x_2, \ldots, x_n]^T$ *is any solution of the companion system of Equation* (5.2.7) *then its first component,* x_1, *is a solution of Equation* (5.2.7) *and, moreover,* $\mathbf{x} = [x_1, x_1^{(1)}, \ldots, x_1^{(n-1)}]^T$.

PROOF We offer a proof for the case $n = 3$. The general situation is completely analogous. Suppose, then, that y solves the equation

$$y^{(3)} + a_2 y^{(2)} + a_1 y^{(1)} + a_0 y = f(t)$$

The argument used to obtain the system (5.2.2) from Equation (5.2.1) proves that $\mathbf{x} = [y, y^{(1)}, y^{(2)}]^T$ is a solution of the companion system. On the other hand, if $\mathbf{x} = [x_1, x_2, x_3]^T$ solves the companion system (5.2.2), then

$$\begin{bmatrix} x_1' \\ x_2' \\ x_3' \end{bmatrix} = \begin{bmatrix} x_2 \\ x_3 \\ -a_0 x_1 - a_1 x_2 - a_2 x_3 + f(t) \end{bmatrix}$$

by a simplification of the right-hand side of Equation (5.2.2). We deduce, therefore, that $x_1^{(1)} = x_2$, $x_2^{(1)} = x_3$, and $x_3^{(1)} = -a_0 x - a_1 x_2 - a_2 x_3 + f(t)$. Then $x_1^{(2)} = x_2^{(1)} = x_3$ and hence $x_1^{(3)} = x_3^{(1)} = -a_0 x_1 - a_1 x_1^{(1)} - a x_1^{(2)} + f(t)$ which shows that $\mathbf{x}$ solves Equation (5.2.1) and that $\mathbf{x} = [x_1, x_1^{(1)}, x_1^{(2)}]^T$.

Having proved this central theorem, we can now bring to bear the entire theory of linear systems on any nth-order equation.

5.2.1. Example

Find a solution of $y^{(2)} - y = 0$.

Solution

The companion system for this equation is

$$\mathbf{x}' = \begin{bmatrix} 0 & 1 \\ 1 & 0 \end{bmatrix} \mathbf{x}$$

We derive with no trouble the solution

$$\mathbf{x}(t) = c_1 \begin{bmatrix} e^t \\ e^t \end{bmatrix} + c_2 \begin{bmatrix} e^{-t} \\ -e^{-t} \end{bmatrix}$$

whose first component, $x_1 = c_1 e^t + c_2 e^{-t}$, is readily verified as a solution of $y^{(2)} - y = 0$.

We may use Theorem 5.2.3 in the other direction, constructing solutions of companion systems from solutions of nth-order equations. For instance, see the following example.

5.2.2. Example

Find a solution of

$$\mathbf{x}' = \begin{bmatrix} 0 & 1 & 0 & 0 \\ 0 & 0 & 1 & 0 \\ 0 & 0 & 0 & 1 \\ 0 & 0 & 1 & 0 \end{bmatrix} \mathbf{x}$$

Solution

The given system is companion to the fourth-order equation

$$y^{(4)} - y^{(2)} = 0 \tag{5.2.8}$$

By inspection $y(t) = t$ is a solution of this equation. Hence $\mathbf{x} = [y, y^{(1)}, y^{(2)}, y^{(3)}]^T = [t, 1, 0, 0]^T$ must solve the companion system. It does.

Our last illustration in this section of the use of Theorem 5.2.3 is an extremely important one. Is it possible for two functions v and z to solve the initial-value problem? Suppose that v and z are solutions of

$$y^{(3)} + a_2 y^{(2)} + a_1 y^{(1)} + a_0 y = f(t)$$

$$y(t_0) = y_1 \qquad y^{(1)}(t_0) = y_2 \qquad y^{(2)}(t_0) = y_3$$

Then $\mathbf{x}_1 = [v, v^{(1)}, v^{(2)}]^T$ and $\mathbf{x}_2 = [z, z^{(1)}, z^{(2)}]^T$ must solve the same initial-value problem,

$$\mathbf{x}' = \begin{bmatrix} 0 & 1 & 0 \\ 0 & 0 & 1 \\ -a_0 & -a_1 & -a_2 \end{bmatrix} \mathbf{x} + f(t) \begin{bmatrix} 0 \\ 0 \\ 1 \end{bmatrix} \qquad \mathbf{x}(t_0) = \begin{bmatrix} y_1 \\ y_2 \\ y_3 \end{bmatrix}$$

Then by Theorem 4.4.3, $\mathbf{x}_1 \equiv \mathbf{x}_2$ and hence $v \equiv z$. Similarly one may prove the following "existence and uniqueness" theorem for the nth-order initial-value problem.

5.2.4. THEOREM *The initial-value problem*

$$\begin{gathered} y^{(n)} + a_{n-1} y^{(n-1)} + \cdots + a_1 y^{(1)} + a_0 y = f(t) \\ y(t_0) = y_1, \quad y^{(1)}(t_0) = y_2, \ldots, y^{(n-1)}(t_0) = y_n \end{gathered} \tag{5.2.9}$$

has a solution and this solution is unique.

It follows immediately from this theorem that $y(t) \equiv 0$ is the sole solution of

$$y^{(n)} + a_{n-1} y^{(n-1)} + \cdots + a_1 y^{(1)} + a_0 y = 0$$

$$y(0) = y^{(1)}(0) = \cdots = y^{(n-1)}(0) = 0$$

For, $y(t) \equiv 0$ is obviously a solution and the theorem guarantees it is the *only* solution.

5.2.3. Example

Solve

$$y^{(2)} + y = 0 \qquad y(0) = 0 \qquad y'(0) = 1$$

Solution

The companion system is

$$\mathbf{x}' = \begin{bmatrix} 0 & 1 \\ -1 & 0 \end{bmatrix} \mathbf{x}$$

with initial condition

$$\mathbf{x}(0) = \begin{bmatrix} y(0) \\ y'(0) \end{bmatrix} = \begin{bmatrix} 0 \\ 1 \end{bmatrix} = \mathbf{x}_0$$

We find

$$\mathbf{x}(t) = \begin{bmatrix} \sin t \\ \cos t \end{bmatrix}$$

is a solution of the companion system and satisfies the initial condition. Hence

$$y(t) = \sin t$$

is the solution of the given problem.

5.2.4. Example

Solve

$$y^{(3)} - y^{(2)} = t \qquad y(0) = 1 \qquad y'(0) = 1 \qquad y^{(2)}(0) = -1$$

Solution

The companion system is

$$\mathbf{x}' = \begin{bmatrix} 0 & 1 & 0 \\ 0 & 0 & 1 \\ 0 & 0 & 1 \end{bmatrix} \mathbf{x} + \begin{bmatrix} 0 \\ 0 \\ t \end{bmatrix}$$

Using variation of parameters, we find a particular solution

$$\mathbf{x}_p(t) = \begin{bmatrix} \dfrac{-t^3}{6} - \dfrac{t^2}{2} - t - 1 \\ -\dfrac{t^2}{2} - t - 1 \\ -t - 1 \end{bmatrix}$$

The general solution is found to be

$$\mathbf{x}(t) = \begin{bmatrix} 1 & t & e^t \\ 0 & 1 & e^t \\ 0 & 0 & e^t \end{bmatrix} \mathbf{c} + \mathbf{x}_p(t)$$

where $\mathbf{c}$ is a vector of arbitrary constants. The initial condition on the given third-order initial-value problem places the initial condition

$$\mathbf{x}(0) = \begin{bmatrix} 1 \\ 1 \\ -1 \end{bmatrix}$$

on the companion system. Therefore $\mathbf{c}$ is determined by

$$\mathbf{x}(0) = \begin{bmatrix} 1 \\ 1 \\ -1 \end{bmatrix} = \begin{bmatrix} 1 & 0 & 1 \\ 0 & 1 & 1 \\ 0 & 0 & 1 \end{bmatrix} \mathbf{c} + \begin{bmatrix} -1 \\ -1 \\ -1 \end{bmatrix}$$

Thus $\mathbf{c} = [2, 2, 0]^T$ and hence

$$\mathbf{x}(t) = \begin{bmatrix} \frac{-t^3}{6} - \frac{t^2}{2} + t + 1 \\ -\frac{t^2}{2} - t + 1 \\ -t - 1 \end{bmatrix}$$

from which

$$x_1(t) = y(t) = -\frac{t^3}{6} - \frac{t^2}{2} + t + 1$$

is obtained as the required solution.

It is convenient to use the companion system to define the general solution of an nth-order equation.

5.2.5. DEFINITION *The general solution to the nth-order equation (5.2.9) is defined as the first entry in the general solution of the corresponding companion system.*

Thus, for example, the function $x_1(t) = c_1 e^t + c_2 e^{-t}$ is the general solution of $y^{(2)} - y = 0$. (See Example 5.2.1.) The companion system may also be used to prove the following "principles of superposition."

5.2.6. THEOREM *If y_1 and y_2 are any solutions of*

$$y^{(n)} + a_{n-1}y^{(n-1)} + \cdots + a_1 y^{(1)} + a_0 y = f_1(t)$$

and

$$y^{(n)} + a_{n-1}y^{(n-1)} + \cdots + a_1 y^{(1)} + a_0 y = f_2(t)$$

respectively, then $ay_1 + by_2$ is a solution of

$$y^{(n)} + a_{n-1}y^{(n-1)} + \cdots + a_1 y^{(1)} + a_0 y = af_1(t) + bf_2(t)$$

If $f_1 = f_2 = 0$ then Theorem 5.2.6 reduces to the important corollary which follows.

5.2.7. COROLLARY *If y_1 and y_2 are any two solutions of the homogeneous equation*

$$y^{(n)} + a_{n-1}y^{(n-1)} + \cdots + a_1 y^{(1)} + a_0 y = 0$$

then so is $ay_1 + by_2$.

Exercise 5.2

Part 1

Write the companion systems for the equations given in Problems 1–12.

1. $y^{(2)} - y^{(1)} = 0$
2. $y^{(2)} + y^{(1)} = 0$
3. $y^{(2)} - y = 0$
4. $y^{(2)} + y = 0$

5. $y^{(2)} = 0$
6. $y^{(2)} - y^{(1)} - y = 1$
7. $y^{(4)} - y^{(3)} = 0$
8. $y^{(4)} - y^{(1)} = 0$
9. $y^{(4)} - y = 0$
10. $y^{(3)} - y^{(2)} + y^{(1)} - y = 1$
11. $y^{(3)} - (\sin t)y = f(t)$
12. $y^{(2)} + a_1(t)y^{(1)} + a_0(t)y = 0$

In Problems 13–23 below you are given the system's matrix for the companion system of which nth-order equation?

13. $\begin{bmatrix} 0 & 1 \\ 0 & 0 \end{bmatrix}$
14. $\begin{bmatrix} 0 & 1 \\ 1 & 0 \end{bmatrix}$
15. $\begin{bmatrix} 0 & 1 \\ 0 & 1 \end{bmatrix}$
16. $\begin{bmatrix} 0 & 1 \\ 1 & 1 \end{bmatrix}$
17. $\begin{bmatrix} 0 & 1 \\ -1 & 0 \end{bmatrix}$
18. $\begin{bmatrix} 0 & 1 \\ 0 & -1 \end{bmatrix}$
19. $\begin{bmatrix} 0 & 1 & 0 \\ 0 & 0 & 1 \\ -1 & 0 & 1 \end{bmatrix}$
20. $\begin{bmatrix} 0 & 1 & 0 \\ 0 & 0 & 1 \\ 0 & -1 & 1 \end{bmatrix}$
21. $\begin{bmatrix} 0 & 1 & 0 \\ 0 & 0 & 1 \\ -1 & 1 & 0 \end{bmatrix}$
22. $\begin{bmatrix} 0 & 1 & 0 \\ 0 & 0 & 1 \\ 1 & -1 & 0 \end{bmatrix}$

Part 2

1. Verify that $y = c_1 t + c_2$ is a solution of Equation (5.2.8). What is the corresponding solution of the companion system? Verify.
2. Suppose y solves $y^{(2)} + a_1 y^{(1)} + a_0 y = 0$. If y is tangent to the t-axis, prove that $y \equiv 0$.
3. Suppose y_1 and y_2 are solutions of $y^{(2)} + a_1 y^{(1)} + a_0 y = f(t)$. Prove that $y_1 \equiv y_2$ if the graphs of these functions are mutually tangent at some point $(t_0, y(t_0))$. Hint: What initial-value problem does $y_1 - y_2$ solve?
4. Write out the steps of the proof of Theorem 5.2.3 for the general case.
5. Find a fundamental matrix for the companion system of Equation (5.2.8) and hence write down a solution of this fourth-order equation with four arbitrary constants.
6. (The Euler Equation) The equation

$$t^2 y^{(2)} + a_1 t y^{(1)} + a_0 y = 0$$

is the second-order case of the family of equations known as Euler's Equations. Its solutions are easy to obtain and indicate the kind of phenomena one can expect near $t = 0$, a point at which the coefficient of $y^{(2)}$ vanishes.

In the Euler Equation above, let $x_1 = y$, $x_2 = ty'$. Find x_1', x_2' in terms of x_1 and x_2 and t. Show that

$$\mathbf{x}' = \begin{bmatrix} x_1' \\ x_2' \end{bmatrix} = \frac{1}{t}\begin{bmatrix} 0 & 1 \\ -a_0 & 1-a_1 \end{bmatrix}\mathbf{x}$$

7. Replace the independent variable t by s according to the substitution $t = e^s$ in the system of Problem 6 above. Solve the resulting system and thus find a general solution of $t^2 y^{(2)} - 2ty^{(1)} + 2y = 0$.
8. Prove Theorem 5.2.4 for the general case.

9. (Principle of superposition) If y_1 and y_2 are both solutions of the same nth-order homogeneous equation, show by direct substitution that $c_1y_1+c_2y_2$ is also a solution. Offer another proof by reference to the companion system.
10. (Principle of superposition) If y_1 satisfies $y^{(n)}+a_{n-1}y^{(n-1)}+\cdots+a_0y=f_1$ and y_2 satisfies $y^{(n)}+\cdots+a_0y=f_2$ show that ay_1+by_2 satisfies $y^{(n)}+\cdots+a_0y=af_1+bf_2$ by direct substitution and by reference to the companion systems.
11. (a) Prove that the general solution of any nth-order homogeneous equation contains n arbitrary constants.
(b) Prove that every initial-value problem for an nth-order equation can be solved by proper selection of the n constants in the general solution.

5.3. THE AUXILIARY EQUATION: DISTINCT ROOTS

In principle there is no need to explore the nth-order equation any further. The companion system is, after all, nothing but a special case of a linear system with constant coefficients and its complete solution may therefore be obtained by the techniques already developed. As we have seen, we then take the first component of the general solution of the companion system as the general solution of the nth-order equation. However, the nth-order equation occurs so often that it is inefficient to involve the entire theory of linear systems whenever we wish its solution. Because of the simple structure of the companion matrix it is possible to avoid the companion system entirely in answering some questions. It is the purpose of this and the next section to indicate how this is done. We might add here that many authors solve nth-order equations without any reference at all to the theory of linear systems.

5.3.1. DEFINITION *The algebraic equation*

$$C(\lambda)=\lambda^n+a_{n-1}\lambda^{n-1}+\cdots+a_1\lambda+a_0=0 \tag{5.3.1}$$

is called the auxiliary equation of

$$y^{(n)}+a_{n-1}y^{(n-1)}+\cdots+a_1y^{(1)}+a_0y=0 \tag{5.3.2}$$

It is interesting to observe that $(-1)^nC(\lambda)$ is the characteristic polynomial of the companion system to Equation (5.3.2). We leave the demonstration to the reader as an exercise. It is certainly easy to verify this in the cases $n=2$ and 3. Since we do not need this result we shall not stop for its proof. Instead we establish next a critical identity. Although stated and proved for the case $n=3$, the general case is straightforward.

5.3.2. THEOREM Let $C(\lambda)=\lambda^3+a_2\lambda^2+a_1\lambda+a_0$. *Then for any number* λ

$$\begin{bmatrix}0 & 1 & 0\\ 0 & 0 & 1\\ -a_0 & -a_1 & -a_2\end{bmatrix}\begin{bmatrix}1\\ \lambda\\ \lambda^2\end{bmatrix}=\begin{bmatrix}\lambda\\ \lambda^2\\ \lambda^3\end{bmatrix}-C(\lambda)\begin{bmatrix}0\\ 0\\ 1\end{bmatrix} \tag{5.3.3}$$

PROOF The multiplication of the left-hand side of Equation (5.3.3) yields

$$\begin{bmatrix} \lambda \\ \lambda^2 \\ -a_0 \quad -a_1\lambda \quad -a_2\lambda^2 \end{bmatrix} = \begin{bmatrix} \lambda \\ \lambda^2 \\ \lambda^3 \quad -\lambda^3 \quad -a_2\lambda^2 \quad -a_1\lambda \quad -a_0 \end{bmatrix} = \begin{bmatrix} \lambda \\ \lambda^2 \\ \lambda^3 \end{bmatrix} - C(\lambda)\begin{bmatrix} 0 \\ 0 \\ 1 \end{bmatrix}$$

which completes the proof.

From the identity expressed in this theorem we can see that any root of $C(\lambda)$ will be an eigenvalue of A_c with $[1, \lambda, \lambda^2]^T$ as a corresponding eigenvector. We state this as a theorem.

5.3.3. THEOREM *Let λ_0 be a zero of* $C(\lambda) = \lambda^3 + a_2\lambda^2 + a_1\lambda + a_0$. *(That is,* $C(\lambda_0) = 0$.*) Then λ_0 is an eigenvalue of A_c and $[1, \lambda_0, \lambda_0^2]^T$ is a corresponding eigenvector.*

PROOF Put $\lambda = \lambda_0$ in Equation (5.3.3) and find

$$\begin{bmatrix} 0 & 1 & 0 \\ 0 & 0 & 1 \\ -a_0 & -a_1 & -a_2 \end{bmatrix}\begin{bmatrix} 1 \\ \lambda_0 \\ \lambda_0^2 \end{bmatrix} = \begin{bmatrix} \lambda_0 \\ \lambda_0^2 \\ \lambda_0^3 \end{bmatrix} = \lambda_0 \begin{bmatrix} 1 \\ \lambda_0 \\ \lambda_0^2 \end{bmatrix}$$

For $n = 2$ the appropriate identity is

$$\begin{bmatrix} 0 & 1 \\ -a_0 & -a_1 \end{bmatrix}\begin{bmatrix} 1 \\ \lambda \end{bmatrix} = \begin{bmatrix} \lambda \\ \lambda^2 \end{bmatrix} - C(\lambda)\begin{bmatrix} 0 \\ 1 \end{bmatrix}$$

where $C(\lambda) = \lambda^2 + a_1\lambda + a_0$. Hence if λ_0 is a zero of $C(\lambda)$ in this case then λ_0 is an eigenvalue of

$$\begin{bmatrix} 0 & 1 \\ -a_0 & -a_1 \end{bmatrix} = A_c$$

and $[1, \lambda_0]^T$ is a corresponding eigenvector.

In view of the above, the companion system always has solutions of the form

$$\begin{bmatrix} 1 \\ \lambda_0 \end{bmatrix} e^{\lambda_0 t} \quad \text{or} \quad \begin{bmatrix} 1 \\ \lambda_0 \\ \lambda_0^2 \end{bmatrix} e^{\lambda_0 t}$$

depending upon its order, and therefore $x(t) = e^{\lambda_0 t}$ is always a solution of the second- or third-order equation whenever λ_0 is a root of the auxiliary equation. This result is true for the general case as well.

5.3.4. THEOREM *The equation*

$$y^{(n)} + a_{n-1}y^{(n-1)} + \cdots + a_1 y^{(1)} + a_0 y = 0$$

has the solution $y(t) = ke^{\lambda_0 t}$ if and only if λ_0 is a root of its auxiliary equation

$$\lambda^n + a_{n-1}\lambda^{n-1} + \cdots + a_1\lambda + a_0 = 0$$

5.3.1. Example

Solve $y^{(2)} - 3y^{(1)} + 2y = 0$.

Solution

We read off the auxiliary equation as $\lambda^2 - 3\lambda + 2 = 0$ and from its zeros, $\lambda_1 = 1$ and $\lambda_2 = 2$, we have the two solutions $y_1(t) = c_1 e^t$ and $y_2(t) = c_2 e^{2t}$. It is not difficult to see that $y(t) = c_1 e^t + c_2 e^{2t}$ is a general solution. (Why?)

5.3.2. Example

Solve $y^{(3)} - y^{(1)} = 0$.

Solution

The roots of the auxiliary equation $C(\lambda) = \lambda^3 - \lambda = 0$ are $\lambda_1 = 0$, $\lambda_2 = 1$, $\lambda_3 = -1$ and hence $y_1(t) = c_1$, $y_2(t) = c_2 e^t$, and $y_3(t) = c_3 e^{-t}$ are each solutions. As before, $y(t) = c_1 + c_2 e^t + c_3 e^{-t}$ is also a solution by superposition; in fact it is a general solution.

In the two examples worked above no explicit reference has been made to the companion system. In dealing with initial-value problems or inhomogeneous equations the fundamental matrix of the companion system is very convenient. Fortunately, the construction of a fundamental matrix from solutions of the nth-order equation is very simple. Refer for a moment back to Theorem 5.2.3 and Example 5.2.2. We see there that solutions of the companion system are simply vectors whose first entry is a solution of the nth-order equation and whose succeeding entries are successive derivatives of the first entry. Thus the first row of a fundamental matrix consists of n solutions of the nth-order equation; the next row is the derivative, term by term, of the first row, the third row is the derivative of the second, and so on. It is, of course, necessary to show that the determinant of the resulting matrix is not zero in order to insure that we really have a fundamental matrix.

5.3.3. Example

Find a fundamental matrix for the companion system of $y^{(3)} - y^{(1)} = 0$.

Solution

From what has just been said and from the work done in Example 5.3.2 we have

$$\Phi(t) = \begin{bmatrix} 1 & e^t & e^{-t} \\ 0 & e^t & -e^{-t} \\ 0 & e^t & e^{-t} \end{bmatrix}$$

and since $\det \Phi = 2$, the above matrix is a fundamental matrix. Note the relationship between successive rows!

To see how fundamental matrices are useful in initial-value problems consider the next example.

5.3.4. Example

Find a solution of $y^{(3)} - y^{(1)} = 0$, $y(0) = 0$, $y^{(1)}(0) = 1$, and $y^{(2)}(0) = 0$.

Solution

The fundamental matrix of the companion system for the given differential equation was derived in the previous example. The general solution of this companion system is, therefore,

$$\mathbf{x}(t) = \begin{bmatrix} 1 & e^t & e^{-t} \\ 0 & e^t & -e^{-t} \\ 0 & e^t & e^{-t} \end{bmatrix} \mathbf{c} \tag{5.3.4}$$

We now select $\mathbf{c}$ so that

$$\mathbf{x}(0) = \begin{bmatrix} 1 & 1 & 1 \\ 0 & 1 & -1 \\ 0 & 1 & 1 \end{bmatrix} \mathbf{c} = \begin{bmatrix} 0 \\ 1 \\ 0 \end{bmatrix} \tag{5.3.5}$$

Thus $\mathbf{c} = [0, \frac{1}{2}, -\frac{1}{2}]^{\mathrm{T}}$. When this value of $\mathbf{c}$ is used in Equation (5.3.4) we obtain a solution to the initial-value problem companion to the given equation. This solution's first entry is the required solution, to wit,

$$y(t) = \frac{e^t}{2} - \frac{e^{-t}}{2}$$

There is another way of obtaining this answer. If we start from $y(t) = c_1 + c_2 e^t + c_3 e^{-t}$, the general solution of the homogeneous equation, we determine c_1, c_2, and c_3 so that y meets the initial conditions. That is,

$$\begin{aligned} y(0) &= c_1 + c_2 + c_3 = 0 \\ y^{(1)}(0) &= \quad\;\; c_2 - c_3 = 1 \\ y^{(2)}(0) &= \quad\;\; c_2 + c_3 = 0 \end{aligned}$$

Note that this is precisely the set of equations represented in matrix-vector form by Equation (5.3.5) and that these equations are obtained in almost identical fashion.

The importance of the fundamental matrix is even more apparent in the attempt to obtain particular solutions of the inhomogeneous equation.

5.3.5. Example

Solve $y^{(2)} - 3y^{(1)} + 2y = e^t$.

Solution

A fundamental matrix of this equation is

$$\Phi(t) = \begin{bmatrix} e^t & e^{2t} \\ e^t & 2e^{2t} \end{bmatrix}$$

By variation of parameters, we find the particular solution

$$\mathbf{x}_p(t) = \begin{bmatrix} e^t & e^{2t} \\ e^t & 2e^{2t} \end{bmatrix} \begin{bmatrix} -t \\ -e^{-t} \end{bmatrix} = \begin{bmatrix} -(t+1)e^t \\ -(t+2)e^t \end{bmatrix}$$

The general solution of the companion system is therefore

$$\mathbf{x}(t) = \Phi(t)\mathbf{k} + \mathbf{x}_p(t) = \begin{bmatrix} k_1e^t + k_2e^{2t} - (t+1)e^t \\ k_1e^t + 2k_2e^{2t} - (t+2)e^t \end{bmatrix}$$

and hence $y(t) = k_1e^t + k_2e^{2t} - (t+1)e^t$ is the general solution of $y^{(2)} - 3y^{(1)} + 2y = e^t$.

To construct real solutions of the nth-order equations via Theorem 5.3.4 when the roots of the auxiliary equation are complex we need to refer once again to that remarkable formula of Euler:

$$e^{a+ib} = e^a \cdot e^{ib} = e^a(\cos b + i \sin b)$$

With this formula and an appeal to the companion system we assert the following extension of Theorem 5.3.4.

5.3.5. THEOREM *The equation*

$$y^{(n)} + a_{n-1}y^{(n-1)} + \cdots + a_1y^{(1)} + a_0y = 0$$

has the solutions $y_1(t) = k_1e^{at} \cos bt$ *and* $y_2(t) = k_2e^{at} \sin bt$ *if and only if* $\lambda = a + ib$ *is a root of the auxiliary equation.*

5.3.6. Example

Find solutions of $y^{(2)} + y^{(1)} + y = 0$.

Solution

The roots of the auxiliary equation are $\lambda_1 = -\frac{1}{2} + (i\sqrt{3}/2)$, $\lambda_2 = -\frac{1}{2} - i(\sqrt{3}/2)$. From Theorem 5.3.5 (or for that matter, from a study of the corresponding companion system) we find the general solution:

$$y(t) = k_1e^{-t/2}\left(\cos \frac{\sqrt{3}t}{2}\right) + k_2e^{-t/2}\left(\sin \frac{\sqrt{3}t}{2}\right)$$

Exercise 5.3

Part 1

Find the general solution of the equations given in Problems 1–15.

1. $y^{(2)} - y^{(1)} = 1$
2. $y^{(4)} - y = 0$
3. $y^{(2)} - 7y^{(1)} + 12y = 0$
4. $y^{(4)} - y = \sin t$
5. $y^{(2)} + y^{(1)} = e^{-t} - 1$
6. $y^{(4)} - y = e^t + t$
7. $y^{(2)} + 4y = 0$
8. $y^{(3)} - 3y^{(2)} + 4y^{(1)} = 0$
9. $y^{(4)} - 2y^{(3)} - y^{(2)} + 2y^{(1)} = 0$
10. $y^{(2)} + y = 1$
11. $y^{(2)} - 2y^{(1)} + 2y = 1 - t$
12. $y^{(3)} + y^{(1)} = 0$
13. $y^{(2)} + 3y^{(1)} + 2y = \cos 2t$
14. $y^{(2)} - y = e^{t^2}$
15. $y^{(2)} + (a+b)y^{(1)} + aby = 0 \quad a \neq b$

Solve the following 12 initial-value problems.

16. $y^{(2)} - y^{(1)} = 0 \quad y(0) = 0 \quad y^{(1)}(0) = 1$
17. $y^{(2)} + 9y = \sin 3t \quad y(0) = 1 \quad y'(0) = 0$
18. $y^{(3)} - y^{(1)} = 0 \quad y(0) = y'(0) = y^{(2)}(0) = -1$
19. $y^{(2)} - y = e^t \quad y(0) = -1 \quad y'(0) = 0$
20. $y^{(2)} + y = 1 \quad y(0) = 0 \quad y'(0) = 0$
21. $y^{(4)} - y = 0 \quad y(0) = -1 \quad y^{(1)}(0) = 1 \quad y^{(2)}(0) = y^{(3)}(0) = 0$
22. $y^{(3)} + y^{1)} = 0 \quad y(0) = y^{(1)}(0) = y^{(2)}(0) = 1$
23. $y^{(2)} - 2y^{(1)} + 2y = 1 - t \quad y(0) = y^{(1)}(0) = 0$
24. $y^{(2)} + y^{(1)} = e^{-t} - 1 \quad y(0) = 1 \quad y^{(1)}(0) = 0$
25. $y^{(2)} + y^{(1)} = 1 \quad y(0) = y^{(1)}(0) = 0$
26. $y^{(2)} + 4y = \sin 2t \quad y(\pi) = 0 \quad y^{(1)}(\pi) = 0$
27. $y^{(2)} = e^t \sin t \quad y(0) = 0 \quad y^{(1)}(0) = 0$

28. The functions 1, e^{-t}, e^t are solutions of what third-order, homogeneous equation?
29. The functions $e^{-t} \sin 3t$, $e^{-t} \cos 3t$ are solutions of what second-order, homogeneous equation?
30. Find a fourth-order, homogeneous equation for which the functions $\cos t$ and $\sin 2t$ are solutions. Is it possible for these functions to be solutions of the same second-order homogeneous equation with real, constant coefficients? If so, find such an equation. If not, why not?
31. Find the general solution to the companion system for the equation of Example 5.3.2. Verify that $y(t) = c_1 + c_2 e^t + c_3 e^{-t}$ is therefore the general solution of the given third-order equation.
32. Complete the steps which prove that the function $\mathbf{x}_p(t)$ of Example 5.3.5 is the result of applying the method of variation of parameters to the companion system of the given equation.

Part 2

1. Find the general solution of the equation

$$y^{(2)} + (\alpha + \beta) y^{(1)} + \alpha\beta y = e^{-\alpha t} \qquad \alpha \neq \beta$$

2. Find the general solution of the equation

$$y^{(2)} + (\alpha + \beta) y^{(1)} + \alpha\beta y = te^{-\alpha t} \qquad \alpha \neq \beta$$

3. Prove Theorem 5.3.4 by substituting $y(t) = ke^{\lambda_0 t}$ into the given differential equation.
4. Suppose the auxiliary Equation (5.3.1) has n distinct roots, $\lambda_1, \lambda_2, \ldots, \lambda_n$. Find a fundamental matrix of the companion system of Equation (5.3.2) and compute its determinant.
5. (a) Find a fundamental matrix for the system companion to $y^{(2)} - 3y^{(1)} + 2y = 0$ and normalize at $t = 0$.
   (b) Use Equation (4.4.10) of Corollary 4.4.2 to find a particular solution of the companion system of $y^{(2)} - 3y^{(1)} + 2y = f(t)$.
   (c) Hence show that $y_p(t) = \int_0^t f(s)(-e^{t-s} + e^{2(t-s)})\, ds$ is a particular solution of $y^{(2)} - 3y^{(1)} + 2y = f(t)$. Verify this formula for $f(t) = e^t$.
6. Same as Problem 5 above, except $y^{(2)} + y = f(t)$.

5.4. THE AUXILIARY EQUATION: REPEATED ROOTS

Theorems 5.3.4 and 5.3.5 enable us to construct as many "essentially different" solutions as there are distinct roots of the auxiliary equation. For each complex root we obtain two solutions; but the complex roots come in conjugate pairs and the solutions obtained for the complex conjugate of $a+ib$ are just multiples of the solutions obtained from $a-ib$. If the auxiliary equation has n distinct roots, a fundamental matrix of the companion system may be constructed as illustrated in Example 5.3.3. The point here is that we can prove $\det \Phi \neq 0$ under these circumstances. We save this discussion for the general theory presented in Section 5.5. In that section we also make precise the term "essentially different" solutions. The reader may have already anticipated our approach, but for the time being we choose not to enter into a discussion of these ideas. Instead, let us study the case where the auxiliary equation has repeated roots.

We have remarked earlier that the characteristic equation and the auxiliary equation are constant (-1 or $+1$) multiples of each other. They must, therefore, have the same roots with the same multiplicities. A root repeated k times in one is repeated k times in the other. The companion system, in this case, may have (and, as we shall see, always does have) root vectors of order equal to the number of times a root is repeated. The first components of the solutions constructed from root vectors of a companion system will be seen to be the functions given by Equation (5.4.1). The proof of this result is not difficult but again we defer the argument to Section 5.5. Let us, however, state the above result as follows.

5.4.1. THEOREM *If λ_0 is a k-fold root of the auxiliary equation, then*

$$y_1(t) = e^{\lambda_0 t} \qquad y_2(t) = te^{\lambda_0 t}, \ldots, y_k(t) = t^{k-1}e^{\lambda_0 t} \tag{5.4.1}$$

are each solutions of the corresponding differential equation. If the root $\lambda_0 = a+ib$ is complex, then the solutions are

$$\begin{array}{ll} y_1(t) = e^{at}\cos bt & y_2(t) = e^{at}\sin bt \\ \vdots & \vdots \\ y_{2k-1}(t) = t^{k-1}e^{at}\cos bt & y_{2k}(t) = t^{k-1}e^{at}\sin bt \end{array} \tag{5.4.2}$$

There are no complications except some tedious arithmetic in applying this theorem. We study a number of examples to illustrate this point.

5.4.1. Example

Solve $y^{(2)} - 2ay^{(1)} + a^2y = 0$.

Solution

The roots of the auxiliary equations are $\lambda_1 = \lambda_2 = a$. From the theorem, $y_1(t) = e^{at}$ and $y_2(t) = te^{at}$ are two solutions. We know by superposition that

$$\begin{aligned} y(t) &= c_1 e^{at} + c_2 t e^{at} \\ &= (c_1 + c_2 t) e^{at} \end{aligned}$$

is also a solution and it is not difficult to see that it is the general solution. For, accepting Theorem 5.4.1, then

$$\Phi(t) = \begin{bmatrix} e^{at} & te^{at} \\ ae^{at} & (1+at)e^{at} \end{bmatrix}$$

is a fundamental matrix since $\det \Phi = e^{2at} > 0$. Hence the general solution of the companion system is $\mathbf{x}(t) = \Phi(t)\mathbf{c}$ whose first component is $c_1 e^{at} + c_2 t e^{at}$.

5.4.2. Example

Solve $y^{(2)} - 2ay^{(1)} + a^2 y = e^{bt}$.

Solution

We use variation of parameters on the companion system. (If $b \neq a$, the function ke^{bt} is a particular solution for the correct choice of k. By substitution into the differential equation we find $k = (b-a)^{-2}$. If $b = a$, however, the appropriate trial function is not nearly so obvious.) From the fundamental matrix computed in the previous example and the formula $\Phi\mathbf{u}' = \mathbf{f}$, which defines the method of variation of parameters, we have

$$\begin{bmatrix} e^{at} & te^{at} \\ ae^{at} & (1+at)e^{at} \end{bmatrix} \begin{bmatrix} u_1' \\ u_2' \end{bmatrix} = \begin{bmatrix} 0 \\ e^{bt} \end{bmatrix} \tag{5.4.3}$$

Row reductions on the system (5.4.3) leads to the simpler equivalent system

$$\begin{bmatrix} 1 & 0 \\ 0 & 1 \end{bmatrix} \begin{bmatrix} u_1' \\ u_2' \end{bmatrix} = \begin{bmatrix} -te^{(b-a)t} \\ e^{(b-a)t} \end{bmatrix}$$

from which we deduce $u_1' = -te^{(b-a)t}$ and $u_2' = e^{(b-a)t}$. Now two cases arise depending on whether $a = b$ or not. Suppose first that $a = b$. Then $u_1 = -t^2/2$ and $u_2 = t$. Thus, a particular solution of the companion system is

$$\begin{bmatrix} e^{bt} & te^{bt} \\ be^{bt} & (1+bt)e^{bt} \end{bmatrix} \begin{bmatrix} -t^2/2 \\ t \end{bmatrix}$$

whose first component is

$$y_p(t) = \frac{-t^2}{2} e^{bt} + t^2 e^{bt} = \frac{t^2}{2} e^{bt} \tag{5.4.4}$$

a solution of $y^{(2)} - 2by^{(1)} + b^2 y = e^{bt}$.

If $a \neq b$ then $u_1 = -(b-a)^{-1} te^{(b-a)t} + (b-a)^{-2} e^{(b-a)t}$, $u_2 = (b-a)^{-1} e^{(b-a)t}$, and the first component of the particular solution of the companion system is the function

$$y_p(t) = \left(\frac{-t}{b-a} + \frac{1}{(b-a)^2}\right)e^{bt} + \frac{t}{b-a}e^{bt} = (b-a)^{-2}e^{bt} \tag{5.4.5}$$

a solution of $y^{(2)} - 2ay^{(1)} + a^2y = e^{bt}$, $a \neq b$.

Remarks

There is nothing in the analysis we have done to prohibit either a or b from being zero. Also, we may even allow b to be complex. Suppose, for instance, $a = \alpha$ and that $b = i\beta$. Then from Equation (5.4.5)

$$y_p(t) = (\alpha - i\beta)^{-2}e^{i\beta t}$$

is a solution of $y^{(2)} - 2\alpha y^{(1)} + \alpha^2 y = e^{i\beta t}$. Now the real part of $y_p(t)$ is a solution of the equation

$$\text{Re}\,[y^{(2)} - 2\alpha y^{(1)} + \alpha^2 y] = y^{(2)} - 2\alpha y^{(1)} + \alpha^2 y = \text{Re}\; e^{i\beta t} = \cos \beta t$$

But

$$\text{Re}\,\{(\alpha - i\beta)^{-2}e^{i\beta}\} = \text{Re}\left\{\frac{(\alpha + i\beta)^2(\cos\beta t + i \sin\beta t)}{(\alpha^2+\beta^2)^2}\right\}$$

$$= \frac{(\alpha^2-\beta^2)\cos\beta t - 2\alpha\beta \sin\beta t}{(\alpha^2+\beta^2)^2}$$

The reader is asked to verify this result in Exercise 5.4, Part 1, Problem 24.

Exercise 5.4

Part 1

Find the general solution of the equation given in the following 10 problems.

1. $y^{(3)} - y^{(2)} = 0$
2. $y^{(3)} - y^{(2)} = 1 + t + \sin 2t$
3. $y^{(4)} + 2y^{(2)} + y = 0$
4. $y^{(3)} - 4y^{(2)} + 4y^{(1)} = 0$
5. $y^{(3)} - 4y^{(2)} + 4y^{(1)} = e^{2t}$
6. $y^{(6)} - y^{(4)} = 0$
7. $y^{(4)} + 2y^{(2)} + y = \cosh t$
8. $y^{(4)} + 2y^{(2)} + y = t \cos t$
9. $y^{(3)} = 0$
10. $y^{(2)} - 6y^{(1)} + 9y = t$

Solve the following 12 initial-value problems.

11. $y^{(3)} - y^{(2)} = 0$
    $y(0) = 1 \qquad y^{(1)}(0) = y^{(2)}(0) = 0$
12. $y^{(3)} - y^{(2)} = 0$
    $y(0) = y^{(1)}(0) = 0 \qquad y^{(2)}(0) = 1$
13. $y^{(2)} + 2y^{(1)} + 2y = 0$
    $y(0) = 1 \qquad y^{(1)}(0) = 0$
14. $y^{(2)} + 2y^{(1)} + y = 0$
    $y(0) = 0 \qquad y^{(1)}(0) = 1$
15. $y^{(2)} + 2y^{(1)} + y = 0$
    $y(0) = \alpha \qquad y^{(1)}(0) = \beta$
16. $y^{(2)} - 6y^{(1)} + 9y = t$
    $y(0) = y^{(1)}(0) = 0$
17. $y^{(3)} - 6y^{(2)} - 9y^{(1)} = 0$
    $y(0) = 1 \qquad y^{(1)}(0) = y^{(2)}(0) = 0$
18. $y^{(2)} + 2ay^{(1)} + a^2y = e^{at}$
    $y(0) = 0 \qquad y^{(1)}(0) = 0$

19. $y^{(2)}+2ay^{(1)}+a^2y=e^{bt}$
$b \neq a \qquad y(0)=y^{(1)}(0)=0$

20. $y^{(6)}-y^{(4)}=0$
$y(0)=y^{(3)}(0)=y^{(5)}(0)=-1$
$y^{(1)}(0)=y^{(2)}(0)=1$

21. $y^{(3)}+3y^{(2)}+3y^{(1)}+y=e^{-t}$
$y(0)=y^{(1)}(0)=y^{(2)}(0)=0$

22. $y^{(3)}-3y^{(2)}+4y=0$
$y(0)=2 \qquad y^{(1)}(0)=-1$
$y^{(2)}(0)=0$

23. Find the lowest-order homogeneous equation for which the following sets of functions are solutions.

(a) $1, t, t^2$ (b) e^{-t}, te^{-t}, t^2e^t

(c) $\sin\frac{\sqrt{2}}{2}t, t\sin 2t$ (d) te^{2t}

24. Verify that

$$y(t)=\frac{1}{(\alpha^2+\beta^2)^2}\{(\alpha^2-\beta^2)\cos\beta t-2\alpha\beta\sin\beta t\}$$

is a solution of $y^{(2)}-2\alpha y^{(1)}+\alpha^2y=\cos\beta t$.

25. Show by substitution into $y^{(2)}-2\alpha y^{(1)}+\alpha^2y=0$ that $y(t)=(C_1+C_2t)e^{\alpha t}$ is a solution.

Part 2

1. Write the companion matrix for the equation $y^{(2)}-2\alpha y^{(1)}+\alpha^2y=0$ and verify that $[0, 1]^T, [1, \alpha]^T$ is a chain. (See Definition 3.2.3.)
2. Same as Problem 1 above except $y^{(3)}-3\alpha y^{(2)}+3\alpha^2y^{(1)}-y=0$ and $[0, 0, 1]^T$, $[0, 1, \alpha]^T, [1, \alpha, \alpha^2]^T$ is a chain. (See Definition 3.2.3.)
3. Solve $y^{(2)}-2\alpha y^{(1)}+\alpha^2y=f(t)$ by applying the method of variation of parameters to the companion system.
Hint: Follow the steps used in Example 5.4.2 to arrive at

$$\begin{bmatrix}1 & 0\\0 & 1\end{bmatrix}\begin{bmatrix}u_1'\\u_i\end{bmatrix}=\begin{bmatrix}-te^{-\alpha t}f(t)\\e^{-\alpha t}f(t)\end{bmatrix}$$

4. Solve Problem 3 above by use of Corollary 4.4.2, Equation (4.4.10) and compare answers.

5.5.* THE GENERAL THEORY OF THE *n*th-ORDER LINEAR EQUATION WITH CONSTANT COEFFICIENTS

This section is entirely theoretical. We propose to prove all the results we have previously used without proof or whose proof we have offered in special cases. We shall also state and prove a few new theorems which have interest in their own right and are often included in the general theory of the linear equation.

Throughout this section let us agree to the following notational conventions:

We write

$$C(\lambda)=\lambda^n+a_{n-1}\lambda^{n-1}+\cdots+a_1\lambda+a_0 \tag{5.5.1}$$

for the auxiliary equation of

$$y^{(n)}+a_{n-1}y^{(n-1)}+\cdots+a_1y^{(1)}+a_0y=0 \tag{5.5.2}$$

and A_c as the systems matrix in the companion system

$$\mathbf{x}'=\begin{bmatrix} 0 & 1 & 0 \cdots & 0 \\ 0 & 0 & 1 \cdots & 0 \\ \cdot & \cdot & \cdot & \cdot \\ \cdot & \cdot & \cdot & \cdot \\ \cdot & \cdot & \cdot & \cdot \\ 0 & 0 & 0 \cdots & 1 \\ -a_0 & -a_1 & -a_2 \cdots & -a_{n-1} \end{bmatrix}\mathbf{x} \tag{5.5.3}$$

5.5.1. THEOREM *For each* λ,

$$A_c\begin{bmatrix} 1 \\ \lambda \\ \cdot \\ \cdot \\ \cdot \\ \lambda^{n-1} \end{bmatrix}=\lambda\begin{bmatrix} 1 \\ \lambda \\ \cdot \\ \cdot \\ \cdot \\ \lambda^{n-1} \end{bmatrix}-C(\lambda)\begin{bmatrix} 0 \\ \cdot \\ \cdot \\ \cdot \\ 0 \\ 1 \end{bmatrix} \tag{5.5.4}$$

PROOF As in the proof of the special case given by Theorem 5.3.1, simply compute

$$A_c\begin{bmatrix} 1 \\ \lambda \\ \cdot \\ \cdot \\ \cdot \\ \lambda^{n-1} \end{bmatrix}=\begin{bmatrix} \lambda \\ \cdot \\ \cdot \\ \cdot \\ \lambda^{n} \\ \lambda^n-\lambda^n-a_{n-1}\lambda^{n-1}-\cdots-a_0 \end{bmatrix}$$

$$=\lambda\begin{bmatrix} 1 \\ \lambda \\ \cdot \\ \cdot \\ \cdot \\ \lambda^{n-1} \end{bmatrix}-C(\lambda)\begin{bmatrix} 0 \\ \cdot \\ \cdot \\ \cdot \\ 0 \\ 1 \end{bmatrix}$$

5.5.2. COROLLARY *If λ_0 is a zero of $C(\lambda)$ then λ_0 is an eigenvalue of A_c and $[1, \lambda_0, \ldots, \lambda_0^{n-1}]^T$ is a corresponding eigenvector. Hence if λ_0 is real, $y(t) = ke^{\lambda_0 t}$ is a solution of Equation (5.5.2). If λ_0 is complex, say $\lambda_0 = a+ib$, then $y_1(t) = k_1 \cos bt$, $y_2(t) = k_2 \sin bt$ are solutions.*

Next we consider the matrix B(λ) defined by

$$B(\lambda)=\left[\begin{array}{c|ccc} 1 & 0 & \cdots & 0 \\ \hline \lambda & & & \\ \lambda^2 & & & \\ \cdot & & I & \\ \cdot & & & \\ \lambda^{n-1} & & & \end{array}\right]$$

where I is the $(n-1)(n-1)$ identity.

Because $\det B(\lambda)=1$, we have

$$\det(A_c-\lambda I)=\det\{(A_c-\lambda I)B(\lambda)\}$$

But

$$(A_c-\lambda I)B(\lambda)=\begin{bmatrix} 0 & 1 & 0 \cdots & 0 & 0 \\ 0 & -\lambda & 1 \cdots & 0 & 0 \\ \cdot & \cdot & \cdot & \cdot & \cdot \\ \cdot & \cdot & \cdot & \cdot & \cdot \\ \cdot & \cdot & \cdot & \cdot & \cdot \\ 0 & 0 & 0 \cdots & -\lambda & 1 \\ -c(\lambda) & -a_1 & -a_2 \cdots & -a_{n-2} & -a_{n-1}-\lambda \end{bmatrix}$$

Hence,

$$\det(A_c-\lambda I)=(-1)^n C(\lambda)\det\begin{bmatrix} 1 & 0 & 0\cdots & 0 & 0 \\ -\lambda & 1 & 0\cdots & 0 & 0 \\ 0 & -\lambda & 1\cdots & 0 & 0 \\ \cdot & \cdot & \cdot & \cdot & \cdot \\ \cdot & \cdot & \cdot & \cdot & \cdot \\ \cdot & \cdot & \cdot & \cdot & \cdot \\ 0 & 0 & 0\cdots & 1 & 0 \\ 0 & 0 & 0\cdots & -\lambda & 1 \end{bmatrix}$$

$$=(-1)^n C(\lambda)$$

This provides a connection between the auxiliary polynomial of the nth-order equation and the characteristic polynomial of its companion matrix.

5.5.3. THEOREM *The characteristic polynomial of* A_c *is given by* $(-1)^n C(\lambda)$.

Let us reconsider the identity given by Equation (5.5.4). Set $u=[1,\lambda,\ldots,\lambda^{n-1}]^T$ and write this identity as

$$(A_c-\lambda I)\mathbf{u}=-C(\lambda)\begin{bmatrix}0\\ \cdot\\ \cdot\\ 0\\ 1\end{bmatrix} \qquad (5.5.5)$$

The derivative of an identity is itself an identity. Thus differentiate Equation (5.5.5) with respect to λ and obtain the identity

$$\frac{d}{d\lambda}(\mathrm{A}_c-\lambda\mathrm{I})\mathbf{u}=(\mathrm{A}_c-\lambda\mathrm{I})\mathbf{u}^{(1)}-\mathbf{u}=-\mathrm{C}^{(1)}(\lambda)\begin{bmatrix}0\\ \cdot\\ \cdot\\ \cdot\\ 0\\ 1\end{bmatrix} \tag{5.5.6}$$

Another differentiation leads to the identity

$$\frac{d^2}{d\lambda^2}(\mathrm{A}_c-\lambda\mathrm{I})\mathbf{u}=(\mathrm{A}_c-\lambda\mathrm{I})\mathbf{u}^{(2)}-2\mathbf{u}^{(1)}=-\mathrm{C}^{(2)}(\lambda)\begin{bmatrix}0\\ \cdot\\ \cdot\\ \cdot\\ 0\\ 1\end{bmatrix} \tag{5.5.7}$$

In general then,

$$\frac{d^k}{d\lambda^k}(\mathrm{A}_c-\lambda\mathrm{I})\mathbf{u}=(\mathrm{A}_c-\lambda\mathrm{I})\mathbf{u}^{(k)}-k\mathbf{u}^{(k-1)}=-\mathrm{C}^{(k)}(\lambda)\begin{bmatrix}0\\ \cdot\\ \cdot\\ \cdot\\ 0\\ 1\end{bmatrix} \tag{5.5.8}$$

a result whose proof can be effected by induction. We summarize these identities in the following theorem.

5.5.4. THEOREM *For each nonnegative integer, k, and every number λ we have*

$$(\mathrm{A}_c-\lambda\mathrm{I})\mathbf{u}^{(k)}=k\mathbf{u}^{(k-1)}-\mathrm{C}^{(k)}(\lambda)\begin{bmatrix}0\\ \cdot\\ \cdot\\ \cdot\\ 0\\ 1\end{bmatrix} \tag{5.5.9}$$

5.5.5. COROLLARY *If* $\lambda=\lambda_0$ *is a k-fold root of* C(λ) *the k functions* $y_1=e^{\lambda_0 t}$, $y_2=te^{\lambda_0 t},\ldots,y_k=t^{k-1}k^{\lambda_0 t}$ *are solutions of the* nth-*order equation* (5.5.2).

PROOF If λ_0 is a k-fold root of $\mathrm{C}(\lambda)$ then

$$\mathrm{C}(\lambda_0)=\mathrm{C}^{(1)}(\lambda_0)=\cdots=\mathrm{C}^{(k-1)}(\lambda_0)=0$$

If we use this fact in Equations (5.5.9) we deduce, after division by $k!$

$$(\mathrm{A}_c-\lambda_0\mathrm{I})\left(\frac{\mathbf{u}^{(1)}(\lambda_0)}{1!}\right)=\mathbf{u}(\lambda_0)$$

$$(\mathrm{A}_c-\lambda_0\mathrm{I})\left(\frac{\mathbf{u}^{(2)}(\lambda_0)}{2!}\right)=\frac{\mathbf{u}^{(1)}(\lambda_0)}{1!}$$

$$(\mathrm{A}_c-\lambda_0\mathrm{I})\left(\frac{\mathbf{u}^{(3)}(\lambda_0)}{3!}\right)=\frac{\mathbf{u}^{(2)}(\lambda_0)}{2!}$$

$$\vdots$$

$$(\mathrm{A}_c-\lambda_0\mathrm{I})\left(\frac{\mathbf{u}^{(k-1)}(\lambda_0)}{(k-1)!}\right)=\frac{\mathbf{u}^{(k-2)}(\lambda_0)}{(k-2)!}$$

Since $\mathbf{u}(\lambda_0)$ is an eigenvector, as we know from Corollary 5.5.2, these relationships prove that

$$\frac{\mathbf{u}^{(k-1)}(\lambda_0)}{(k-1)!},\frac{\mathbf{u}^{(k-2)}(\lambda_0)}{(k-2)!},\ldots,\frac{\mathbf{u}^{(1)}(\lambda_0)}{1!},\mathbf{u}(\lambda_0) \tag{5.5.10}$$

is a chain of root vectors of length k by Theorem 3.2.2. The corresponding solutions of the companion system are given by Theorem 3.3.1. We wish only the first component of these solutions. Note that the first entry in each vector of the chain (5.5.10) is 0 except for the eigenvector whose first entry is 1. Thus the first components of the solutions given in Equations (3.3.1) are given by $t^{k-1}e^{\lambda_0 t}, t^{k-2}e^{\lambda_0 t},\ldots,te^{\lambda_0 t}, e^{\lambda_0 t}$. (We have suppressed the constants.) This proves the corollary. This corollary has been seen before as Theorem 5.4.1. If A is an arbitrary $n\times n$ matrix and λ_0 is a k-fold eigenvalue of A, $k\leq n$, then we cannot conclude, in general, that there is a chain of root vectors corresponding to λ_0 of any length >1. If $\mathrm{A}=\mathrm{I}$, there are no chains of length 2, for example. However, Corollary 5.5.5, besides enabling us to write down immediately the solutions of Equation (5.5.2), shows that companion matrices have chains of length k. Let m_i be the multiplicity of λ_i, an eigenvalue of A_c, $i=1,2,\ldots,k$. Suppose $n=m_1+m_2+\cdots+m_k$. Let $\mathbf{u}_{i,m_i},\mathbf{u}_{i,m_i-1},\ldots,\mathbf{u}_{i,1}$ be a chain corresponding to the eigenvalue λ_i. We have the following corollary.

5.5.6. COROLLARY *The n vectors*

$$\mathbf{u}_{1,m_1},\mathbf{u}_{1,m_1-1},\ldots,\mathbf{u}_{1,1}\text{—}\textit{corresponding to } \lambda_1$$
$$\mathbf{u}_{2,m_2},\mathbf{u}_{2,m_2-1},\ldots,\mathbf{u}_{2,1}\text{—}\textit{corresponding to } \lambda_2$$
$$\vdots$$
$$\mathbf{u}_{k,m_k},\mathbf{u}_{k,m_k-1},\ldots,\mathbf{u}_{k,1}\text{—}\textit{corresponding to } \lambda_k$$

span $\mathscr{R}^n$.

PROOF Each chain is a sequence of linearly independent vectors† and no vector from a chain corresponding to *one* eigenvalue is a linear combination of vectors from chains corresponding to other eigenvalues. Hence these n vectors form a linearly independent set and therefore span $\mathscr{R}^n$.

5.5.7. COROLLARY *If* λ_0 *is the sole eigenvalue of* A_c *then Equation* (5.5.2) *has the general solution*

$$y(t) = (c_1 + c_2 t + \cdots + c_n t^{n-1}) e^{\lambda_0 t}$$

5.5.8. DEFINITION *The solutions* $y_1, y_2, \ldots, y_n$ *of Equation* (5.5.2) *are called linearly independent if the vectors*

$$\begin{bmatrix} y_1 \\ y_1^{(1)} \\ \cdot \\ \cdot \\ \cdot \\ y_1^{(n-1)} \end{bmatrix} \begin{bmatrix} y_2 \\ y_2^{(1)} \\ \cdot \\ \cdot \\ \cdot \\ y_2^{(n-1)} \end{bmatrix} \quad \cdots \quad \begin{bmatrix} y_n \\ y_n^{(1)} \\ \cdot \\ \cdot \\ \cdot \\ y_n^{(n-1)} \end{bmatrix} \tag{5.5.11}$$

are linearly independent at any t_0.

Since the vectors in (5.5.11) are solutions of the system companion to Equation (5.5.2) they are independent for all t or for no t by Corollary 4.2.3. Thus we need not specify a point at which to evaluate the vectors in (5.5.11) to check their independence. Furthermore, the matrix Φ,

$$\Phi = \begin{bmatrix} y_1 & y_2 & \cdots & y_n \\ y_1^{(1)} & y_2^{(1)} & & y_n^{(1)} \\ \cdot & \cdot & & \cdot \\ \cdot & \cdot & & \cdot \\ \cdot & \cdot & & \cdot \\ y_1^{(n-1)} & y_2^{(n-1)} & & y_n^{(n-1)} \end{bmatrix}$$

is a fundamental matrix of the companion system if and only if the solutions $y_1, y_2, \ldots, y_n$ are linearly independent. But for Φ to be a fundamental matrix $\det \Phi \neq 0$ is necessary and sufficient. The function $W(t) = \det \Phi(t)$ is known as the *Wronskian* of the solutions. In many works on differential equations the Wronskian plays a central role. We have proved the following theorem.

5.5.9. THEOREM *The solutions* $y_1, y_2, \ldots, y_n$ *are linearly independent if and only if their Wronskian*

$$W(t) = \det \Phi(t) \tag{5.5.12}$$

is never zero.

† See Exercise 3.3, Part 2, Problem 4.

Because of the way we have defined a general solution of Equation (5.5.2) and linear independence, Theorem 5.5.9 supplies an alternative means for constructing a general solution to (5.5.2).

5.5.10. THEOREM *If $y_1, y_2, \ldots, y_n$ are linearly independent solutions of Equation* (5.5.2) *then*

$$y(t) = c_1 y_1(t) + c_2 y_2(t) + \cdots + c_n y_n(t)$$

is a general solution.

Finally it should be remarked that a definition of linear independence for an arbitrary sequence of functions, $v_1, v_2, \ldots, v_m$, is usually given as:

$v_1, v_2, \ldots, v_m$ are linearly independent if $c_1 v_1 + c_2 v_2 + \cdots + c_m v_m \equiv 0$
This implies $c_1 = c_2 = \cdots = c_m = 0$.

This definition does not require that any of the functions given be differentiable. If these are solutions of (5.5.2) then they are linearly independent in the sense of Definition 5.5.8 if and only if they are linearly independent in the latter sense. The proof of this is found in Exercise 5.5, Problems 6 and 7.

Exercise 5.5

1. Write out the steps in the proof of Theorem 5.5.3 in the 2×2 and 3×3 cases.
2. Prove Theorem 5.5.4 by induction.
3. What are the solutions of the 3×3 companion system for which Equation (5.5.10) is a chain assuming $k = 3$? What are the first components of these solutions? The second components?
4. Suppose A_c is 2×2 with eigenvalues $\lambda_0 = \alpha + i\beta$, $\bar{\lambda}_0 = \alpha - i\beta$. What is the general solution of Equation (5.5.2)? Suppose A_c is $2r \times 2r$ with $\lambda_0 = \alpha + i\beta$, $\bar{\lambda}_0 = \alpha - i\beta$ as the only distinct eigenvalues. What is the general solution of Equation (5.5.2)? Hint: Study Corollary 5.5.7.
5. Prove Theorem 5.5.10 explicitly.
6. If $y_1, y_2, \ldots, y_n$ are solutions of (5.5.2) and are linearly independent according to definition 5.5.8, prove that they are linearly independent in the sense of the definition given after Theorem 5.5.10. Hint: Suppose $c_1 y_1(t) + \cdots + c_n y_n(t) \equiv 0$. Set $\mathbf{c} = [c_1, c_2, \ldots, c_n]^T$. Why is

   $$\Phi(t)\mathbf{c} \equiv \mathbf{0}$$

   Hence, why is $\mathbf{c} = \mathbf{0}$?
7. If $y_1, y_2, \ldots, y_n$ are solutions of (5.5.2) linearly independent in the sense of the definition given after Theorem 5.5.10, prove they are linearly independent in the sense of Definition 5.5.8. Hint: Suppose the contrary. Then there exist constants $c_1, c_2, \ldots, c_n$ such that $\mathbf{c} = [c_1, c_2, \ldots, c_n]^T$ implies

   $$\Phi(t_0)\mathbf{c} = \mathbf{0} \quad \text{but} \quad \mathbf{c} \neq \mathbf{0}$$

   Why? But $\mathbf{x}(t) = \Phi(t)\mathbf{c}$ is a solution of the companion system which vanishes at t_0. Hence $\Phi(t)\mathbf{c} \equiv \mathbf{0}$, $\mathbf{c} \neq \mathbf{0}$. Thus, $c_1 y_1(t) + \cdots + c_n y_n(t) \equiv 0$ (why?) and yet not all the c's are zero.

5.6.* AN EXAMPLE FROM MECHANICS

As a motivation for the succeeding sections, we present an example of a type of problem which arises frequently in engineering and physical applications.

Figure 5.6.1

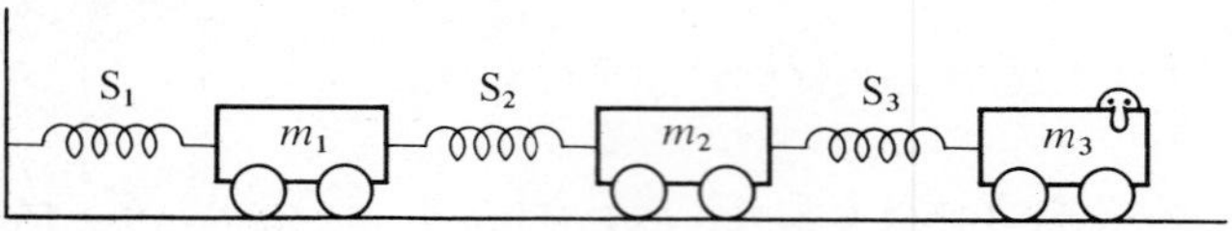

Consider the system of springs and masses in Figure 5.6.1. We assume that the only forces acting result from the extension and compression of the springs and follow Hooke's Law. Thus the tension in each spring is

$$\begin{aligned} T_1 &= k_1 x_1 \\ T_2 &= k_2(x_2 - x_1) \\ T_3 &= k_3(x_3 - x_2) \end{aligned}$$

where $k_i > 0$ is the spring constant of the spring S_i and x_i represents the displacement to the right of the mass m_i from its rest position so that $x_1 = x_2 = x_3 = 0$ when all the springs are neither stretched nor compressed. (A negative value for x_1, for instance, means that m_1 is displaced to the left of its rest position.)

When T_i is positive it tends to accelerate m_i to the left and m_{i-1} to the right according to Newton's Second Law of Motion. We thus have

$$\begin{aligned} m_1 x_1'' &= -T_1 + T_2 \\ m_2 x_2'' &= -T_2 + T_3 \\ m_3 x_3'' &= -T_3 \end{aligned}$$

or

$$\begin{bmatrix} m_1 & 0 & 0 \\ 0 & m_2 & 0 \\ 0 & 0 & m_3 \end{bmatrix} \mathbf{x}'' = \begin{bmatrix} -k_1 x_1 + k_2(x_2 - x_1) \\ -k_2(x_2 - x_1) + k_3(x_3 - x_2) \\ -k_3(x_3 - x_2) \end{bmatrix}$$

$$= \begin{bmatrix} -(k_1 + k_2) & k_2 & 0 \\ k_2 & -(k_2 + k_3) & k_3 \\ 0 & k_3 & -k_3 \end{bmatrix} \begin{bmatrix} x_1 \\ x_2 \\ x_3 \end{bmatrix}$$

or

$$\mathbf{M}\mathbf{x}'' = \mathbf{S}\mathbf{x} \tag{5.6.1}$$

Note that S is a symmetric matrix. This corresponds to the fact that the force acting on m_i due to the position of m_j is equal to that acting on m_j due to the position of m_i since the springs transmit these forces symmetrically. It should be pointed out that the symmetric matrices occur very frequently in physical problems.

Exercise 5.6

1. In Figure 5.6.1 remove S_3 and m_3 and derive the following equations of motion

$$\begin{bmatrix} m_1 & 0 \\ 0 & m_2 \end{bmatrix}\mathbf{x}'' = \begin{bmatrix} -(k_1+k_2) & k_2 \\ k_2 & -k_2 \end{bmatrix}\mathbf{x}$$

2. Write the equations of motion for the spring system illustrated in Figure 5.6.2.

Figure 5.6.2

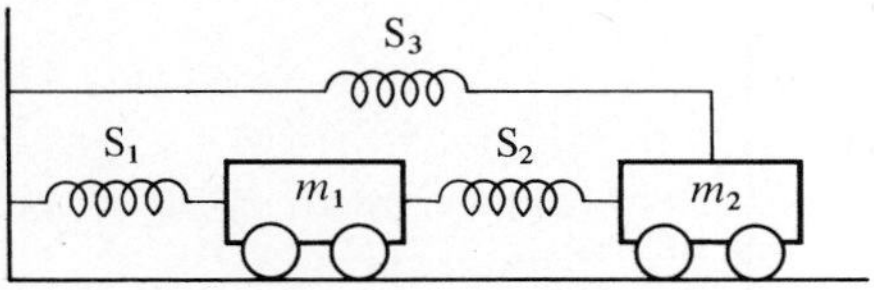

3. Write the equations of motion for the spring system illustrated in Figure 5.6.3.

Figure 5.6.3

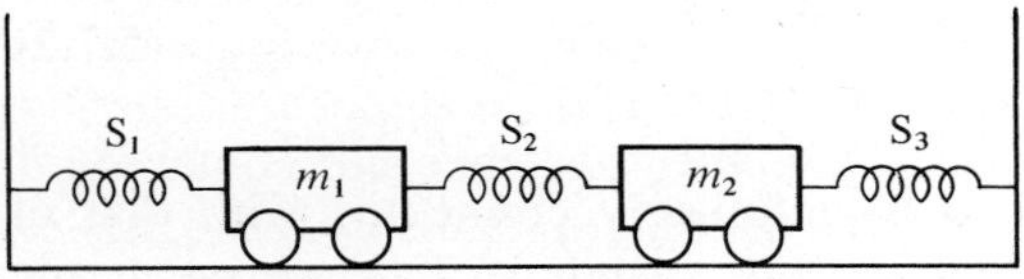

4. Another very common occurrence of systems of second order is found in electrical engineering. Using Kirchhoff's Laws (review Section 3.7) show that

$$\begin{bmatrix} i_1 \\ i_2 \end{bmatrix}'' + \begin{bmatrix} 0 & 0 \\ 0 & \dfrac{R}{L} \end{bmatrix}\begin{bmatrix} i_1 \\ i_2 \end{bmatrix}' + \begin{bmatrix} \dfrac{1}{LC} & -\dfrac{1}{LC} \\ -\dfrac{1}{LC} & \dfrac{1}{LC} \end{bmatrix}\begin{bmatrix} i_1 \\ i_2 \end{bmatrix} = \begin{bmatrix} \dfrac{V(t)}{L} \\ 0 \end{bmatrix}$$

represents the currents in Figure 5.6.4.

Figure 5.6.4

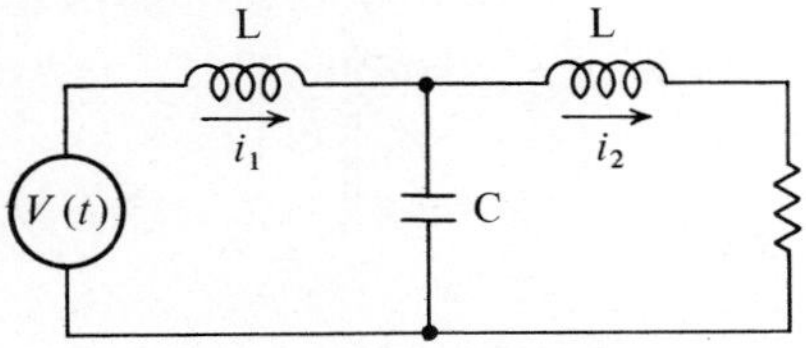

5.7.* THE GENERAL SYSTEM OF SECOND-ORDER EQUATIONS

Prompted by our experience in Section 5.6, we devote this section to the study of the general second-order linear system

$$\mathbf{y}^{(2)} + \mathbf{A}\mathbf{y}^{(1)} + \mathbf{B}\mathbf{y} = \mathbf{f} \tag{5.7.1}$$

where A and B are constant $n \times n$ matrices, with real entries,

$$A = \begin{bmatrix} a_{11} & \cdots & a_{1n} \\ \cdot & & \cdot \\ \cdot & & \cdot \\ \cdot & & \cdot \\ a_{n1} & \cdots & a_{nn} \end{bmatrix} \qquad B = \begin{bmatrix} b_{11} & \cdots & b_{1n} \\ \cdot & & \\ \cdot & & \\ \cdot & & \\ b_{n1} & \cdots & b_{nn} \end{bmatrix}$$

and **f** is a real-valued, continuous vector function of t,

$$\mathbf{f}(t) = \begin{bmatrix} f_1(t) \\ \cdot \\ \cdot \\ \cdot \\ f_n(t) \end{bmatrix}$$

Corresponding to the system (5.7.1) is the initial-value problem

$$\begin{gathered} \mathbf{y}^{(2)} + \mathrm{A}\mathbf{y}^{(1)} + \mathrm{B}\mathbf{y} = \mathbf{f} \\ \mathbf{y}(t_0) = \mathbf{y}_0 \qquad \mathbf{y}^{(1)}(t_0) = \mathbf{y}_1 \end{gathered} \tag{5.7.2}$$

whose theory may be expected to be intimately connected with that of (5.7.1). By a solution of the inhomogeneous system (5.7.1), we mean any twice differentiable function $\mathbf{y}(t)$ such that

$$\mathbf{y}^{(2)}(t) + \mathrm{A}\mathbf{y}^{(1)} + \mathrm{B}\mathbf{y}(t) = \mathbf{f}(t) \quad \text{for all } t$$

If $\mathbf{y}(t)$ also meets the conditions at t_0 described by Equation (5.7.2) it is called a solution of the initial-value problem. We might proceed to solve (5.7.1) after the fashion developed in Chapters 3 and 4. This would involve the substitution of various "trial" functions such as $\mathbf{u}e^{\lambda t}$ into the given equations to deduce special solutions. Instead of this approach, we choose to mirror the ideas developed in this chapter. We begin by supposing $\mathbf{y}(t)$ is a solution of the initial-value problem (5.7.2) and by setting $\mathbf{x}(t) = \mathbf{y}^{(1)}(t)$. From Equation (5.7.1) it follows that

$$\begin{aligned} \mathbf{y}^{(1)} &= \mathrm{O}\mathbf{y} + \mathrm{I}\mathbf{x} \\ \mathbf{x}^{(1)} &= -\mathrm{B}\mathbf{y} - \mathrm{A}\mathbf{x} + \mathbf{f} \end{aligned} \tag{5.7.3}$$

(Note O is the $n \times n$ zero matrix and I the $n \times n$ identity.) Now define **z** as the $2n$-dimensional vector whose first n components are the components of y and whose last n components are the components of **x**. That is,

$$\mathbf{z} = [y_1, y_2, \ldots, y_n, y_1^{(1)}, y_2^{(1)}, \ldots, y_n^{(1)}]^{\mathrm{T}}$$

If we study Equations (5.7.3) we can see that $\mathbf{z}(t)$ is a solution of the first-order, $2n$-dimensional system

$$\mathbf{z}' = \mathrm{A}_c\mathbf{z} + \begin{bmatrix} 0 \\ \cdot \\ \cdot \\ \cdot \\ 0 \\ f_1 \\ \cdot \\ \cdot \\ \cdot \\ f_n \end{bmatrix} \quad \text{where} \quad \mathrm{A}_c = \begin{bmatrix} 0 & \cdots & 0 & 1 & 0 & \cdots & 0 \\ 0 & \cdots & 0 & 0 & 1 & \cdots & 0 \\ \cdot & & \cdot & \cdot & \cdot & & \cdot \\ \cdot & & \cdot & \cdot & \cdot & & \cdot \\ \cdot & & \cdot & \cdot & \cdot & & \cdot \\ 0 & \cdots & 0 & 0 & 0 & \cdots & 1 \\ -b_{11} & \cdots & -b_{1n} & -a_{11} & -a_{12} & \cdots & -a_{1n} \\ \cdot & & \cdot & \cdot & \cdot & & \cdot \\ \cdot & & \cdot & \cdot & \cdot & & \cdot \\ \cdot & & \cdot & \cdot & \cdot & & \cdot \\ -b_{n1} & \cdots & -b_{nn} & -a_{n1} & -a_{n2} & \cdots & -a_{nn} \end{bmatrix} \tag{5.7.4}$$

Conversely, if $\mathbf{z}(t)$ solves Equation (5.7.4) it is easy to verify that its first n components form an n-dimensional vector form satisfying Equation (5.7.1) and its last n components form its derivative. Also, if we adjoin to the system (5.7.4) the initial condition

$$\mathbf{z}(t_0) = [y_1(t_0), y_2(t_0), \ldots, y_n(t_0), y_1^{(1)}(t_0), y_2^{(1)}(t_0), \ldots, y_n^{(1)}(t_0)]^{\mathrm{T}} \tag{5.7.5}$$

then the first n components of $\mathbf{z}(t)$ form a vector which is a solution of the initial-value problem (5.7.2). Let us make the following convenient abbreviations suggested by the definition of $\mathbf{z}$ and the matrix of the system (5.7.4):

$$\mathrm{A}_c = \begin{bmatrix} \mathrm{O} & \mathrm{I} \\ -\mathrm{B} & -\mathrm{A} \end{bmatrix} \qquad \mathbf{z} = \begin{bmatrix} \mathbf{y} \\ \mathbf{y}^{(1)} \end{bmatrix} \qquad \mathbf{z}(t_0) = \begin{bmatrix} \mathbf{y}_0 \\ \mathbf{y}_1 \end{bmatrix} = \begin{bmatrix} \mathbf{y}(t_0) \\ \mathbf{y}^{(1)}(t_0) \end{bmatrix}$$

and

$$\begin{bmatrix} \mathbf{0} \\ \mathbf{f} \end{bmatrix} = \begin{bmatrix} 0 \\ \cdot \\ \cdot \\ \cdot \\ 0 \\ f_1 \\ \cdot \\ \cdot \\ \cdot \\ f_n \end{bmatrix}$$

We may now express the conclusions just deduced in a form reminiscent of the earlier section of this chapter.

5.7.1. THEOREM *Every solution* $\mathbf{z}(t) = \begin{bmatrix} \mathbf{y}(t) \\ \mathbf{x}(t) \end{bmatrix}$ *of*

$$\mathbf{z}' = \begin{bmatrix} \mathrm{O} & \mathrm{I} \\ -\mathrm{B} & -\mathrm{A} \end{bmatrix} \mathbf{z} + \begin{bmatrix} \mathbf{0} \\ \mathbf{f} \end{bmatrix} \qquad \mathbf{z}(t_0) = \begin{bmatrix} \mathbf{y}_0 \\ \mathbf{y}_1 \end{bmatrix} \tag{5.7.6}$$

has the property that $\mathbf{y}(t)$ *satisfies*

$$\mathbf{y}^{(2)} + \mathbf{A}\mathbf{y}^{(1)} + \mathbf{B}\mathbf{y} = \mathbf{f}(t) \qquad \mathbf{y}(t_0) = \mathbf{y}_0 \qquad \mathbf{y}^{(1)}(t_0) = \mathbf{y}_1 \tag{5.7.7}$$

and $\mathbf{y}^{(1)}(t) = \mathbf{x}(t)$. *Conversely, every solution* $\mathbf{y}(t)$ *of Equation* (5.7.7) *defines*

$$\mathbf{z}(t) = \begin{bmatrix} \mathbf{y}(t) \\ \mathbf{y}'(t) \end{bmatrix}$$

a solution of Equation (5.7.6).

We have, therefore, replaced the system of n second-order equations with a "companion system" of $2n$ first-order equations. Since the latter initial-value problem always has a unique solution so does the former and indeed Theorem 5.7.1 tells us how to obtain it. The reader is invited to complete the theory by paralleling the ideas presented in Sections 5.1 through 5.5. We conclude this section with a corollary and an example.

5.7.2. COROLLARY *The dimension of the solution space of the homogeneous system*

$$\mathbf{y}^{(2)} + \mathbf{A}\mathbf{y}^{(1)} + \mathbf{B}\mathbf{y} = \mathbf{0}$$

is $2n$.

PROOF Suppose $\mathbf{y}_1, \mathbf{y}_2, \ldots, \mathbf{y}_m$ are solutions of Equation (5.7.7). Then

$$\mathbf{z}_1 = \begin{bmatrix} \mathbf{y}_1 \\ \mathbf{y}_1' \end{bmatrix}, \mathbf{z}_2 = \begin{bmatrix} \mathbf{y}_2 \\ \mathbf{y}_2' \end{bmatrix}, \ldots, \mathbf{z}_m = \begin{bmatrix} \mathbf{y}_m \\ \mathbf{y}_m' \end{bmatrix}$$

are solutions of

$$\mathbf{z}' = \begin{bmatrix} \mathbf{O} & \mathbf{I} \\ -\mathbf{B} & -\mathbf{A} \end{bmatrix} \mathbf{z} \tag{5.7.8}$$

But the dimension of the solution space of Equation (5.7.8) is $2n$. Hence $m \leqslant 2n$. Now suppose $\mathbf{z}_1, \mathbf{z}_2, \ldots, \mathbf{z}_{2n}$ are any linearly independent solutions of the system (5.7.8). Define $\mathbf{y}_i$ as the vector whose entries are the first n entries of $\mathbf{z}_i$, $i = 1, 2, \ldots, 2n$. If

$$\alpha_1 \mathbf{y}_1 + \alpha_2 \mathbf{y}_2 + \cdots + \alpha_{2n} \mathbf{y}_{2n} = \mathbf{0}$$

then by differentiation,

$$\alpha_1 \mathbf{y}_1' + \alpha_2 \mathbf{y}_2' + \cdots + \alpha_{2n} \mathbf{y}_{2n}' = \mathbf{0}$$

From these last two results,

$$\alpha_1 \mathbf{z}_1 + \alpha_2 \mathbf{z}_2 + \cdots + \alpha_{2n} \mathbf{z}_{2n} = \mathbf{0}$$

which, from the linear independence of $\mathbf{z}_1, \mathbf{z}_2, \ldots, \mathbf{z}_{2n}$, we deduce $\alpha_1 = \alpha_2 = \cdots = \alpha_{2n} = 0$. Thus $\mathbf{y}_1, \mathbf{y}_2, \ldots, \mathbf{y}_{2n}$ is linearly independent and the dimension of the solution space is at least $2n$. But $m \leqslant 2n$ shows that it is at most $2n$. This completes the proof.

5.7.1. Example

Find the unique solution of

$$\mathbf{y}^{(2)}+\begin{bmatrix} 1 & 1 \\ 0 & 1 \end{bmatrix}\mathbf{y}=-\begin{bmatrix} 1 \\ 1 \end{bmatrix} \qquad \mathbf{y}(0)=\mathbf{0} \qquad \mathbf{y}'(0)=\mathbf{0} \tag{5.7.9}$$

Solution

We study

$$\mathbf{z}'=\begin{bmatrix} 0 & 0 & 1 & 0 \\ 0 & 0 & 0 & 1 \\ 1 & 1 & 0 & 0 \\ 0 & -1 & 0 & 0 \end{bmatrix}\mathbf{z}+\begin{bmatrix} 0 \\ 0 \\ -1 \\ -1 \end{bmatrix} \qquad \mathbf{z}(0)=\mathbf{0} \tag{5.7.10}$$

The standard analysis leads to the following four linearly independent solutions of the homogeneous equation

$$\mathbf{z}_1(t)=\begin{bmatrix} 1 \\ 0 \\ 1 \\ 0 \end{bmatrix}e^t \quad \mathbf{z}_2(t)=\begin{bmatrix} 1 \\ 0 \\ -1 \\ 0 \end{bmatrix}e^{-t} \quad \mathbf{z}_3(t)=\begin{bmatrix} \sin t \\ -2\sin t \\ \cos t \\ -2\cos t \end{bmatrix} \quad \mathbf{z}_4(t)=\begin{bmatrix} -\cos t \\ 2\cos t \\ \sin t \\ -2\sin t \end{bmatrix}$$

The constant vector

$$\mathbf{z}_p(t)=\begin{bmatrix} 2 \\ -1 \\ 0 \\ 0 \end{bmatrix}$$

is readily verified as a solution of the inhomogeneous equation. From the general solution of the companion system we obtain the unique solution of the companion initial-value system, Equation (5.7.10), namely,

$$\mathbf{z}(t)=-\frac{3}{4}\begin{bmatrix} 1 \\ 0 \\ 1 \\ 0 \end{bmatrix}e^t-\frac{3}{4}\begin{bmatrix} 1 \\ 0 \\ -1 \\ 0 \end{bmatrix}e^{-t}+\frac{1}{2}\begin{bmatrix} -\cos t \\ 2\cos t \\ \sin t \\ -2\sin t \end{bmatrix}+\begin{bmatrix} 2 \\ -1 \\ 0 \\ 0 \end{bmatrix}$$

Therefore,

$$\mathbf{u}(t)=-\frac{3}{4}\begin{bmatrix} 1 \\ 0 \end{bmatrix}e^t-\frac{3}{4}\begin{bmatrix} 1 \\ 0 \end{bmatrix}e^{-t}+\frac{1}{2}\begin{bmatrix} -\cos t \\ 2\cos t \end{bmatrix}+\begin{bmatrix} 2 \\ -1 \end{bmatrix} \tag{5.7.11}$$

is the required solution of Equation (5.7.9).

Exercise 5.7

Part 1

1. Find the companion system for

$$\begin{bmatrix} 1 & 0 \\ 0 & 2 \end{bmatrix}\mathbf{y}'' - \begin{bmatrix} -2 & 1 \\ 1 & -1 \end{bmatrix}\mathbf{y} = \mathbf{0}$$

2. Find the companion system for

$$\mathbf{y}'' = \begin{bmatrix} -1 & 0 \\ 0 & -4 \end{bmatrix}\mathbf{y}$$

3. Find solutions of the companion system to the equation given in Problem 2.
4. Substitute $\mathbf{y} = \mathbf{u}e^{\lambda t}$ into the system given in Problem 2 and hence determine λ and $\mathbf{u}$ so that $\mathbf{u}e^{\lambda t}$ is a solution. Compare with the results of Problem 3.
5. Verify by substitution that the vector given in Equation (5.7.11) solves the system (5.7.9).
6. Find a general solution of the system (5.7.10), ignoring the restraint $\mathbf{z}(0) = \mathbf{0}$.
7. Following the method described in this section, solve

$$\mathbf{y}^{(2)} + \begin{bmatrix} 2 & 1 \\ 2 & 3 \end{bmatrix}\mathbf{y} = \mathbf{0} \qquad \mathbf{y}(0) = \begin{bmatrix} 1 \\ 1 \end{bmatrix} \qquad \mathbf{y}^{(1)}(0) = \begin{bmatrix} 0 \\ 1 \end{bmatrix}$$

Part 2

1. Extend Theorem 5.7.1 to systems of mth-order equations.
2. Define what is meant by a general solution of

$$\mathbf{y}^{(2)} + \mathrm{A}\mathbf{y}^{(1)} + \mathrm{B}\mathbf{y} = \mathbf{0}$$

3. Show that $\mathbf{u}e^{\lambda t}$ is a nontrivial solution of

$$\mathbf{y}^{(2)} + \mathrm{A}\mathbf{y}^{(1)} + \mathrm{B}\mathbf{y} = \mathbf{0}$$

if and only if

$$[\lambda^2\mathrm{I} + \lambda\mathrm{A} + \mathrm{B}]\mathbf{u} = \mathbf{0} \qquad \mathbf{u} \neq 0$$

4. Show that λ is an eigenvalue of C

$$\mathrm{C} = \begin{bmatrix} \mathrm{O} & \mathrm{I} \\ -\mathrm{B} & -\mathrm{A} \end{bmatrix}$$

and $\boldsymbol{\omega} = \begin{bmatrix} \mathbf{u} \\ \lambda\mathbf{u} \end{bmatrix}$ is a corresponding eigenvector if and only if

$$[\lambda^2\mathrm{I} + \lambda\mathrm{A} + \mathrm{B}]\mathbf{u} = \mathbf{0} \qquad \mathbf{u} \neq \mathbf{0}$$

Hint: Compute $\mathrm{C}\boldsymbol{\omega}$.

5. Refer to Problem 4. Construct matrices A and B such that $\mathbf{u}$ is not an eigenvector of A or B but $\boldsymbol{\omega}$ is an eigenvector of C.
6. Refer to Problem 4. Show that

$$\det[\lambda^2\mathrm{I} + \lambda\mathrm{A} + \mathrm{B}] = 0$$

is a necessary and sufficient condition for λ to be an eigenvalue of C.

7. Using the results of Problem 4, show that the eigenvalues of

$$C = \begin{bmatrix} O & I \\ E & O \end{bmatrix}$$

are the square roots of the eigenvalues of E. Furthermore, in spite of Problem 5, show that if $\mathbf{u}$ is an eigenvector of E corresponding to the eigenvalue $\mu = \lambda^2$ of E, then

$$\boldsymbol{\omega} = \begin{bmatrix} \mathbf{u} \\ \sqrt{\mu}\mathbf{u} \end{bmatrix}$$

is an eigenvector of C. Is the converse true?

8.* In what way are the eigenvalues of M and N related to the eigenvalues of

$$C = \begin{bmatrix} O & I \\ -MN & M+N \end{bmatrix}$$

Hint: Use Problem 6.

9.* (a) Refer to Problem 8 for notation. If $\mathbf{u}$ is an eigenvector of N, is

$$\boldsymbol{\omega} = \begin{bmatrix} \mathbf{u} \\ \lambda\mathbf{u} \end{bmatrix}$$

an eigenvector of C for some values of λ?
Hint: Use Problem 4.
(b) In part (a) above replace N with M. What conclusions can you draw?
(c) Test your answer to (a) and (b) against the example

$$M = \begin{bmatrix} 0 & -1 \\ -1 & 0 \end{bmatrix} \qquad N = \begin{bmatrix} 0 & 1 \\ -6 & 5 \end{bmatrix}$$

10.* In what way are the eigenvalues and eigenvectors of F related to the eigenvalues and eigenvectors of

$$\begin{bmatrix} O & I \\ -F^2 & 2F \end{bmatrix}$$

11. Show that $\mathbf{u} \neq \mathbf{0}$ and μ can be found so that $\mathbf{y}(t) = e^{\mu t}\mathbf{u}$ is a solution of

$$\mathbf{y}^{(2)} + A\mathbf{y}^{(1)} + B\mathbf{y} = e^{\mu t}\mathbf{v}_0$$

where $\mathbf{v}_0$ is a constant, provided

$$\det[\mu^2 I + \mu A + B] = 0$$

Contrast this with Problem 3.

5.8.* AN IMPORTANT SPECIAL CASE

The system

$$M\mathbf{x}'' = T\mathbf{x} + \mathbf{f}_0 \qquad (5.8.1)$$

is a special case of the general system of second-order equations. If M and T are suitably restricted, it is a system of some interest for two reasons. It is simple to solve and it is a mathematical model for a variety of phenomena including the spring coupled masses of Section 5.6 as well as certain items from atomic and molecular theory. The restriction on M is that it be a diagonal matrix with positive entries on the diagonal. The restrictions on T are more complicated. For the present we will assume that T be symmetric and have negative eigenvalues only.

If we normalize Equation (5.8.1) by multiplying through by M^{-1} (which exists because of the simple nature of M) the coefficient of $\mathbf{y}$ becomes $M^{-1}T$. Unfortunately, $M^{-1}T$ is not necessarily symmetric. Since symmetric matrices always have a full complement of eigenvectors (the eigenvectors span $\mathscr{R}^n$), this loss is serious. We therefore transform the equation in a manner which is not so straightforward.

Letting

$$M = \begin{bmatrix} m_{11} & & O \\ & \ddots & \\ O & & m_{nn} \end{bmatrix}$$

we denote

$$M^{1/2} = \begin{bmatrix} m_{11}^{1/2} & & O \\ & \ddots & \\ O & & m_{nn}^{1/2} \end{bmatrix}$$

Then, if $\mathbf{x}(t)$ is a solution of (5.8.1) and we set

$$\mathbf{y}(t) = M^{1/2}\mathbf{x}(t)$$

it follows that

$$\begin{aligned} M\mathbf{x}''(t) &= M^{1/2}\mathbf{y}''(t) = T\mathbf{x}(t) + \mathbf{f}_0 \\ &= TM^{-1/2}\mathbf{y}(t) + \mathbf{f}_0 \end{aligned}$$

so that $\mathbf{y}(t)$ satisfies

$$\mathbf{y}'' = S\mathbf{y} + \mathbf{F}_0 \tag{5.8.2}$$

where

$$S = M^{-1/2}TM^{-1/2} \quad \text{and} \quad \mathbf{F}_0 = M^{-1/2}\mathbf{f}_0$$

Conversely, if $\mathbf{y}(t)$ satisfies (5.8.2) then $\mathbf{x}(t) = M^{-1/2}\mathbf{y}(t)$ satisfies (5.8.1); furthermore, we can prove that S is symmetric and has only negative eigenvalues (see Exercise 5.8, Part 2, Problem 4) which justifies the extra effort in obtaining (5.8.2).

To begin this study, assume $\mathbf{F}_0 = \mathbf{0}$. The system (5.8.2) is then

$$\mathbf{y}'' = S\mathbf{y} \tag{5.8.3}$$

with companion system

$$\mathbf{z}' = \begin{bmatrix} \mathrm{O} & \mathrm{I} \\ \mathrm{S} & \mathrm{O} \end{bmatrix} \mathbf{z} \tag{5.8.4}$$

$$= \mathrm{S}_c \mathbf{z}$$

Consider the vector $\boldsymbol{\omega}$, defined by

$$\boldsymbol{\omega} = \begin{bmatrix} \mathbf{u} \\ \lambda^{1/2}\mathbf{u} \end{bmatrix}$$

where $\mathbf{u}$ is an eigenvector of S and λ a corresponding eigenvalue. Then

$$\mathrm{S}_c\boldsymbol{\omega} = \begin{bmatrix} \mathrm{O} & \mathrm{I} \\ \mathrm{S} & \mathrm{O} \end{bmatrix} \begin{bmatrix} \mathbf{u} \\ \lambda^{1/2}\mathbf{u} \end{bmatrix} = \begin{bmatrix} \lambda^{1/2}\mathbf{u} \\ \mathrm{S}\mathbf{u} \end{bmatrix}$$

$$= \begin{bmatrix} \lambda^{1/2}\mathbf{u} \\ \lambda\mathbf{u} \end{bmatrix} = \lambda^{1/2} \begin{bmatrix} \mathbf{u} \\ \lambda^{1/2}\mathbf{u} \end{bmatrix} = \lambda^{1/2}\boldsymbol{\omega}$$

Hence $\boldsymbol{\omega}$ is an eigenvector of S_c corresponding to the eigenvalue $\lambda^{1/2}$. From Theorem 5.7.1, it follows, therefore, that

$$\mathbf{y}(t) = \mathbf{u}e^{\sqrt{\lambda}t}$$

is a solution of (5.8.3). Since S is symmetric $\mathbf{u}$ may be selected real (Theorem 2.5.4). Since λ is negative, $\lambda^{1/2}$ is pure imaginary and $e^{\sqrt{\lambda}t}$ represents trigonometric functions. In view of this, set $\omega^2 = -\lambda$, then the real and imaginary parts of $\mathbf{u}e^{\sqrt{\lambda}t}$ are solutions; that is,

$$\mathbf{u}\cos \omega t \quad \text{and} \quad \mathbf{u}\sin \omega t$$

Let us review these facts by working an example.

5.8.1. Example

Solve

$$\mathbf{y}'' = \mathrm{S}\mathbf{y}$$

where

$$\mathrm{S} = \begin{bmatrix} -5 & 2 \\ 2 & -2 \end{bmatrix}$$

Solution

For S, $\lambda_1 = -1$, $\lambda_2 = -6$, and $\mathbf{x}_1 = \begin{bmatrix} 1 \\ 2 \end{bmatrix}$, $\mathbf{x}_2 = \begin{bmatrix} -2 \\ 1 \end{bmatrix}$. The reader is urged to verify that

$$\begin{bmatrix} \mathbf{x}_1 \\ i\sqrt{6}\mathbf{x}_1 \end{bmatrix} \quad \text{and} \quad \begin{bmatrix} \mathbf{x}_2 \\ i\sqrt{6}\mathbf{x}_2 \end{bmatrix}$$

solve the companion system. We construct

$$\begin{bmatrix}1\\2\end{bmatrix}\cos t \qquad \begin{bmatrix}1\\2\end{bmatrix}\sin t \qquad \begin{bmatrix}-2\\1\end{bmatrix}\cos 6^{1/2}t \qquad \begin{bmatrix}-2\\1\end{bmatrix}\sin 6^{1/2}t$$

and easily verify that these functions are solutions of the given problem. The general solution is then

$$y(t) = \begin{bmatrix}1\\2\end{bmatrix}(a\cos t + b\sin t) + \begin{bmatrix}-2\\1\end{bmatrix}(c\cos 6^{1/2}t + d\sin 6^{1/2}t)$$

For reference, we state the following theorem.

5.8.1. THEOREM *If* $-\lambda_1^2, -\lambda_2^2, \ldots, -\lambda_n^2$ *are the eigenvalues of a symmetric matrix* S *(with possible duplications) and* $\mathbf{x}_1, \mathbf{x}_2, \ldots, \mathbf{x}_n$ *are corresponding linearly independent real eigenvectors then the arbitrary linear combination of*

$$\begin{gathered}\mathbf{x}_1\cos\lambda_1 t, \mathbf{x}_2\cos\lambda_2 t, \ldots, \mathbf{x}_n\cos\lambda_n t\\ \mathbf{x}_1\sin\lambda_1 t, \mathbf{x}_2\sin\lambda_2 t, \ldots, \mathbf{x}_n\sin\lambda_n t\end{gathered} \tag{5.8.5}$$

is the general solution of

$$\mathbf{y}'' = \mathbf{S}\mathbf{y} \tag{5.8.6}$$

PROOF The proof is basically the argument given before Example 5.8.1 combined with the material of Section 5.7 and the observation that symmetric matrices always have n linearly independent real eigenvectors.†

To solve

$$\mathbf{y}'' = \mathbf{S}\mathbf{y} + \mathbf{F}_0 \tag{5.8.7}$$

we note that $\mathbf{y}_p(t) = -\mathrm{S}^{-1}\mathrm{F}_0$ is surely a solution when $\mathbf{F}_0$ is constant. Suppose $\mathbf{y}_c(t)$ is a general solution of (5.8.6).

By superposition,

$$\mathbf{y}(t) = \mathbf{y}_c(t) - \mathrm{S}^{-1}\mathbf{F}_0$$

is a general solution of (5.8.7). In terms of $\mathbf{x}$, the original variable, we refer to Equation (5.8.2) and write

$$\begin{aligned}\mathrm{M}^{1/2}\mathbf{x}(t) &= \mathbf{y}_c(t) - (\mathrm{M}^{-1/2}\mathrm{T}\mathrm{M}^{-1/2})^{-1}\mathrm{M}^{-1/2}\mathbf{f}_0\\ &= \mathbf{y}_c(t) - (\mathrm{M}^{1/2}\mathrm{T}^{-1}\mathrm{M}^{1/2})\mathrm{M}^{-1/2}\mathbf{f}_0\\ &= \mathbf{y}_c(t) - \mathrm{M}^{1/2}\mathrm{T}^{-1}\mathbf{f}_0\end{aligned}$$

Therefore

$$\mathbf{x}(t) = \mathrm{M}^{-1/2}\mathbf{y}_c(t) - \mathrm{T}^{-1}\mathbf{f}_0$$

solves Equation (5.8.1).

† See Theorem 2.5.4 and Section 3.6 for a proof of this result.

Exercise 5.8

Part 1

1. Find the solutions of

$$\begin{bmatrix} 9 & 0 \\ 0 & 4 \end{bmatrix} \mathbf{y}'' = \begin{bmatrix} -45 & 12 \\ 12 & -8 \end{bmatrix} \mathbf{y} + \begin{bmatrix} 3 \\ 4 \end{bmatrix}$$

2. Find the solutions of

$$\mathbf{y}'' = \begin{bmatrix} -4 & 1 & 2 \\ 1 & -3 & 1 \\ 2 & 1 & -4 \end{bmatrix} \mathbf{y}$$

3. If B is symmetric show that
   (a) ABA^T is symmetric.
   (b) ABA is symmetric if A is also symmetric.
4. (a) Show that $T = \begin{bmatrix} -2 & 1 \\ 1 & -1 \end{bmatrix}$ has negative eigenvalues.

   (b) If $M = \begin{bmatrix} \frac{1}{4} & 0 \\ 0 & 1 \end{bmatrix}$ show that $M^{-1}T$ has negative eigenvalues, but is not symmetric.

Part 2

1. Let $\mathbf{x}_1, \mathbf{x}_2, \ldots, \mathbf{x}_n$ be n real linearly independent eigenvectors of the symmetric matrix S. Show that

$$S\mathbf{x} \cdot \mathbf{x} = \sum_{i=1}^{n} \lambda_i \alpha_i^2 \|\mathbf{x}_i\|^2$$

   where

   (a) $\mathbf{x} = \sum_{i=1}^{n} \alpha_i \mathbf{x}_i$

   (b) $\lambda_1, \lambda_2, \ldots, \lambda_n$ are eigenvalues of S.
2. Under the conditions of Problem 1, prove:

   $S\mathbf{x} \cdot \mathbf{x} < 0$ for all $\mathbf{x} \neq 0$ if and only if the eigenvalues of S are all negative.

3.* Assume A is nonsingular and S is symmetric with only negative eigenvalues. Using the results of Problem 2, prove:

   "A^TSA has only negative eigenvalues."

   Hint: Start with $(A^TSA)\mathbf{x} \cdot \mathbf{x} = SA\mathbf{x} \cdot A\mathbf{x}$ (why?).
4. Prove that $M^{-1/2}\, TM^{-1/2}$ has only negative eigenvalues and is symmetric. Hint: Use Problem 3, Part 1 and Problem 3, Part 2.

Chapter 6

AN INTRODUCTION TO THE THEORY OF ANALYTIC DIFFERENTIAL SYSTEMS

6.1. INTRODUCTION

In more advanced works† it is proved that a linear system,

$$\mathbf{x}' = \mathrm{A}(t)\mathbf{x} \qquad a < t < b \tag{6.1.1}$$

with a real-valued, continuous $n \times n$ matrix, $\mathrm{A}(t)$, always has n linearly independent solutions on $a < t < b$. If $\mathrm{A}(t)$ is neither constant nor one-dimensional, the actual construction of even one of these solutions is usually an imposing task. (The one-dimensional problem is easy and is given as Exercise 6.1, Problem 10.) With this major exception, the theory of (6.1.1) mirrors the theory of the constant coefficient system in remarkable detail. Indeed, a careful perusal of the proofs in Chapter 4 will show that the constancy of A was irrelevant in many of the theorems.

The starting point of the theory is the construction of a matrix $\Phi(t)$ from the solutions of $\mathbf{x}' = \mathrm{A}(t)\mathbf{x}$ so normalized at $t = t_0$ that for $a < t < b$ and $a < t_0 < b$,

(1) $\Phi'(t) \equiv \mathrm{A}(t)\Phi(t) \qquad \Phi(t_0) = \mathrm{I}$
(2) $\Phi(t)\mathbf{x}_0$ is the unique solution of $\mathbf{x}' = \mathrm{A}(t)\mathbf{x} \qquad \mathbf{x}(t_0) = \mathbf{x}_0$
(3) $\Phi(t)\mathbf{x}_0 + \Phi(t)\int_{t_0}^{t} \Phi^{-1}(s)\mathbf{f}(s)\,ds$ is the unique solution of

$$\mathbf{x}' = \mathrm{A}(t)\mathbf{x} + \mathbf{f} \qquad \mathbf{x}(t_0) = \mathbf{x}_0$$

We assume $\mathbf{f}$ continuous on (a, b). The matrix, Φ, constructed to satisfy (1) is called the *normalized fundamental matrix* of $\mathbf{x}' = \mathrm{A}(t)\mathbf{x}$. Properties (2)

† For instance, W. Hurewicz, *Lectures on Ordinary Differential Equations*, The Technology Press, Cambridge, Mass., 1958.

and (3) and many others follow from (1) in the same way that their special cases were proved in Chapter 4. In the following exercise set are listed the more important theorems.

Having briefly described the similarities between the constant and non-constant cases of linear systems, let us turn to an equally brief discussion of their major difference—the construction of solutions. It is the rare system for which some finite combination of known functions are solutions. In fact, the systems themselves are often used to define and analyze new functions. The exponential function may be defined as the solution of $x' = x$, $x(0) = 1$ and its well-known properties deduced by direct analysis of the defining differential equation. The Bessel, Legendre, Laguerre, and Hermite functions are other important examples defined as solutions of certain initial-value problems.

The new feature introduced in this chapter is the possibility of representing solutions of (6.1.1) in the form

$$\mathbf{x}(t) = \mathbf{u}_0 + \mathbf{u}_1(t-t_0) + \cdots + \mathbf{u}_k(t-t_0)^k + \cdots \tag{6.1.2}$$

a power series. This approach has a difficulty quite apart from the obvious problem of determining the infinitely many vectors $\mathbf{u}_0, \mathbf{u}_1, \mathbf{u}_2, \ldots$ for even if these vectors were known, we have no assurance that (6.1.2) makes sense. The series

$$\begin{bmatrix}0\\0\end{bmatrix} + \begin{bmatrix}1\\0\end{bmatrix}t + \begin{bmatrix}2!\\0\end{bmatrix}t^2 + \cdots + \begin{bmatrix}k!\\0\end{bmatrix}t^k + \cdots$$

does not make sense if $t \neq 0$. We must not only find $\mathbf{u}_0, \mathbf{u}_1, \ldots$, if possible, but prove the convergence of (6.1.2) to a solution.

The critical assumption which permits the existence of series solutions to (6.1.1) is the "analyticity" of $A(t)$—a concept defined in the next section. Continuity and even the existence of derivatives of all orders is not sufficient to insure analyticity. In spite of this quite restrictive hypothesis, most of the physically important differential equations and systems meet this requirement. The study of such systems is the goal of this chapter. Although a thorough investigation requires a knowledge of the theory of functions of a complex variable, we present an extensive introduction into the major ideas. The proofs are often omitted.

Exercise 6.1

1. Show that if $\mathbf{x}_1$ and $\mathbf{x}_2$ are solutions of $\mathbf{x}' = A(t)\mathbf{x} + \mathbf{f}(t)$, then $\mathbf{x}_1 - \mathbf{x}_2$ is a solution of $\mathbf{x}' = A(t)\mathbf{x}$.
2. (Superposition) Show that if $\mathbf{x}_1$ is a solution of $\mathbf{x}' = A(t)\mathbf{x} + \mathbf{f}_1(t)$ and $\mathbf{x}_2$ is a solution of $\mathbf{x}' = A(t)\mathbf{x} + \mathbf{f}_2(t)$ then $\mathbf{x}_1 + \mathbf{x}_2$ is a solution of $\mathbf{x}' = A(t)\mathbf{x} + \mathbf{f}_1(t) + \mathbf{f}_2(t)$.
3. Use property (2) in the text to show that $\mathbf{x}' = A(t)\mathbf{x} + \mathbf{f}(t)$ $\mathbf{x}(t_0) = \mathbf{x}_0$ can have no more than one solution.
4. What property is expressed by the conclusion of Problem 2 if $\mathbf{f}_1 = 0$? If $\mathbf{f}_1 = 0 = \mathbf{f}_2$?

5. Use property (1) in the text to prove that $\Phi(t)\mathbf{x}_0$ is a solution of $\mathbf{x}' = A(t)\mathbf{x}$, $\mathbf{x}(t_0) = \mathbf{x}_0$.
6. Which proofs in Chapter 4 depend upon A being constant?
7. Show that $\mathbf{x}(t) \equiv \mathbf{0}$ is the only solution of $\mathbf{x}' = A(t)\mathbf{x}$, $\mathbf{x}(t_0) = \mathbf{0}$.
8. Show that if $\mathbf{x}_1(t), \mathbf{x}_2(t), \ldots, \mathbf{x}_n(t)$ are solutions of $\mathbf{x}' = A(t)\mathbf{x}$ and $\mathbf{x}_1(t_0), \mathbf{x}_2(t_0), \ldots, \mathbf{x}_n(t_0)$ are linearly dependent, then $\mathbf{x}_1(t), \mathbf{x}_2(t), \ldots, \mathbf{x}_n(t)$ are linearly dependent for any other value of t for which $A(t)$ is defined. Hint: Assume the contrary, examine $c_1\mathbf{x}_1(t) + c_2\mathbf{x}_2(t) + \cdots + c_n\mathbf{x}_n(t)$, and use the conclusion of Problem 7.
9. As in Problem 8, show that if $\mathbf{x}_1(t), \mathbf{x}_2(t), \ldots, \mathbf{x}_n(t)$ are linearly independent at $t = t_0$ they are linearly independent for any other value of t for which $A(t)$ is defined.
10. (The integrating factor) The one-dimensional equation $x' = a(t)x + f(t)$ can always be solved by use of a multiplying factor

$$\mu(t) = \exp\left[-\int a(t)\,dt\right]$$

(a) Show that

$$\frac{d}{dt}[\mu x] = \mu x' - \mu a(t)x$$

(b) Thus show that

$$\frac{d}{dt}[\mu x] = \mu f(t)$$

is equivalent to $x' = a(t)x + f(t)$ and hence

$$x(t) = \frac{1}{\mu(t)} \int \mu(t) f(t)\,dt$$

(c) Show that for continuous $a(t)$, $\mu \neq 0$.
(d) Prove that the function given in (b) is a solution of $x' = a(t)x + f(t)$ by direct substitution.

The following eight problems depend upon the conclusions reached in Problem 10 above. In each case find the general solution over the interval stated.

11. $x' = (\cos t)x$, all t
12. $x' = 2tx$, all t
13. $x' = (\cos t)x + \sin t$, $x(\pi) = 4$, all t
14. $x' = (\tan t)x$, $x(0) = 1$, all t
15. $x' = (1-t)x + t^2$, all t
16. $x' = (1/t)x$, all $t > 0$
17. $x' = (1/t)x$, all $t < 0$
18. $x' = |t|x$, $x(0) = 1$, all t

6.2. POWER SERIES EXPANSIONS OF ANALYTIC FUNCTIONS

In the sections to follow we will deal with differential equations of the form

$$\mathbf{x}' = A(t)\mathbf{x}$$

where the entries of the coefficient matrix $A(t)$ are expressed as power series. We will find that a solution of (6.2.1) is a power series whose coefficients can be computed in terms of $A(t)$ and the initial value $\mathbf{x}(t_0) = \mathbf{x}_0$.

6.2.1. DEFINITION *A function, f, of a real variable is said to be analytic at the real number,* t_0 *if there exist*

(1) *a positive number r and*

(2) *a sequence of numbers* $a_0, a_1, a_2, \ldots$

such that for $|t-t_0| < r$

$$f(t) = \lim_{N\to\infty} \sum_{k=0}^{N} a_k(t-t_0)^k = \sum_{k=0}^{\infty} a_k(t-t_0)^k \tag{6.2.1}$$

The following theorem states that one may differentiate an analytic function by *first* differentiating each term $a_k(t-t_0)^k$ to obtain $ka_k(t-t_0)^{k-1}$ and then performing the infinite summation. This formal procedure may seem reasonable but one must remember that we are combining an infinite and an infinitesimal process. The theorem is by no means trivial.

6.2.2. THEOREM *If f is analytic at* t_0 *and for* $|t-t_0| < r$

$$f(t) = \sum_{k=0}^{\infty} a_k(t-t_0)^k$$

then for each positive integer j, the jth derivative is expressible as,

$$f^{(j)}(t) = \sum_{k=0}^{\infty} \frac{(k+j)!}{k!} a_{k+j}(t-t_0)^k \quad \text{for} \quad |t-t_0| < r \tag{6.2.2}$$

so that $f^{(j)}$ *is also analytic at* t_0 *and*

$$f^{(j)}(t_0) = j!a_j \tag{6.2.3}$$

The constants $a_k = (1/k!)f^{(k)}(t_0)$ are known as the *Taylor Coefficients* of f at t_0 and the expression

$$\sum_{k=0}^{\infty} a_k(t-t_0)^k = \sum_{k=0}^{\infty} \frac{1}{k!} f^{(k)}(t_0)(t-t_0)^k$$

is known as the *Taylor Series* of f at t_0.

Analytic functions form a special class of functions for

(i) In general, a function may not have derivatives of all orders.

(ii) A function, f, may have derivatives of all orders but the Taylor Series

$$\sum_{k=0}^{\infty} \frac{1}{k!} f^{(k)}(t_0)(t-t_0)^k$$

may converge only at $t = t_0$.

(iii) A function, f, may have a Taylor Series

$$\sum_{k=0}^{\infty} \frac{1}{k!} f^{(k)}(t_0)(t-t_0)^k$$

which converges when $|t-t_0| < r$ for some $r > 0$ but not to $f(t)$, i.e.,

$$f(t) \neq \sum_{k=0}^{\infty} \frac{1}{k!} f^{(k)}(t_0)(t-t_0)^k$$

unless $t = t_0$.

Taylor Series for some elementary functions are listed in the next example.

6.2.1. Example

By direct application of Equation (6.2.3) we find:

(1) $$e^t = \sum_{k=0}^{\infty} \frac{t^k}{k!} \quad \text{for all } t$$

(2) $$\sin t = \sum_{k=0}^{\infty} \frac{(-1)^k t^{2k+1}}{(2k+1)!} \quad \text{for all } t$$

(3) $$\cos t = \sum_{k=0}^{\infty} \frac{(-1)^k t^{2k}}{(2k)!} \quad \text{for all } t$$

(4) $$\frac{1}{1-t} = \sum_{k=0}^{\infty} t^k \qquad -1 < t < 1$$

(5) $$\frac{1}{1+t} = \sum_{k=0}^{\infty} (-1)^k t^k \qquad -1 < t < 1$$

(6) $$\ln(1+t) = \sum_{k=0}^{\infty} \frac{(-1)^k t^{k+1}}{k+1} \qquad -1 < t \leqslant 1$$

(7) $$\arctan t = \sum_{k=0}^{\infty} \frac{(-1)^k t^{2k+1}}{2k+1} \qquad -1 \leqslant t \leqslant 1$$

6.2.2. Example

Consider the series

$$\sum_{n=0}^{\infty} \frac{(-1)^n t^{2n}}{2^{2n}(n!)^2} \tag{6.2.4}$$

By the ratio test this series converges for all t. Let us name this series by writing

$$J_0(t) = \sum_{n=0}^{\infty} \frac{(-1)^n t^{2n}}{2^{2n}(n!)^2}$$

The function J_0 is analytic and by Theorem 6.2.2, Equation (6.2.3),

$$\frac{J_0^{(2k)}(0)}{(2k)!} = \frac{(-1)^k}{2^{2k}(k!)^2} \qquad k = 0, 1, 2, \ldots$$

$$J_0^{(2k+1)}(0) = 0 \qquad k = 0, 1, 2, \ldots$$

It is important to realize that J_0 is simply a name for the given series; an abbreviation for a complicated limit process. The phrase "f is analytic at t_0" is often used without mention of the interval $-r < t - t_0 < r$ in those instances where we are less concerned about the value of r than the fact that series (6.2.1) converges to f for all t on some interval containing t_0.

Series expansions for functions can often be found by differentiating the series expansions for simpler functions. In Example 6.2.1 the connection between the expansion of $\sin t$ and $\cos t$ and the fact that $(d/dt) \sin t = \cos t$ illustrates this fact. There are other such connections in this list.

Power series may be added, subtracted, multiplied, and divided. This forms still another method for constructing series expansions for function.

6.2.3. THEOREM *If*

$$f(t) = \sum_{k=0}^{\infty} f_k (t - t_0)^k \qquad -r < t - t_0 < r$$

$$g(t) = \sum_{k=0}^{\infty} g_k (t - t_0)^k \qquad -r < t - t_0 < r$$

then

$$f(t) \pm g(t) = \sum_{k=0}^{\infty} (f_k \pm g_k)(t - t_0)^k$$

in $-r < t - t_0 < r$. *Similarly,*

$$f(t) g(t) = \sum_{k=0}^{\infty} \sum_{i=0}^{k} f_i g_{k-i} (t - t_0)^k$$

for $-r < t - t_0 < r$.

Next is one last theorem which we will use extensively.

6.2.4. THEOREM *If*

$$\sum_{k=0}^{\infty} f_k (t - t_0)^k \equiv 0 \qquad -r < t - t_0 < r$$

then $f_0 = f_1 = \cdots = 0$

We actually have enough theorems now to prove this. We set this as a problem in Exercise 6.4, Part 2, Problem 4. With this last theorem stated, we now turn to the analogous definitions and theorems for analytic matrices.

6.2.5. DEFINITION *The $n\times m$ matrix $A(t)$ is said to be analytic at $t=t_0$ if each of its entries is analytic at $t=t_0$.*

In this definition it is possible for each of the entries to have a series convergent on a different interval about t_0. The smallest of these intervals is an interval of convergence common to all the entries and is taken as the interval of convergence for $A(t)$. Each of the above theorems can now be reworded for analytic $A(t)$ by simply requiring the hypotheses to hold simultaneously for each entry of $A(t)$ and the conclusion interpreted similarly. For instance,

$$A(t)=\sum_{k=0}^{\infty} A_k(t-t_0)^k \qquad -r<t-t_0<r$$

is simply an abbreviation for $n\cdot m$ expansions, one for each entry of $A(t)$.

6.2.3. Example

Find A_k such that

$$\begin{bmatrix} \cos t & \sin t \\ e^t & \dfrac{1}{1-t} \end{bmatrix} = \sum_{k=0}^{\infty} A_k\, t^k$$

Solution

Referring to Example 6.2.2 for the relevant expansions, we find

$$A_{2i}=\begin{bmatrix} \dfrac{(-1)^i}{(2i)!} & 0 \\ \dfrac{1}{(2i)!} & 1 \end{bmatrix} \qquad i=0,1,2\ldots$$

and

$$A_{2i+1}=\begin{bmatrix} 0 & \dfrac{(-1)^i}{(2i+1)!} \\ \dfrac{1}{(2i+1)!} & 1 \end{bmatrix} \qquad i=0,1,2\ldots$$

The interval of convergence is $-1<t<1$ and this shows that A is analytic at $t_0=0$.

All in all, the properties of analytic matrices are made to depend upon the properties of their analytic entries.

Since column vectors are $m\times 1$ matrices and row vectors are $1\times m$ matrices all we have said applies to vectors as special cases.

6.2.4. Example

Find $\mathbf{x}_k$ such that

$$\mathbf{x}(t) = \begin{bmatrix} e^t \\ e^{2t} \end{bmatrix} = \sum_{k=0}^{\infty} \mathbf{x}_k t^k$$

Solution

Since

$$e^t = \sum_{k=0}^{\infty} \frac{t^k}{k!}$$

we have

$$e^{2t} = \sum_{k=0}^{\infty} \frac{2^k t^k}{k!}$$

and hence

$$\mathbf{x}_k = \begin{bmatrix} \dfrac{1}{k!} \\ \dfrac{2^k}{k!} \end{bmatrix} = \frac{1}{k!} \begin{bmatrix} 1 \\ 2^k \end{bmatrix}$$

and therefore

$$\mathbf{x}(t) = \sum_{k=0}^{\infty} \begin{bmatrix} 1 \\ 2^k \end{bmatrix} \frac{t^k}{k!}$$

Exercise 6.2

Part 1

In each of the following 10 problems you may assume that f is analytic at t_0. Use Equations (6.2.3) to find the Taylor Series for these functions at t_0.

1. e^t at $t_0 = 1$
2. e^t at $t_0 = -1$
3. $\dfrac{1}{1-t^2}$ at $t_0 = 0$
4. $\dfrac{1}{(1-t)^2}$ at $t_0 = 0$
5. $\ln(1-t)$ at $t_0 = 0$
6. $\dfrac{1+t}{1-t}$ at $t_0 = 0$
7. $\ln\dfrac{1+t}{1-t}$ at $t_0 = 0$
8. $(1-t)^{-\alpha}$ at $t_0 = 0$ $\quad \alpha > 0$
9. $\cos^2 t$ at $t_0 = 0$

10. $\dfrac{e^t}{1-t}$ at $t_0 = 0$
11. Show that the expansion of $1/(1-t^2)$ may be obtained from $1/(1-t)$ by substituting t^2 for t (Problem 3).
12. Show that the expansion for $1/[(1-t)^2]$ may be obtained from the expansion for $1/(1-t)$ by differentiation (Problem 4).
13. Same as Problem 12 for ln $(1-t)$ (Problem 5).
14. Show that the expansion for $(1+t)/(1-t)$ may be obtained by multiplying $(1+t)$ by the expansion for $(1-t)^{-1}$ (Problem 6).
15. Compare the expansion of Problem 8 with the binomial theorem.
16. Expand $\cos^2 t$ by observing $\cos^2 t = \frac{1}{2} + \frac{1}{2}\cos 2t$ (Problem 9).
17. Find the Taylor Series for

(a) $\mathbf{x}(t) = \begin{bmatrix} 1 \\ 1 \\ e^t \end{bmatrix} \qquad t_0 = 0$

(b) $\mathbf{x}(t) = \begin{bmatrix} e^t \\ e^t \\ e^t \end{bmatrix} \qquad t_0 = 0$

(c) $\mathbf{x}(t) = \begin{bmatrix} \sin t \\ 1+t \\ t^2 \end{bmatrix} \qquad t_0 = 0$

18. Find the Taylor Series at t_0 for the following matrices

(a) $A(t) = \begin{bmatrix} 1 & 1 \\ 1 & \dfrac{1}{1-t} \end{bmatrix}$

(b) $A(t) = \begin{bmatrix} e^t & 0 \\ 0 & 1 \end{bmatrix}$

(c) $A(t) = \begin{bmatrix} \dfrac{1}{1-t} & \dfrac{1}{1+t} \\ t+1 & t-1 \end{bmatrix}$

Part 2

1. Find three nonzero terms in the Taylor Series expansion of $\tan t$ at $t_0 = 0$.
2. Show that the series in (6.2.4) converges for all t.
3. Show that $\Sigma_{n=0}^{\infty} t^n$ converges for all $t, -1 < t < 1$. How do you know it converges to $1/(1-t)$?
4. Prove Theorem 6.2.4.
5. Are polynomials analytic functions? Explain!
6. Are rational functions analytic functions? Explain! (Hint:

$$\frac{1}{(t-t_0)(t-t_1)} = \frac{A}{t-t_0} + \frac{B}{t-t_1} \quad \text{if} \quad t_0 \neq t_1$$

but $A/(t-\tilde{t})$ is analytic except where $t = \tilde{t}$.)

6.3. SERIES SOLUTIONS OF $\mathbf{x}' = A(t)\mathbf{x}$

The initial-value problem

$$\mathbf{x}' = A(t)\mathbf{x} \qquad \mathbf{x}(t_0) = \mathbf{x}_0 \tag{6.3.1}$$

is now studied under the stringent hypothesis that $A(t)$ is analytic at t_0. Such points, t_0, are called *ordinary points* of the differential system (6.3.1). If $A(t) = A$, a constant matrix, then A is analytic for all t and thus every point is an ordinary point. Thus the results we establish for $A(t)$ analytic at t_0 hold for systems with constant coefficients at every point (see Section 6.5). We accept without proof the fundamental result that there exists a unique solution of (6.3.1) analytic at t_0 and hence with power series representation, say†

$$\mathbf{x}(t) = \sum_{k=0}^{\infty} \mathbf{x}_k (t-t_0)^k \tag{6.3.2}$$

The question we propose to answer is: How are the constant vectors $\mathbf{x}_k$ related to $A(t)$ and $\mathbf{x}(t_0)$? We know from Taylor's Theorem that

$$\mathbf{x}_k = \frac{\mathbf{x}^{(k)}(t_0)}{k!} \qquad k = 0, 1, 2, \ldots \tag{6.3.3}$$

so to find $\mathbf{x}_k$ we must differentiate (6.3.1) $(k-1)$ times and evaluate the result at t_0. A useful formula for this purpose is known as *Leibnitz's Rule*. (See Exercise 6.3, Part 2, Problem 1.)

$$\frac{d^k}{dt^k}[A(t)\mathbf{x}(t)] = \sum_{i=0}^{k} \binom{k}{i} A^{(i)}(t)\mathbf{x}^{(k-i)}(t)$$

where $\binom{k}{i} = (k!/i!(k-i)!)$ are the standard binomial coefficients. From (6.3.3) we have

$$\mathbf{x}_{k+1} = \frac{\mathbf{x}^{(k+1)}(t_0)}{(k+1)!} = \frac{1}{(k+1)!}\left[\frac{d^k}{dt^k}\mathbf{x}'(t)\right]_{t=t_0}$$

$$= \frac{1}{(k+1)!}\left[\frac{d^k}{dt^k}A(t)\mathbf{x}(t)\right]_{t=t_0}$$

Using Leibnitz's Rule this implies

$$\mathbf{x}_{k+1} = \frac{1}{(k+1)!}\sum_{i=0}^{k} \binom{k}{i} A^{(i)}(t_0)\mathbf{x}^{(k-i)}(t_0) \tag{6.3.4}$$

From Taylor's Theorem

$$A(t) = \sum_{k=0}^{\infty} \frac{A^{(k)}(t_0)}{k!}(t-t_0)^k$$

† For a proof of this theorem see H. Hochstadt, *Differential Equations: A Modern Approach*, Holt, Rinehart & Winston, New York, 1964.

Setting $\mathbf{A}_k = (\mathbf{A}^{(k)}(t_0)/k!)$, and writing $\mathbf{x}_k = (\mathbf{x}^{(k)}(t_0)/k!)$ as in Equation (6.3.3), Equation (6.3.4) may be written as

$$\mathbf{x}_{k+1} = \frac{k!}{(k+1)!} \sum_{i=0}^{k} \mathbf{A}_i \mathbf{x}_{k-i}$$

Therefore,

$$\mathbf{x}_{k+1} = \frac{1}{k+1} \sum_{i=0}^{k} \mathbf{A}_i \mathbf{x}_{k-i} \tag{6.3.5}$$

for $k = 0, 1, \ldots$ and

$$\mathbf{x}_0 = \mathbf{x}(t_0)$$

by hypothesis. Because a unique solution of (6.3.1) analytic at t_0 exists we can assert the following.

6.3.1. THEOREM *Let* $\mathbf{A}(t)$ *be analytic at* t_0 *and have the expansion*

$$\mathbf{A}(t) = \sum_{k=0}^{\infty} \mathbf{A}_k (t-t_0)^k \qquad -r < t-t_0 < r$$

and let $\mathbf{x}_0$ *be any vector. Then the expressions*

$$\begin{aligned} \mathbf{x}_1 &= \mathbf{A}_0 \mathbf{x}_0 \\ \mathbf{x}_2 &= \tfrac{1}{2}\{\mathbf{A}_0 \mathbf{x}_1 + \mathbf{A}_1 \mathbf{x}_0\} \\ &\;\cdot \\ &\;\cdot \\ \mathbf{x}_{k+1} &= \frac{1}{k+1}\{\mathbf{A}_0 \mathbf{x}_k + \mathbf{A}_1 \mathbf{x}_{k-1} + \cdots + \mathbf{A}_k \mathbf{x}_0\} \end{aligned} \tag{6.3.6}$$

determine $\mathbf{x}_1, \mathbf{x}_2, \ldots$ *uniquely and*

$$\mathbf{x}(t) = \sum_{k=0}^{\infty} \mathbf{x}_k (t-t_0)^k \qquad -r < t-t_0 < r$$

is a solution of (6.3.1), *analytic at* t_0.

We see that each Taylor Coefficient of the solution $\mathbf{x}_k$ can (eventually) be computed in terms of $\mathbf{x}_0$, the Taylor Coefficients of $\mathbf{A}(t)$, and $\mathbf{x}_1, \ldots, \mathbf{x}_{k-1}$ which have been computed previously.

6.3.1. Example

$$\mathbf{x}' = \begin{bmatrix} 0 & 1 \\ 0 & 3 \end{bmatrix} \mathbf{x}$$

Solution

A general solution for this system is

$$\mathbf{x}(t) = c_1 \begin{bmatrix} 1 \\ 0 \end{bmatrix} + c_2 \begin{bmatrix} 1 \\ 3 \end{bmatrix} e^{3t}$$

as found by the usual "eigenvalue–eigenvector" method. It is interesting to contrast this with the method of power series, which in this case is particularly simple since

$$A_k = O \qquad k \geq 1$$

Hence (6.3.6) becomes, for $n = k+1$,

$$\begin{aligned}\mathbf{x}_n &= \frac{1}{n}\begin{bmatrix}0 & 1\\0 & 3\end{bmatrix}\mathbf{x}_{n-1}\\ &= \frac{1}{n}\cdot\frac{1}{n-1}\begin{bmatrix}0 & 1\\0 & 3\end{bmatrix}^2\mathbf{x}_{n-2}\\ &= \frac{1}{n!}\begin{bmatrix}0 & 1\\0 & 3\end{bmatrix}^n\mathbf{x}_0\end{aligned}$$

It is obvious and easy to prove by induction that

$$\begin{bmatrix}0 & 1\\0 & 3\end{bmatrix}^n = \begin{bmatrix}0 & 3^{n-1}\\0 & 3^n\end{bmatrix}$$

and hence, setting $\mathbf{x}_0 = \begin{bmatrix}a\\b\end{bmatrix}$,

$$\begin{aligned}\mathbf{x}_n &= \frac{1}{n!}\begin{bmatrix}0 & 3^{n-1}\\0 & 3^n\end{bmatrix}\begin{bmatrix}a\\b\end{bmatrix}\\ &= \frac{3^n}{n!}\begin{bmatrix}\frac{b}{3}\\b\end{bmatrix} \qquad n > 1\end{aligned}$$

Thus,

$$\begin{aligned}\mathbf{x}(t) &= \begin{bmatrix}a\\b\end{bmatrix} + \sum_{m=1}^{\infty}\begin{bmatrix}\frac{b}{3}\\b\end{bmatrix}\frac{3^n t^n}{n!}\\ &= \begin{bmatrix}a\\b\end{bmatrix} + \begin{bmatrix}\frac{b}{3}\\b\end{bmatrix}\sum_{n=1}^{\infty}\frac{(3t)^n}{n!}\end{aligned}$$

Since $e^{3t} - 1 = \sum_{n=1}^{\infty} (3t)^n/n!$,

$$\begin{aligned}\mathbf{x}(t) &= \begin{bmatrix}a\\b\end{bmatrix} + \begin{bmatrix}\frac{b}{3}\\b\end{bmatrix}(e^{3t}-1)\\ &= \begin{bmatrix}a - \frac{b}{3}\\0\end{bmatrix} + \begin{bmatrix}\frac{b}{3}\\b\end{bmatrix}e^{3t}\end{aligned}$$

The choices $a - (b/3) = c_1$ and $b = 3c_2$ reduce this expression to the more familiar form given by the eigenvalue–eigenvector technique.

6.3.2. Example

Consider the problem

$$\mathbf{x}' = \begin{bmatrix} 0 & 5t & 0 \\ 0 & 0 & 3t^2 \\ -1 & t^2 & 0 \end{bmatrix} \mathbf{x} \qquad \mathbf{x}(0) = \begin{bmatrix} 0 \\ 0 \\ 1 \end{bmatrix}$$

Here we find

$$\mathrm{A}(t) = \begin{bmatrix} 0 & 0 & 0 \\ 0 & 0 & 0 \\ -1 & 0 & 0 \end{bmatrix} + t \begin{bmatrix} 0 & 5 & 0 \\ 0 & 0 & 0 \\ 0 & 0 & 0 \end{bmatrix} + t^2 \begin{bmatrix} 0 & 0 & 0 \\ 0 & 0 & 3 \\ 0 & 1 & 0 \end{bmatrix}$$

We set

$$\mathbf{x}(t) = \sum_{k=0}^{\infty} \mathbf{x}_k t^k$$

and determine the coefficients, $\mathbf{x}_k$, as follows:

$$\mathbf{x}_0 = \mathbf{x}(0) = \begin{bmatrix} 0 \\ 0 \\ 1 \end{bmatrix}$$

$$\mathbf{x}_1 = \mathbf{x}'(0) = \mathrm{A}(0)\mathbf{x}(0) = \mathrm{A}_0\mathbf{x}_0$$

$$= \begin{bmatrix} 0 & 0 & 0 \\ 0 & 0 & 0 \\ -1 & 0 & 0 \end{bmatrix} \begin{bmatrix} 0 \\ 0 \\ 1 \end{bmatrix} = \mathbf{0}$$

$$\mathbf{x}_2 = \tfrac{1}{2}\{\mathrm{A}_0\mathbf{x}_1 + \mathrm{A}_1\mathbf{x}_0\}$$

$$= \mathbf{0} + \begin{bmatrix} 0 & 5 & 0 \\ 0 & 0 & 0 \\ 0 & 0 & 0 \end{bmatrix} \begin{bmatrix} 0 \\ 0 \\ 1 \end{bmatrix} = \mathbf{0}$$

$$\mathbf{x}_3 = \tfrac{1}{3}\{\mathrm{A}_0\mathbf{x}_2 + \mathrm{A}_1\mathbf{x}_1 + \mathrm{A}_2\mathbf{x}_0\}$$

$$= \mathbf{0} + \mathbf{0} + \tfrac{1}{3} \begin{bmatrix} 0 & 0 & 0 \\ 0 & 0 & 3 \\ 0 & 1 & 0 \end{bmatrix} \begin{bmatrix} 0 \\ 0 \\ 1 \end{bmatrix} = \begin{bmatrix} 0 \\ 1 \\ 0 \end{bmatrix}$$

Noting that $\mathrm{A}_k = 0$ for $k \geqslant 3$, we have

$$\mathbf{x}_4 = \tfrac{1}{4}\{\mathrm{A}_0\mathbf{x}_3 + \mathrm{A}_1\mathbf{x}_2 + \mathrm{A}_2\mathbf{x}_1\}$$

$$= \tfrac{1}{4} \begin{bmatrix} 0 & 0 & 0 \\ 0 & 0 & 0 \\ -1 & 0 & 0 \end{bmatrix} \begin{bmatrix} 0 \\ 1 \\ 0 \end{bmatrix} = \mathbf{0}$$

$$\mathbf{x}_5 = \tfrac{1}{5}\{\mathrm{A}_0\mathbf{x}_4 + \mathrm{A}_1\mathbf{x}_3 + \mathrm{A}_2\mathbf{x}_2\}$$

$$= \tfrac{1}{5} \begin{bmatrix} 0 & 5 & 0 \\ 0 & 0 & 0 \\ 0 & 0 & 0 \end{bmatrix} \begin{bmatrix} 0 \\ 1 \\ 0 \end{bmatrix} = \begin{bmatrix} 1 \\ 0 \\ 0 \end{bmatrix}$$

Continuing in this manner we find

$$\mathbf{x}_6 = \mathbf{x}_7 = \mathbf{x}_8 = \mathbf{0}$$

Thus

$$\begin{aligned} \mathbf{x}_9 &= \tfrac{1}{9}\{A_0\mathbf{x}_8 + A_1\mathbf{x}_7 + A_2\mathbf{x}_6\} = 0 \\ \mathbf{x}_{10} &= \tfrac{1}{10}\{A_1\mathbf{x}_9 + A_2\mathbf{x}_8 + A_3\mathbf{x}_8\} = 0 \end{aligned}$$

and having established

$$\mathbf{x}_6 = \mathbf{x}_7 = \mathbf{x}_8 = \cdots = \mathbf{x}_N = \mathbf{0}$$

we see that

$$\mathbf{x}_{N+1} = \frac{1}{N+1}\{A_0\mathbf{x}_N + A_1\mathbf{x}_{N-1} + A_2\mathbf{x}_{N-2}\} = \mathbf{0}$$

Thus for $k \geqq 6$, $\mathbf{x}_k = \mathbf{0}$.

Note that even in this concrete example we have used mathematical induction to obtain the answer.

We now have

$$\begin{aligned} \mathbf{x}(t) &= \mathbf{x}_0 + t\mathbf{x}_1 + t^2\mathbf{x}_2 + t^3\mathbf{x}_3 + t^4\mathbf{x}_4 + t^5\mathbf{x}_5 \\ &= \begin{bmatrix} t^5 \\ t^3 \\ 1 \end{bmatrix} \end{aligned}$$

The reader should verify that $\mathbf{x}(t)$ is a solution of the initial-value problem.

6.3.3. Example

We consider the same differential equation with a different initial condition:

$$\mathbf{x}' = \begin{bmatrix} 0 & 5t & 0 \\ 0 & 0 & 3t^2 \\ -1 & t^2 & 0 \end{bmatrix} \mathbf{x} \qquad \mathbf{x}(0) = \begin{bmatrix} 1 \\ 1 \\ 1 \end{bmatrix}$$

Now

$$\mathbf{x}_0 = \begin{bmatrix} 1 \\ 1 \\ 1 \end{bmatrix}$$

$$\mathbf{x}_1 = \begin{bmatrix} 0 & 0 & 0 \\ 0 & 0 & 0 \\ -1 & 0 & 0 \end{bmatrix} \begin{bmatrix} 1 \\ 1 \\ 1 \end{bmatrix} = \begin{bmatrix} 0 \\ 0 \\ -1 \end{bmatrix}$$

$$\begin{aligned} \mathbf{x}_2 &= \tfrac{1}{2}\left\{ \begin{bmatrix} 0 & 0 & 0 \\ 0 & 0 & 0 \\ -1 & 0 & 0 \end{bmatrix} \begin{bmatrix} 0 \\ 0 \\ -1 \end{bmatrix} + \begin{bmatrix} 0 & 5 & 0 \\ 0 & 0 & 0 \\ 0 & 0 & 0 \end{bmatrix} \begin{bmatrix} 1 \\ 1 \\ 1 \end{bmatrix} \right\} \\ &= \begin{bmatrix} \frac{5}{2} \\ 0 \\ 0 \end{bmatrix} \end{aligned}$$

$$\mathbf{x}_3 = \tfrac{1}{3}\{A_0\mathbf{x}_2 + A_1\mathbf{x}_1 + A_2\mathbf{x}_0\}$$

$$= \begin{bmatrix} 0 \\ 1 \\ -\frac{1}{2} \end{bmatrix}$$

$$\mathbf{x}_4 = \tfrac{1}{4}\{A_0\mathbf{x}_3 + A_1\mathbf{x}_2 + A_2\mathbf{x}_1\}$$

$$= \begin{bmatrix} 0 \\ -\frac{3}{4} \\ 0 \end{bmatrix}$$

$$\mathbf{x}_5 = \tfrac{1}{5}\{A_0\mathbf{x}_4 + A_1\mathbf{x}_3 + A_2\mathbf{x}_2\}$$

$$= \begin{bmatrix} 1 \\ 0 \\ 0 \end{bmatrix}$$

$$\mathbf{x}_6 = \tfrac{1}{6}\{A_0\mathbf{x}_5 + A_1\mathbf{x}_4 + A_2\mathbf{x}_3\}$$

$$= \begin{bmatrix} -\frac{5}{8} \\ -\frac{1}{4} \\ 0 \end{bmatrix}$$

At this stage, no particular pattern is obvious. We might, by continuing the process and applying some ingenuity, discover a pattern; or we might be satisfied with the statement

$$\mathbf{x}(t) = \begin{bmatrix} 1 \\ 1 \\ 1 \end{bmatrix} + t\begin{bmatrix} 0 \\ 0 \\ 1 \end{bmatrix} + t^2\begin{bmatrix} \frac{5}{2} \\ 0 \\ 0 \end{bmatrix} + t^3\begin{bmatrix} 0 \\ 1 \\ -\frac{1}{2} \end{bmatrix}$$

$$+ t^4\begin{bmatrix} 0 \\ -\frac{3}{4} \\ 0 \end{bmatrix} + t^5\begin{bmatrix} 1 \\ 0 \\ 0 \end{bmatrix} + t^6\begin{bmatrix} -\frac{5}{8} \\ -\frac{1}{4} \\ 0 \end{bmatrix}$$

$$+ \text{terms containing at least } t^7$$

Exercise 6.3

Part 1

Find $\mathbf{x}_0, \mathbf{x}_1, \mathbf{x}_2$, and $\mathbf{x}_3$ for the solutions $\mathbf{x}(t) = \Sigma_{k=0}^{\infty} \mathbf{x}_k (t - t_0)^k$ of the following:

1. $\mathbf{x}' = \begin{bmatrix} 1 & t \\ t & 1 \end{bmatrix}\mathbf{x} \qquad \mathbf{x}(0) = \begin{bmatrix} 1 \\ 0 \end{bmatrix}$

2. $\mathbf{x}' = \begin{bmatrix} 1 & t \\ t & 1 \end{bmatrix}\mathbf{x} \qquad \mathbf{x}(0) = \begin{bmatrix} 0 \\ 1 \end{bmatrix}$

3. $\mathbf{x}' = \begin{bmatrix} 1 & t \\ t & 1 \end{bmatrix}\mathbf{x} \qquad \mathbf{x}(0) = \begin{bmatrix} a \\ b \end{bmatrix}$

4. $\mathbf{x}' = \begin{bmatrix} 1 & t^2 \\ t^2 & 1 \end{bmatrix}\mathbf{x} \qquad \mathbf{x}(0) = \begin{bmatrix} a \\ b \end{bmatrix}$

5. $\mathbf{x}' = \begin{bmatrix} 1 & t & 0 \\ 0 & 0 & 0 \\ 1 & 0 & t \end{bmatrix}\mathbf{x} \qquad \mathbf{x}(0) = \begin{bmatrix} 1 \\ 0 \\ 0 \end{bmatrix}$

6. $\mathbf{x}' = \begin{bmatrix} 1 & t & 0 \\ 0 & 0 & 0 \\ 1 & 0 & t \end{bmatrix}\mathbf{x} \qquad \mathbf{x}(0) = \begin{bmatrix} 0 \\ 1 \\ 0 \end{bmatrix}$

7. $\mathbf{x}' = (1-t)\begin{bmatrix} 0 & 1 & 0 \\ 0 & 0 & 1 \\ 0 & 1 & -1 \end{bmatrix}\mathbf{x} \qquad \mathbf{x}(0) = \begin{bmatrix} 1 \\ 1 \\ 1 \end{bmatrix}$

In the next three problems find an expression for $\mathbf{x}_k$.

8. $\mathbf{x}' = \begin{bmatrix} 0 & 1 & 0 \\ -1 & 0 & 0 \\ 0 & 0 & 1 \end{bmatrix}\mathbf{x} \qquad \mathbf{x}(0) = \begin{bmatrix} 1 \\ 0 \\ 1 \end{bmatrix}$

9. $x' = x \qquad x(0) = 1$

10. $x'' + x = 0 \qquad x(0) = 1 \qquad x'(0) = 0$

(Use the companion system.)

Part 2

1. Let f and g be given functions of t, each infinitely differentiable. Prove, by induction:

$$\frac{d^n}{dt^n}(f \cdot g) = \sum_{k=0}^{n} \binom{n}{k} f^{(n)} g^{(n-k)}$$

2. Differentiate (6.3.2) and substitute into (6.3.1). Assuming

$$A(t) = \sum_{n=0}^{\infty} A_k (t-t_0)^k$$

derive (6.3.5) by equating coefficients of like powers of $(t-t_0)$.

3. Given $\mathbf{x}' = A(t)\mathbf{x}$, $\mathbf{x}(t_0) = \mathbf{x}_0$, let $\mathbf{x}_1, \mathbf{x}_2, \ldots, \mathbf{x}_N$ be computed according to (6.3.5). Show: If

$$\mathbf{z}(t) = \mathbf{x}_0 + (t-t_0)\mathbf{x}_1 + \cdots + (t-t_0)^N \mathbf{x}_N$$

then

$$\mathbf{z}'(t) - A(t)\mathbf{z}(t) = (t-t_0)^N \mathbf{q}(t)$$

for some function $\mathbf{q}(t)$ analytic at t_0.

4. Apply (6.3.5) to the case $A(t) = A$, a constant matrix and thus derive

$$\mathbf{x}(t) = \sum_{k=0}^{\infty} \frac{(t-t_0)^k}{k!} A^k \mathbf{x}_0$$

solves $\mathbf{x}' = A\mathbf{x}, \mathbf{x}(t_0) = \mathbf{x}_0$.

Let $\mathbf{z}' = A(t)\mathbf{z}$, where

$$A(t) = \begin{bmatrix} 0 & 1 \\ -\sum_{k=0}^{\infty} q_k t^k & -\sum_{k=0}^{\infty} p_k t^k \end{bmatrix}$$

and

$$\mathbf{z}_k = \begin{bmatrix} y_k \\ x_k \end{bmatrix}$$

and

$$\mathbf{z}(t) = \sum_{k=0}^{\infty} \mathbf{z}_k t^k$$

be standard notation for the next three problems.

5. Use Equations (6.3.6) to derive

$$\mathbf{z}_{k+1} = \begin{bmatrix} y_{k+1} \\ x_{k+1} \end{bmatrix} = \frac{1}{k+1} \begin{bmatrix} x_k \\ -\sum_{i=0}^{k} (q_{k-i} y_i + p_{k-i} x_i) \end{bmatrix}$$

$k = 0, 1, 2, \ldots$

6. Use the result of Problem 5 to show

$$y_{k+1} = -\frac{1}{k(k+1)} \sum_{i=0}^{k} (q_{k-1-i} + k p_{k-i}) y_{k-i}$$

where $q_{-1} = 0$ and $k = 0, 1, 2, \ldots$.

7. Use the result of Problem 6 to show that

$$y(t) = \sum_{k=0}^{\infty} y_k t^k$$

solves $y'' + p(t) y' + q(t) y = 0$ implies

$$y_{k+1} = -\frac{1}{k(k+1)} \sum_{i=0}^{k} (q_{k-1-i} + k i p_{k-i}) y_i$$

$q_{-1} = 0$, $k = 0, 1, 2, \ldots$. Hint: Look at the companion system and write $p(t) = \Sigma_{k=0}^{\infty} p_k t^k$, $q_k(t) = \Sigma_{k=0}^{\infty} q_k t^k$.

8. Suppose

$$y(t) = \sum_{k=0}^{\infty} y_k t^k$$

solves

$$y'' - t y' - y = 0 \qquad y(0) = 1 \qquad y'(0) = 0$$

Find the companion matrix and solve the companion system by the method of this section. Hence show that

$$y(t) = 1 + \sum_{k=0}^{\infty} \frac{t^{2k}}{2^n k!}$$

9. Find two solutions, one satisfying $y(0) = 1$, $y'(0) = 0$ and the other satisfying $y(0) = 0$, $y'(0) = 1$ for the following equations:

(a) $y'' - t^2 y = 0$ (b) $y'' - ty = 0$
(c) $y'' = 0$ (d) $y'' + e^t y = 0$

Use the method of this section on the companion system to these second-order equations.

6.4. THE LEGENDRE EQUATION: AN EXAMPLE OF THE ORDINARY POINT FOR ONE-DIMENSIONAL EQUATIONS

What we have described until now is a fairly general approach to analytic systems of differential equations at ordinary points. Frequently special equations are of special interest and are amenable to special techniques. Many one-dimensional second-order equations fit this category. We discuss below one such example as an illustration of the basic ideas.

The so-called *Legendre Equation* is actually a family of equations parametrized by the variable α and defined for each value of α as

$$(1-t^2)y'' - 2ty' + \alpha(\alpha+1)y = 0 \tag{6.4.1}$$

We could attack the Legendre Equation by studying its companion system via the techniques of Section 6.3. However, it is often more convenient to study specific one-dimensional problems directly. We use the Legendre Equation as an example to illustrate this alternative method. Let

$$y(t) = \sum_{k=0}^{\infty} y_k t^k \tag{6.4.2}$$

be a solution of the Legendre Equation (6.4.1). Differentiating Equation (6.4.2) twice we obtain

$$y'(t) = \sum_{k=0}^{\infty} k y_k t^{k-1}$$

$$y''(t) = \sum_{k=0}^{\infty} k(k-1) y_k t^{k-2}$$

so that upon substitution into equation 6.4.1 we obtain

$$\sum_{k=0}^{\infty} k(k-1)y_k t^{k-2} + \sum_{k=0}^{\infty} -k(k-1)y_k t^k + \sum_{k=0}^{\infty} -2k y_k t^k + \sum_{k=0}^{\infty} \alpha(\alpha+1) y_k t^k \equiv 0$$

Combining the latter three series leads to

$$\sum_{k=0} k(k-1)y_k t^{k-2} - \sum_{k=0}^{\infty} \{k^2 + k - \alpha(\alpha+1)\} y_k\, t^k \equiv 0 \tag{6.4.3}$$

Also

$$\sum_{k=0}^{\infty} k(k-1)y_k t^{k-2} = \sum_{k=2}^{\infty} k(k-1) y_k t^{k-2} = \sum_{k=0}^{\infty} (k+2)(k+1) y_{k+2} t^k$$

the first equality holds since we have removed from the sum only terms with value zero, the second follows by replacement of k by $k+2$. Thus (6.4.3) becomes

$$\sum_{k=0}^{\infty} \{(k+2)(k+1)y_{k+2} - [k^2+k-\alpha(\alpha+1)]y_k\}t^k \equiv 0$$

which is possible if and only if

$$(k+2)(k+1)\,y_{k+2} = [k^2+k-\alpha(\alpha+1)]\,y_k$$

for $k = 0, 1, 2, \ldots$.
Or in neater form,

$$y_{k+2} = \frac{(k-\alpha)(k+\alpha+1)}{(k+2)(k+1)}\,y_k \tag{6.4.4}$$

for $k = 0, 1, 2, \ldots$.
Let us consider the case where $\alpha = 2r$, an even integer, and the solution, $y(t)$, satisfies the initial conditions $y(0) = 1$, $y'(0) = 0$. Now $y_1 = y'(0) = 0$ and thus from Equation (6.4.4), $y_3 = y_1 = 0$, which in turn implies $y_5 = y_3 = 0$ and so on for all odd-subscripted y_k. For the even-subscripted y_k we see that when $k = 2r$, $y_{2r+2} = 0$ (recall $\alpha = 2r$) and hence $y_{2r+2} = y_{2r+4} = \cdots = 0$. The remaining y_k, namely, $y_0, y_2, \ldots, y_{2r}$ are all unequal to zero and therefore $y(t)$ is a polynomial of degree $2r = \alpha$. Similarly, if $\alpha = 2r+1$, an odd integer, and $y(t)$ is a solution of (6.4.1) with $y(0) = 0$, $y'(0) = 1$ then, as above, $y(t)$ will be a polynomial of degree $\alpha = 2r+1$. We can therefore assert; *the Legendre Equation always has a polynomial solution of degree α whenever α is a nonnegative integer.* These polynomial solutions to Legendre's Equation are constant multiples of the so-called *Legendre Polynomials*

$$P_k(t) = \frac{1}{2^k k!}\frac{d^k}{dt^k}\{(t^2-1)^k\}$$

Showing this will be the aim of some of the problems in Part 2 of the following exercise set.

Exercise 6.4

Part 1

1. Assume $\alpha = 2r+1$ in Equation (6.4.4) and $y(0) = 0$, $y'(0) = 1$. Show the following:

   (a) $y_0 = y_2 = y_4 = \cdots 0$
   (b) $y_{2r+3} = y_{2r+5} = \cdots = 0$
   (c) $y_1, y_3, y_5 \cdots y_{2r+1}$

   are each not zero.

In the next seven problems assume a power series solution and obtain its coefficients by following the procedure illustrated in the text for Legendre's Equation.

2. $y''+y=0 \qquad y(0)=1 \qquad y(0)=0$
   (Compare with Exercise 6.3, Part 1, Problem 9.)
3. $y''-ty'-y=0 \qquad y(0)=1 \qquad y'(0)=0$
   (Compare with Exercise 6.3, Part 2, Problem 8.)
4. (a) $y''-t^2y=0$
   (b) $y''-ty=0$
   (Compare with Exercise 6.3, Part 2, Problem 9.)
5. Use Equation (6.4.4) to express y_{2k} in terms of y_0, assuming α is not an integer.
6. Same as Problem 5 except express y_{2r+1} in terms of y_1.
7. The equation

$$y''-2ty'+2\alpha y=0$$

   is called Hermite's Equation. If α is a positive integer there is a polynomial solution. Find it!
8. If α is not an integer there are two linearly independent solutions of Hermite's Equation as given in Problem 7. Find them!

Part 2

1. Let $u(t)=(t^2-1)^k$. Show that $u(t)$ is a solution of $(t^2-1)y'-2kty=0$.
2. Using Leibnitz' Formula, show that

$$\frac{d^{k+1}}{dt^{k+1}}\{(t^2-1)u'-2ktu\}=(t^2-1)u^{(k+2)}+2tu^{(k+1)}-k(k+1)u^{(k)}$$

3. Setting $z(t)=u^{(k)}(t)$, use Problem 2 to show that $z(t)$ satisfies (6.4.1) for $\alpha=k$.
4. Use Leibnitz' Formula to show that

$$z(t)=\frac{d^k}{dt^k}u(t)=\frac{d^k}{dt^k}\{(t-1)^k(t+1)^k\}=k!\sum_{j=0}^{k}\binom{k}{j}^2(t-1)^{k-j}(t+1)^j$$

5. Use Problem 4 to show that $z(1)=k!\,2^k$, and thus $P_k(t)$ satisfies $P_k(1)=1$ as well as (6.4.1).
6. Show that if $\alpha=k$ is a nonnegative integer, (6.4.1) has a solution which is *not* a polynomial.
7. Show that for $\alpha=n$, if $y(t)$ is a polynomial solution of (6.4.1) then $x(t)=KP_k(t)$ for some constant K.

6.5. THE EXPONENTIAL OF A MATRIX

The special case $A(t)=A$, A a constant, $n\times n$ matrix, reduces Equations (6.3.6) to the particularly simple statements:

$$\mathbf{x}_1=\mathbf{A}\mathbf{x}_0$$
$$\mathbf{x}_2=\tfrac{1}{2}\mathbf{A}\mathbf{x}_1$$
$$\vdots$$
$$\mathbf{x}_k=\frac{1}{k}\mathbf{A}\mathbf{x}_{k-1}$$

By repeated "back" substitution $\mathbf{x}_k$ may be written explicitly in terms of $\mathbf{x}_0$, to wit:

$$\mathbf{x}_k = \frac{1}{k}\mathrm{A}\mathbf{x}_{k-1} = \frac{1}{k}\mathrm{A}\left[\frac{1}{k-1}\mathrm{A}\mathbf{x}_{k-2}\right]$$

$$= \frac{1}{k(k-1)}\mathrm{A}^2\mathbf{x}_{k-2}$$

$$= \cdots = \frac{1}{k!}\mathrm{A}^k\mathbf{x}_0$$

Hence,

$$\mathbf{x}(t) = \sum_{k=0}^{\infty} \frac{(t-t_0)^k}{k!}\mathrm{A}^k\mathbf{x}_0 \tag{6.5.1}$$

is the unique solution of

$$\mathbf{x}' = \mathrm{A}\mathbf{x} \qquad \mathbf{x}(t_0) = \mathbf{x}_0 \tag{6.5.2}$$

6.5.1. Example

Use (6.5.1) to solve

$$\mathbf{x}' = \begin{bmatrix} 0 & 1 \\ 1 & 0 \end{bmatrix}\mathbf{x} \qquad \mathbf{x}(0) = \begin{bmatrix} 1 \\ 1 \end{bmatrix}$$

Solution

We set

$$\mathrm{A} = \begin{bmatrix} 0 & 1 \\ 1 & 0 \end{bmatrix}$$

and compute

$$\mathrm{A}^2 = \mathrm{I}, \quad \mathrm{A}^3 = \mathrm{A}_0, \quad \mathrm{A}^4 = \mathrm{I}, \quad \ldots$$

Thus,

$$\mathrm{A}^{2k}\mathbf{x}_0 = \mathbf{I}\mathbf{x}_0 = \begin{bmatrix} 1 \\ 1 \end{bmatrix}$$

while

$$\mathrm{A}^{2k+1}\mathbf{x}_0 = \mathrm{A}\mathbf{x}_0 = \begin{bmatrix} 1 \\ 1 \end{bmatrix}$$

Substituting into (6.5.1), we find

$$\mathbf{x}(t) = \sum_{k=0}^{\infty} \frac{t^k}{k!}\begin{bmatrix} 1 \\ 1 \end{bmatrix} = \begin{bmatrix} 1 \\ 1 \end{bmatrix} e^t$$

because

$$e^t = \sum_{k=0}^{\infty} \frac{t^k}{k!} \tag{6.5.3}$$

The power series method is thus seen to be an alternative to the eigenvalue–eigenvector method used in the earlier chapters. Like the eigenvalue–eigenvector method, its ease of application is closely connected to the simplicity of A.

There is another aspect of the power series approach to the constant coefficient problem which is of some interest. Set $t_0 = 0$ in Equations (6.5.1) and (6.5.2) so that

$$\mathbf{x}(t) = \sum_{k=0}^{\infty} \frac{t^k}{k!} \mathrm{A}^k \mathbf{x}_0 \tag{6.5.4}$$

is the unique solution of

$$\mathbf{x}' = \mathrm{A}\mathbf{x} \qquad \mathbf{x}(0) = \mathbf{x}_0 \tag{6.5.5}$$

A comparison of Equations (6.5.4) and (6.5.3) suggests a definition for $e^{t\mathrm{A}}$; specifically,

$$e^{t\mathrm{A}}\mathbf{u} = \sum_{n=0}^{\infty} \frac{t^k}{k!} \mathrm{A}^k \mathbf{u} \tag{6.5.6}$$

If the left-hand side were to be viewed solely as an abbreviation for the right-hand side, this definition would be a convenient mnemonic device for remembering the solution of (6.5.5) as $e^{t\mathrm{A}}\mathbf{x}_0$. What makes it particularly exciting is that it suggests studying functions of a matrix. (In fact, polynomial functions of A have already been introduced in Section 3.5.) Does $e^{t\mathrm{A}}$ behave analogously to e^{ta}? Can we make sense of $e^{t\mathrm{A}} \cdot e^{t\mathrm{B}}$, $(e^{t\mathrm{A}})^{-1}$, etc? Instead of delving deeply into these questions, we only remark that by Theorem 4.2.5 and Theorem 4.2.8,

$$\mathbf{x}(t) = \Phi(t)\mathbf{x}_0$$

is also the unique solution of $\mathbf{x}' = \mathrm{A}\mathbf{x}$, $\mathbf{x}(0) = \mathbf{x}_0$ if Φ is a fundamental matrix of A normalized at $t = 0$. Therefore,

$$e^{t\mathrm{A}}\mathbf{x}_0 \equiv \sum_{k=0}^{\infty} \frac{t^k}{k!} \mathrm{A}^k \mathbf{x}_0 = \Phi(t)\mathbf{x}_0$$

An alternative definition of $e^{t\mathrm{A}}$ is thus

$$e^{t\mathrm{A}} \equiv \Phi(t) \tag{6.5.7}$$

where Φ is the normalized fundamental matrix of A at $t = 0$. Equation (6.5.7) is meant to be interpreted as

$$e^{tA}\mathbf{u} = \Phi(t)\mathbf{u}$$

for each constant vector **u** with n entries.

6.5.2. Example

Show that

$$e^{(t+s)A} = e^{tA} \cdot e^{sA} \tag{6.5.8}$$

Solution

The proposed equality should be understood in the following sense: Given any constant vector **u**,

$$e^{(t+s)A}\mathbf{u} = \mathbf{v}$$

is a vector defined by (6.5.6) with $t+s$ replacing t. Similarly $e^{sA}\mathbf{u} = \mathbf{w}$ is interpreted via (6.5.6) with s replacing t and

$$e^{tA}\mathbf{w} \equiv e^{tA}[e^{sA}\mathbf{u}]$$

Then Equation (6.5.8) means: for every **u**

$$e^{tA}\mathbf{w} = \mathbf{v}$$

We have remarked above that

$$e^{tA}\mathbf{u} = \Phi(t)\mathbf{u}$$

so that

$$\begin{aligned} e^{(t+s)A}\mathbf{u} &= \Phi(t+s)\mathbf{u} \\ &= \Phi(t)\Phi(s)\mathbf{u} \end{aligned}$$

by Theorem 4.3.1. But

$$\begin{aligned} \Phi(t)\Phi(s)\mathbf{u} &= \Phi(t)\,[\Phi(s)\mathbf{u}] \\ &= \Phi(t)\,[e^{sA}\mathbf{u}] \\ &= e^{tA}[e^{sA}\mathbf{u}] \end{aligned}$$

which completes the proof.

Exercise 6.5

Part 1

In Problems 1–5 use the power series to solve the given systems as outlined in Example 6.5.1.

1. $\mathbf{x}' = \begin{bmatrix} 1 & 0 \\ 0 & 1 \end{bmatrix}\mathbf{x} \qquad \mathbf{x}(0) = \begin{bmatrix} 1 \\ 0 \end{bmatrix}$

2. $\mathbf{x}' = \begin{bmatrix} 0 & 1 & 0 \\ 0 & 0 & 1 \\ 0 & 0 & 0 \end{bmatrix}\mathbf{x} \qquad \mathbf{x}(0) = \begin{bmatrix} 1 \\ 0 \\ 1 \end{bmatrix}$

3. $y''+y=0 \qquad y(0)=0 \qquad y'(0)=1$

4. $$\mathbf{x}' = \begin{bmatrix} 1 & 1 & 0 \\ 0 & 1 & 0 \\ 0 & 0 & 0 \end{bmatrix} \mathbf{x} \qquad \mathbf{x}(0) = \begin{bmatrix} 1 \\ 1 \\ 0 \end{bmatrix}$$

5. $y' = 2y \qquad y(0) = -1$

Part 2

1. Show that the solution of

$$\mathbf{x}' = \begin{bmatrix} 0 & 1 \\ 1 & 0 \end{bmatrix} \mathbf{x} \qquad \mathbf{x}(0) = \mathbf{x}_0$$

may be written as

$$\mathbf{x}(t) = \begin{bmatrix} \sinh t & \cosh t \\ \cosh t & \sinh t \end{bmatrix} \mathbf{x}_0$$

by using the power series method.

2. Suppose A is diagonal

$$\mathrm{A} = \begin{bmatrix} d_{11} & 0 & \cdots 0 \\ 0 & d_{22} & \cdots 0 \\ \cdot & \cdot & \cdot \\ \cdot & \cdot & \cdot \\ \cdot & \cdot & \cdot \\ 0 & 0 & \cdots d_{nn} \end{bmatrix}$$

Evaluate $e^{\mathrm{A}t}$ and verify that $\mathbf{x}(t) = e^{\mathrm{A}t}\mathbf{x}_0$ solves

$$\mathbf{x}' = \mathrm{A}\mathbf{x} \qquad \mathbf{x}(0) = \mathbf{x}_0$$

6.6.* THE REGULAR SINGULAR POINT

If $\mathrm{A}(t)$ is not analytic at t_0 then

$$\mathbf{x}' = \mathrm{A}(t)\mathbf{x}$$

is said to have a *singular* point at t_0. In our work we usually deal with matrices $\mathrm{A}(t)$ which are analytic at all but a few exceptional points where $\mathrm{A}(t)$ fails to be analytic. In the previous sections we have confined our attention to series solutions about ordinary points—points where $\mathrm{A}(t)$ is analytic. In this section we investigate the possibility of a series solutions at certain kinds of singular points. These special singular points are selected on the basis that a form of series solution succeeds in yielding at least one solution. We confine ourselves to singular points at $t_0 = 0$. Singular points occurring elsewhere may be transformed to the origin by the change of variables $t - t_0 = \mathrm{T}$.

6.6.1. DEFINITION *The point* $t = 0$ *is a regular singular point of* $\mathbf{x}' = \mathrm{A}(t)\mathbf{x}$ *if*

$$A(t) = \frac{1}{t} \sum_{k=0}^{\infty} B_k t^k \qquad B_0 \neq 0 \tag{6.6.1}$$

where

$$B(t) \equiv \sum_{k=0}^{\infty} B_k t^k \tag{6.6.2}$$

is analytic at $t = 0$.

The restriction, $B_0 \neq 0$, is made to insure that $A(t)$ is not analytic at $t = 0$.

The regular singular point is selected from all possible singular points for two reasons: (1) we have a complete theory of $\mathbf{x}' = A(t)\mathbf{x}$ at a regular singular point—although we touch only the highlights here, and (2) they occur for some of the most important differential equations of mathematical physics.

6.6.1. Example

The following systems and equations have a regular singular point at $t = 0$:

(1) $y' = \frac{c}{t} y$

(2) $\mathbf{x}' = \frac{1}{t} \mathbf{B}\,\mathbf{x}$ B a constant $n \times n$ matrix

(3) $x' = \frac{1}{t}[I + C(t)]\mathbf{x}$ $C(t)$ an $n \times n$ matrix analytic at $t = 0$

The equation $y' = (c/t)y$, whose solution is $y(t) = kt^c$, shows that we cannot expect a solution of the form $\Sigma \mathbf{x}_k t^k$ at all regular singular points. We must generalize the Taylor Series somewhat. Specifically, suppose

$$\mathbf{x}(t) = t^c \sum_{k=0}^{\infty} \mathbf{x}_k t^k \qquad \mathbf{x}_0 \neq \mathbf{0} \tag{6.6.3}$$

where $\Sigma\, \mathbf{x}_k t^k$ converges in some interval, $-r < t < r$. Such a series is called a *Frobenius Series* after the mathematician who did much to clarify the study of the regular singular point. If we assume $\mathbf{x}(t)$ as given by (6.6.3) is a solution of

$$\mathbf{x}' = \frac{1}{t} B(t)\mathbf{x} \tag{6.6.4}$$

then substitution into (6.6.4) yields

$$\sum_{k=0}^{\infty} (k+c)\mathbf{x}_k t^{k+c-1} \equiv \frac{1}{t}\left(\sum_{k=0}^{\infty} B_k t^k\right) \times \left(\sum_{i=0}^{\infty} \mathbf{x}_i t^i\right) t^c$$

$$\equiv \sum_{k=0}^{\infty} \left(\sum_{i=0}^{k} B_i \mathbf{x}_{k-i}\right) t^{k+c-1}$$

If we identify the coefficients of these two series we obtain

$$\begin{aligned} c\mathbf{x}_0 - \mathrm{B}_0\mathbf{x}_0 &= \mathbf{0} \quad \text{when} \quad k=0 \\ (1+c)\mathbf{x}_1 - \mathrm{B}_0\mathbf{x}_1 &= \mathrm{B}_1\mathbf{x}_0 \quad \text{when} \quad k=1 \\ (2+c)\mathbf{x}_2 - \mathrm{B}_0\mathbf{x}_2 &= \mathrm{B}_1\mathbf{x}_1 + \mathrm{B}_2\mathbf{x}_0 \quad \text{when} \quad k=2 \\ (k+c)\mathbf{x}_k - \mathrm{B}_0\mathbf{x}_k &= \sum_{i=1}^{k} \mathrm{B}_i\mathbf{x}_{k-i} \quad \text{any} \quad k \end{aligned} \tag{6.6.5}$$

The first equation

$$(\mathrm{B}_0 - c\mathrm{I})\mathbf{x}_0 = \mathbf{0}$$

is called the *indicial equation* of $\mathbf{x}' = (1/t)\mathrm{B}(t)\mathbf{x}$ at the regular singular point, $t=0$. Since $\mathbf{x}_0 \neq \mathbf{0}$, the indicial equation is solved for any eigenvalue c and corresponding eigenvector $\mathbf{x}_0$. Once c and $\mathbf{x}_0$ are specified the equations

$$(\mathrm{B}_0 - (c+k)\mathrm{I})\mathbf{x}_k = -\sum_{i=1}^{k} \mathrm{B}_i\mathbf{x}_{k-i} \qquad k \geqslant 1 \tag{6.6.6}$$

determines $\mathbf{x}_k$ in terms of $\mathbf{x}_0, \mathbf{x}_1, \ldots, \mathbf{x}_{k-1}$ *unless* $\mathrm{B}_0 - (c+k)\mathrm{I}$ is singular. Indeed, $\mathrm{B}_0 - (c+k)\mathrm{I}$ will be singular if B_0 has an eigenvalue $\hat{c}$ and $-c+\hat{c} = \mathrm{N}$, N a positive integer. For, when $k=\mathrm{N}$, $\mathrm{B}_0 - (c+\mathrm{N})\mathrm{I} = \mathrm{B}_0 - \hat{c}\mathrm{I}$. In this case

$$(\mathrm{B}_0 - \hat{c}\mathrm{I})\mathbf{x}_\mathrm{N} = -\sum_{i=1}^{\mathrm{N}} \mathrm{B}_i\mathbf{x}_{\mathrm{N}-i}$$

may be inconsistent. We can always avoid this difficulty for at least one choice of the eigenvalue c. To wit, suppose the eigenvalues of the $n \times n$ matrix B_0 are $c_1, c_2, \ldots, c_n$, allowing repetitions. We select c from this set so that for each i, $i = 1, 2, \ldots, n$, $c_i - c$ is not a positive integer and hence $\mathrm{B}_0 - (c+k)\mathrm{I}$ is invertible† for each $k \geqslant 1$. It is known that the resulting series

$$\mathbf{x}(t) = \sum_{k=0}^{\infty} \mathbf{x}_k t^{k+c}$$

will represent a solution of $\mathbf{x}' = (1/t)\mathrm{B}(t)\mathbf{x}$ analytic at $t=0$. We accept this theorem without proof. The following examples illustrate the preceding ideas.

6.6.2. Example

Find a series solution of

$$\mathbf{x}' = \frac{1}{t}\begin{bmatrix} 0 & 1 \\ t & 0 \end{bmatrix}\mathbf{x} \qquad \text{about} \quad t=0$$

† See Exercise 6.6, Part 2, Problem 4.

Solution

The point $t=0$ is a regular singular point and

$$\begin{aligned} B(t) &= \begin{bmatrix} 0 & 1 \\ 0 & 0 \end{bmatrix} + t\begin{bmatrix} 0 & 0 \\ 1 & 0 \end{bmatrix} \\ &= B_0 + tB_1 \end{aligned}$$

The eigenvalues of B_0 are both zero and the one-dimensional eigenspace is $\alpha\begin{bmatrix} 1 \\ 0 \end{bmatrix}$. We have, therefore, $c=0$ and $\mathbf{x}_0 = \begin{bmatrix} 1 \\ 0 \end{bmatrix}$. From Equations (6.6.6) we obtain

$$\begin{bmatrix} -k & 1 \\ 0 & -k \end{bmatrix}\mathbf{x}_k = -\begin{bmatrix} 0 & 0 \\ 1 & 0 \end{bmatrix}\mathbf{x}_{k-1} \qquad k \geqslant 1$$

We solve for $\mathbf{x}_k$ in terms of $\mathbf{x}_0$ via the steps:

$$\begin{aligned} \mathbf{x}_k &= \begin{bmatrix} \frac{1}{k^2} & 0 \\ \frac{1}{k} & 0 \end{bmatrix}\mathbf{x}_{k-1} \\ &= \begin{bmatrix} \frac{1}{k^2} & 0 \\ \frac{1}{k} & 0 \end{bmatrix}\begin{bmatrix} \frac{1}{(k-1)^2} & 0 \\ \frac{1}{k-1} & 0 \end{bmatrix}\mathbf{x}_{k-2} \\ &= \begin{bmatrix} \frac{1}{k^2} & 0 \\ \frac{1}{k} & 0 \end{bmatrix}\begin{bmatrix} \frac{1}{(k-1)^2} & 0 \\ \frac{1}{k-1} & 0 \end{bmatrix}\cdots\begin{bmatrix} 1 & 0 \\ 1 & 0 \end{bmatrix}\mathbf{x}_0 \end{aligned}$$

If we multiply together the matrices forming the coefficient of $\mathbf{x}_0$ we obtain

$$\mathbf{x}_k = \begin{bmatrix} \frac{1}{(k!)^2} & 0 \\ \frac{k}{(k!)^2} & 0 \end{bmatrix}\begin{bmatrix} 1 \\ 0 \end{bmatrix} = \frac{1}{(k!)^2}\begin{bmatrix} 1 \\ k \end{bmatrix} \qquad k \geqslant 1$$

Thus

$$\begin{aligned} \mathbf{x}(t) &= \mathbf{x}_0 + \sum_{k=1}^{\infty} \mathbf{x}_k t^k \\ &= \mathbf{x}_0 + \sum_{k=1}^{\infty} \frac{t^k}{(k!)^2}\begin{bmatrix} 1 \\ k \end{bmatrix} = \sum_{k=0}^{\infty} \frac{t^k}{(k!)^2}\begin{bmatrix} 1 \\ k \end{bmatrix} \end{aligned}$$

This series converges for all t as the ratio test shows. There is no other solution of the type we have been discussing because (1) there is no other eigenvalue, and (2) the eigenvalue, 0, has only a one-parameter family of eigenvectors, $\alpha\begin{bmatrix}1\\0\end{bmatrix}$ associated with it.† Therefore the only possible solutions of this type are $\alpha\mathbf{x}(t)$ where $\mathbf{x}(t)$ is given above.

6.6.3. Example

Find a solution of

$$\mathbf{x}' = \frac{1}{t}\begin{bmatrix}0 & 1\\ t & 1\end{bmatrix}\mathbf{x}$$

about $t = 0$.

Solution

Here

$$\mathrm{B}(t) = \mathrm{B}_0 + t\mathrm{B}_1 = \begin{bmatrix}0 & 1\\ 0 & 1\end{bmatrix} + t\begin{bmatrix}0 & 0\\ 1 & 0\end{bmatrix}$$

and the eigenvalues of B_0 are $c_2 = 0$ and $c_1 = 1$.

The choice $c_1 = 1$ must lead to a solution; we save the computation for Exercise 6.6, Part 1, Problem 1. Let us examine the case $c_2 = 0$. Then

$$\mathbf{x}_0 = \begin{bmatrix}1\\0\end{bmatrix}$$

and

$$\begin{bmatrix}-k & 1\\ 0 & -k+1\end{bmatrix}\mathbf{x}_k = -\begin{bmatrix}0 & 0\\ 1 & 0\end{bmatrix}\mathbf{x}_{k-1} \qquad k \geqslant 1$$

As expected, when $k = 1$, $c_2 + k$ becomes the larger eigenvalue $c_1 = 1$, and the system may be inconsistent. In fact, it is! For,

$$\begin{bmatrix}-1 & 1\\ 0 & 0\end{bmatrix}\mathbf{x}_1 = -\begin{bmatrix}0 & 0\\ 1 & 0\end{bmatrix}\begin{bmatrix}1\\0\end{bmatrix} = \begin{bmatrix}0\\1\end{bmatrix}$$

There is no solution corresponding to $c_2 = 0$, of this type.

6.6.4. Example

Find two solutions of

$$\mathbf{x}' = \frac{1}{t}\begin{bmatrix}t & 1\\ 0 & 1\end{bmatrix}\mathbf{x}$$

Solution

Here

$$\mathrm{B}(t) = \mathrm{B}_0 + t\mathrm{B}_1 = \begin{bmatrix}0 & 1\\ 0 & 1\end{bmatrix} + t\begin{bmatrix}1 & 0\\ 0 & 0\end{bmatrix}$$

† For a discussion of other types of solutions see W. Wasow, *Asymptotic Expansions for Ordinary Differential Equations*, Wiley (Interscience), New York, 1965.

The eigenvalues of B_0 are 0 and 1 and the case $c = 1$ is given in Exercise 6.6, Part 1, Problem 2. When $c = 0$, $\mathbf{x}_0 = \begin{bmatrix} \alpha \\ 0 \end{bmatrix}$ and, again,

$$\begin{bmatrix} -k & 1 \\ 0 & 1-k \end{bmatrix} \mathbf{x}_k = -\begin{bmatrix} 1 & 0 \\ 0 & 0 \end{bmatrix} \mathbf{x}_{k-1} \qquad k \geqslant 1$$

In this case, however, when $k = 1$,

$$\begin{bmatrix} -1 & 1 \\ 0 & 0 \end{bmatrix} \mathbf{x}_1 = -\begin{bmatrix} 1 & 0 \\ 0 & 0 \end{bmatrix} \mathbf{x}_0 = -\begin{bmatrix} \alpha \\ 0 \end{bmatrix}$$

for which

$$\mathbf{x}_1 = \begin{bmatrix} \beta \\ \beta - \alpha \end{bmatrix}$$

are solutions for each α and β. Once $k \geqslant 2$ the matrix

$$\begin{bmatrix} -k & 1 \\ 0 & 1-k \end{bmatrix}$$

is nonsingular and hence

$$\mathbf{x}_k = \begin{bmatrix} \dfrac{1}{k} & 0 \\ 0 & 0 \end{bmatrix} \mathbf{x}_{k-1} = \begin{bmatrix} \dfrac{1}{k!} & 0 \\ 0 & 0 \end{bmatrix} \mathbf{x}_1$$

$$= \frac{1}{k!} \begin{bmatrix} \beta \\ 0 \end{bmatrix} = \frac{\beta}{k!} \begin{bmatrix} 1 \\ 0 \end{bmatrix} \qquad k \geqslant 2$$

Therefore, for each choice of α and β we obtain a solution. Selecting $\beta = 0$ and $\alpha = 1$ yields

$$\mathbf{x}(t) = \begin{bmatrix} 1 \\ -t \end{bmatrix}$$

The choice $\alpha = 1, \beta = 1$ yields

$$\mathbf{x}_2 = \begin{bmatrix} 1 \\ 0 \end{bmatrix} + \begin{bmatrix} 1 \\ 0 \end{bmatrix} t + \sum_{n=2}^{\infty} \frac{t^n}{n!} \begin{bmatrix} 1 \\ 0 \end{bmatrix}$$

$$= \begin{bmatrix} 1 \\ 0 \end{bmatrix} e^t$$

Thus the general solution may be written

$$\mathbf{x}(t) = c_1 \mathbf{x}_1(t) + c_2 \mathbf{x}_2(t)$$

$$= c_1 \begin{bmatrix} 1 \\ -t \end{bmatrix} + c_2 \begin{bmatrix} 1 \\ 0 \end{bmatrix} e^t$$

This may readily be verified.

Exercise 6.6

Part 1

1. For the system of Example 6.6.3 show that $c=1$ leads to a solution

$$\mathbf{x}(t)=\sum_{k=0}^{\infty}\frac{t^{k+1}}{k!(k+1)!}\begin{bmatrix}1\\k+1\end{bmatrix}$$

2. For the system of Example 6.6.4 show that the selection $c=1$ leads to the solution

$$\mathbf{x}(t)=\begin{bmatrix}1\\1\end{bmatrix}t+\sum_{k=1}^{\infty}\frac{t^{k+1}}{(k+1)!}\begin{bmatrix}1\\0\end{bmatrix}=\begin{bmatrix}e^t-1\\t\end{bmatrix}$$

Note that the general solution reduces to the above solution for $c_1=-1$, $c_2=1$.

3. If $\mathrm{B}(t)=\mathrm{B}_0$, a constant matrix, show that Equations (6.6.7) reduce to $[\mathrm{B}_0-(k+c)\mathrm{I}]\mathbf{x}_k=\mathbf{0}$. Hence deduce $\mathbf{x}(t)=t^c\mathbf{x}_0$ is always a solution. Then verify by direct substitution.
4. Find the first three terms of a series solution of the following systems.

(a) $\mathbf{x}'=\dfrac{1}{t}\begin{bmatrix}1 & 1+t\\0 & t^2\end{bmatrix}\mathbf{x}$ (c) $\mathbf{x}'=\dfrac{1}{t}\begin{bmatrix}\dfrac{1}{1-t} & 1\\0 & 0\end{bmatrix}\mathbf{x}$

(b) $\mathbf{x}'=\dfrac{1}{t}\begin{bmatrix}\dfrac{1}{1-t} & 1\\0 & 0\end{bmatrix}\mathbf{x}$ (d) $\mathbf{x}'=\dfrac{1}{t}\begin{bmatrix}\sin t & 0\\t & t\end{bmatrix}\mathbf{x}$

5. Solve $y'=(1/t)g(t)y$ by two methods one of which is the power series technique of this chapter:

(a) For $g(t)=1+t$

(b) For $g(t)=\dfrac{1}{1-t}$

(c) For $g(t)=t\tan t$

Part 2

1. Verify

$$\sum_{n=0}^{\infty}a_nt^n\cdot\sum_{k=0}^{\infty}b_kt^k=\sum_{n=0}^{\infty}\left(\sum_{k=0}^{n}a_kb_{n-k}\right)t^n$$

for the functions

(a) $\dfrac{1}{1-t}=1+t+t^2+\cdots+t^n+\cdots$ $\qquad \dfrac{1}{1+t}=1-t+t^2-\cdots$

(b) $e^t=\displaystyle\sum_{n=0}^{\infty}\frac{t^n}{n!}$ $\qquad e^{-t}=\displaystyle\sum_{n=0}^{\infty}\frac{(-1)^nt^n}{n!}$

2. If $\mathbf{x}(t)=t^c\sum_{k=0}^{\infty}\mathbf{x}_kt^k$, $\mathbf{x}_0\neq\mathbf{0}$ is a solution of $\mathbf{x}'=\mathrm{A}(t)\mathbf{x}$, and $t=0$ is an ordinary point, show that $c=0$.

3. A singular point which is not a regular singular point is called an *irregular singular* point. Show that no solution of the form

$$t^c \sum_{k=0}^{\infty} \mathbf{x}_k t^k \qquad \mathbf{x}_0 \neq \mathbf{0}$$

is possible for the system

$$\mathbf{x}' = \frac{1}{t^2}\begin{bmatrix} 1 & 1 \\ 0 & 1 \end{bmatrix}\mathbf{x}$$

which has an irregular singular point at $t = 0$.

4.* (a) Let $c_1, c_2, \ldots, c_n$ be eigenvalues of B_0 not necessarily distinct. Show that the sequence $B_0-(1+c_1)I$, $B_0-(2+c_1)I, \ldots, B_0-(k+c_1)I, \ldots$. contains a singular matrix if and only if $c_j - c_1$ is a positive integer for some j, $1 < j \leqslant n$.
(b) Let the eigenvalues of B_0 be ordered so that $\operatorname{Re} c_1 \geqslant \operatorname{Re} c_i$, $i = 2, 3, \ldots, n$. Deduce from (a) that $B_0 - (k+c_1)T$ is never singular, $k \geqslant 1$.

5. Consider the second-order equation

$$y'' + a(t)y' + b(t)y = 0 \tag{6.6.7}$$

(a) Show: If $y(t)$ is a solution of (6.6.7) then

$$\mathbf{x}(t) = \begin{bmatrix} y(t) \\ ty'(t) \end{bmatrix}$$

satisfies

$$\mathbf{x}'(t) = \frac{1}{t}\begin{bmatrix} 0 & 1 \\ -t^2 b(t) & (1-ta(t)) \end{bmatrix}\mathbf{x} \tag{6.6.8}$$

(b) Show: If

$$\mathbf{x}(t) = \begin{bmatrix} x_1(t) \\ x_2(t) \end{bmatrix}$$

satisfies (6.6.8), $x_1(t)$ is a solution of (6.6.7).

6. Consider the second-order equation

$$y'' + \frac{p(t)}{t}y' + \frac{q(t)}{t^2}y = 0 \tag{6.6.9}$$

(a) If $y(t)$ is a solution of (6.6.9) show that

$$\mathbf{x}(t) = \begin{bmatrix} y(t) \\ ty'(t) \end{bmatrix}$$

is a solution of

$$\mathbf{x}' = \frac{1}{t}\begin{bmatrix} 0 & \\ -q(t) & (1-p(t)) \end{bmatrix}\mathbf{x} \tag{6.6.10}$$

(b) What conditions on p and q would result in the system (6.6.10) having a regular singular point at $t = 0$?

7.* Consider the equation of the nth-order equation

$$y^{(n)}+a_{n-1}(t)y^{(n-1)}+\cdots+a_0(t)y=0 \tag{6.6.11}$$

where $a_0, \ldots, a_{n-1}$ are given functions.
Let $y(t)$ be a solution of (6.6.11) and define

$$\mathbf{x}(t)=\begin{bmatrix} y(t) \\ ty'(t) \\ \cdot \\ \cdot \\ \cdot \\ t^{n-1}y^{(n-1)}(t) \end{bmatrix} \tag{6.6.12}$$

Show that $\mathbf{x}(t)$ satisfies

$$\mathbf{x}'=\frac{1}{t}\begin{bmatrix} 0 & 1 & 0 & 0 & \cdots & 0 & 0 \\ 0 & 1 & 1 & 0 & \cdots & 0 & 0 \\ 0 & 0 & 2 & 1 & \cdots & 0 & 0 \\ \cdot & \cdot & \cdot & \cdot & & \cdot & \cdot \\ \cdot & \cdot & \cdot & \cdot & & \cdot & \cdot \\ \cdot & \cdot & \cdot & \cdot & & \cdot & \cdot \\ 0 & 0 & 0 & 0 & \cdots & n-2 & 1 \\ -t^n a_0 & -t^{n-1}a_1 & -t^{n-2}a_2 & -t^{n-3}a_3 & \cdots & -ta_{n-2} & [n-1-ta_{n-1}] \end{bmatrix}\mathbf{x} \tag{6.6.13}$$

8.* (a) Show that if $\mathbf{x}(t)$ satisfies (6.6.13) then for each component $x_i(t)=t^{i-1}x_1^{(i-1)}(t)$. To do this you may assume the equality is valid for $i=k$ and differentiate, then compare the result with

$$x_i'=\frac{1}{t}((i-1)x_i+x_{i+1})$$

which follows from (6.6.13).
(b) Show that $x_1(t)$ satisfies (6.6.11) by interpreting the last component of (6.6.13) and using Problem 8(a) above.

9.* Let $a_0(t), \ldots, a_{n-1}(t)$ be analytic at $t=0$. Consider the nth-order equation

$$y^{(n)}+\frac{1}{t}a_{n-1}(t)y^{(n-1)}+\cdots+\frac{1}{t^n}a_0(t)y=0 \tag{6.6.14}$$

This equation is said to have a *regular singular point at* $t=0$ if at least one of the coefficients, $(1/t^i)a_{n-i}$, is *not* analytic at $t=0$. To what first-order n-dimensional system with a regular singular point at $t=0$ does (6.6.14) correspond?

6.7.* THE BESSEL EQUATION

The transformation $x_1=y$ and $x_2=ty'$ sends the equation

$$y''+\frac{p(t)}{t}y'+\frac{q(t)}{t^2}y=0$$

into the system

$$\mathbf{x}' = \frac{1}{t}\begin{bmatrix} 0 & 1 \\ -q(t) & (1-p(t)) \end{bmatrix}\mathbf{x} \qquad \mathbf{x} = \begin{bmatrix} x_1 \\ x_2 \end{bmatrix}$$

It is not difficult to see that the system will have a regular singular point at $t = 0$ when $q(t)$ and $p(t)$ are analytic at $t = 0$ and, moreover, the first component of every solution of the system is a solution of the second-order equation. All this and somewhat more is described in Exercise 6.6, Part 2, Problems 5–9. We are led, therefore, to the following definition.

6.7.1. DEFINITION *The point $t = 0$ is a regular singular point of Equation* (6.7.1) *if*

(1) *p and q are analytic at $t = 0$.*
(2) *p_0, q_0, and q_1 are not all zero.*

Here we assume

$$p(t) = p_0 + p_1 t + \cdots + p_k t^k + \cdots$$

and

$$q(t) = q_0 + q_1 t + \cdots + q_k t^k + \cdots$$

Although the above transformation is very useful for general or theoretical studies, for the computation of coefficients of a specific equation it is often unnecessary to convert to a system. We have seen a similar situation in Section 6.4.

We shall study, very briefly, the differential equation

$$t^2 y'' + t y' + (t^2 - \alpha^2) y = 0 \tag{6.7.1}$$

(known as *Bessel's Equation of Index α*) as an illustration of the problems and methods of the direct approach.

Assume a solution, $y(t)$, of (6.7.1) of the form

$$y(t) = \sum_{k=0}^{\infty} y_k t^{k+c} \qquad y_0 \neq 0$$

We compute

$$t y'(t) = \sum_{h=0}^{\infty} (k+c) y_k t^{k+c}$$

$$t^2 y'' = \sum_{k=0}^{\infty} (k+c)(k+c-1) y_k t^{k+c}$$

Upon substituting into (6.7.1), this leads to

$$\sum_{k=0}^{\infty} [(k+c)^2 - \alpha^2] y_k t^{k+c} + \sum_{k=0}^{\infty} y_k t^{k+c+2} \equiv 0 \tag{6.7.2}$$

But

$$\sum_{k=0}^{\infty} y_k t^{k+c+2} \equiv \sum_{k=2}^{\infty} y_{k-2} t^{k+c}$$

and hence (6.7.2) may be written

$$(c^2-\alpha^2)y_0t^c+[(c+1)^2-\alpha^2]y_1t^{c+1}$$
$$+\sum_{k=2}^{\infty}\{[(k+c)^2-\alpha^2]y_k+y_{k-2}\}t^{k+c}\equiv 0$$

This latter identity implies that every coefficient is zero. That is,

$$y_0(c^2-\alpha^2)=0$$
$$[(c+1)^2-\alpha^2]y_1=0$$

and for $k\geqslant 2$,

$$[(k+c)^2-\alpha^2]y_k+y_{k-2}=0 \tag{6.7.3}$$

We select $c=\alpha$ so as to satisfy $c^2-\alpha^2=0$ and $y_1=0$ to satisfy the next equation. Let us suppose $\alpha\geqslant 0$. (If not, choose $c=-\alpha$.) Then

$$(k+\alpha)^2-\alpha^2=k(k+2\alpha)\geqslant 0$$

and the general recurrence relation, (6.7.2), along with $y_1=0$, implies

$$0=y_3=y_5=\cdots$$

Set $k=2i$ in (6.7.3). Then,

$$\begin{aligned}(k+\alpha)^2-\alpha^2&=k(k+2\alpha)\\&=4i(i+\alpha)\qquad i=1,2,\ldots\end{aligned}$$

Thus,

$$\begin{aligned}4(1+\alpha)y_2&=-y_0 &&\text{using}\quad i=1\\4(2)(2+\alpha)y_4&=-y_2 &&\text{using}\quad i=2\\&\ \ \vdots\\4(j)(j+\alpha)y_{2j}&=-y_{2j-2} &&\text{using}\quad i=j\end{aligned}$$

By back-substitution we find,

$$\begin{aligned}y_{2j}&=\frac{-y_{2j-2}}{4j(j+\alpha)}\\&=\frac{(-1)^2y_{2j-4}}{(4j)(4j-4)(j+\alpha)(j-1+\alpha)}\\&\ \ \vdots\\&=\frac{(-1)^jy_0}{4^jj!(1+\alpha)(2+\alpha)\cdots(j+\alpha)}\end{aligned}$$

Therefore, replacing j by n, we find

$$y(t)=t^\alpha\left[y_0+\sum_{n=1}^{\infty}y_0\frac{(-1)^nt^{2n}}{2^{2n}n!(1+\alpha)\cdots(n+\alpha)}\right] \tag{6.7.4}$$

This power series is convergent for all t and represents, for a special choice of y_0, a function known as the *Bessel Function of Index* α, denoted by $J_\alpha(t)$. If $\alpha = v$, v an integer $\geqslant 0$, we choose $y_0 = 1/2^v v!$ and after some rewriting,

$$J_v(t) = \sum_{n=0}^{\infty} \frac{(-1)^n t^{2n+v}}{2^{2n+v} n!(n+v)!} \tag{6.7.5}$$

is a solution of $t^2y'' + ty' + (t^2 - v^2)y = 0$. A second Frobenius Series solution is possible for some choices of α. We leave it for an exercise to establish the solution,

$$y^*(t) = y_0 t^{-\alpha}\left[1 + \sum_{n=1}^{\infty} \frac{(-1)^n t^{2n}}{2^{2n} n!(1-\alpha)(2-\alpha)\cdots(n-\alpha)}\right] \tag{6.7.6}$$

Exercise 6.7

1. Derive the expression for $J_v(t)$ as given by Equation (6.7.5).
2. Derive the solution $y^*(t)$, Equation (6.7.6) of Bessel's Equation, assuming α is not an integer.
3. Why are the solutions of Bessel's Equation (α not an integer) given by Equations (6.7.4) and (6.7.6) linearly independent?
4. Solve

$$t^2y'' + ty' + (t^2 + \alpha^2)y = 0$$

by use of Frobenius Series. Solutions appropriately normalized are often called modified Bessel Functions. (This equation is obtained from Bessel's Equation by replacing t with it.)

5. The equation

$$ty'' + (1-t)y' + \alpha y = 0$$

is called Laguerre's Equation. If $\alpha = v$, v a nonnegative integer, there is a polynomial solution. Find it by applying the techniques of this section.

6. If α is not an integer, find two linearly independent solutions of Laguerre's Equation as given in Problem 5.

Chapter 7
AN INTRODUCTION TO NONLINEAR EQUATIONS

7.1. WHAT IS A NONLINEAR EQUATION?

Consider the following equations:

(a) $x' = x^2$ (b) $x_1' = x_1 x_2$, $x_2' = tx_1 + x_2$ (c) $x_1' = t^3 x_1$, $x_2' = x_1 \sin x_2$

In each case, there is at least one component of the derivative of the unknown function $\mathbf{x}(t)$ that is not expressed as a linear (time varying or constant) combination of the components of $\mathbf{x}(t)$ plus a given function of time. Such equations are called *nonlinear*. In order to *study* them (we can rarely solve them explicitly) we must introduce new techniques. We do find, however, that the theory of $\mathbf{x}' = \mathrm{A}(t)\mathbf{x} + \mathbf{F}(t)$ is still applicable in some cases.

Although we must sacrifice the matrix notation in dealing with nonlinear equations we preserve the convenience of vector notation for systems. The notation

$$\mathbf{x}' = \mathbf{f}(t, \mathbf{x}) \tag{7.1.1}$$

is a shorthand for the system of equations

$$\begin{aligned} x_1' &= f_1(t, x_1, x_2, \ldots, x_n) \equiv f_1(t, \mathbf{x}) \\ x_2' &= f_2(t, x_1, x_2, \ldots, x_n) \equiv f_2(t, \mathbf{x}) \\ &\vdots \\ x_n' &= f_n(t, x_1, x_2, \ldots, x_n) \equiv f_n(t, \mathbf{x}) \end{aligned} \tag{7.1.2}$$

Thus $\mathbf{f}$ is a vector whose entries are functions of the $n+1$ variables, $t, x_1, x_2, \ldots, x_n$. In specific problems it is common to use the expanded form

(7.1.2) rather than the compact (7.1.1). With this said, we note that (7.1.1) is linear if and only if

$$\mathbf{f}(t, \mathbf{x}) = A(t)\mathbf{x} + \mathbf{g}(t) \tag{7.1.3}$$

for some matrix A(t) and vector function $\mathbf{g}(t)$.

7.1.1. DEFINITION *The system* $\mathbf{x}' = f(t, \mathbf{x})$ *is called autonomous if* $\mathbf{f}$ *is independent of* t.

To some extent the word "autonomous" plays the role for (7.1.1) that "homogeneous" plays for the system $\mathbf{x}' = A\mathbf{x} + \mathbf{g}(t)$, A a constant matrix. It is apparent that $\mathbf{x}' = A\mathbf{x}$ is autonomous and $\mathbf{x}' = A\mathbf{x} + \mathbf{u}e^t$ is not. Only the system (a) above is autonomous. It is always possible to rewrite a nonautonomous system in autonomous form by the introduction of a spurious variable. For example, in the nonautonomous equation

$$x' = t^2x - e^t$$

let $x_1 = t$ and $x_2 = x$. Then, $x_1' = 1$ and $x_2' = x' = t^2x - e^t = x_1^2x_2 - e^{x_1}$. In vector form, this pair of equations is

$$\mathbf{x}' = \begin{bmatrix} x_1 \\ x_2 \end{bmatrix}' = \begin{bmatrix} 1 \\ x_1^2x_2 - e^{x_1} \end{bmatrix}$$

in which $\mathbf{f}$ is independent of t. In general, suppose

$$\mathbf{x}' = \mathbf{f}(t, \mathbf{x}) \qquad \mathbf{x} = \begin{bmatrix} x_1 \\ x_2 \\ \cdot \\ \cdot \\ \cdot \\ x_n \end{bmatrix} \qquad \mathbf{f} = \begin{bmatrix} f_1 \\ f_2 \\ \cdot \\ \cdot \\ \cdot \\ f_n \end{bmatrix}$$

Set

$$\mathbf{y} = \begin{bmatrix} t \\ x_1 \\ x_2 \\ \cdot \\ \cdot \\ \cdot \\ x_n \end{bmatrix} \qquad \mathbf{F} = \begin{bmatrix} 1 \\ f_1 \\ f_2 \\ \cdot \\ \cdot \\ \cdot \\ f_n \end{bmatrix}$$

Then

$$\mathbf{y}' = \begin{bmatrix} 1 \\ x_1' \\ x_2' \\ \cdot \\ \cdot \\ x_n' \end{bmatrix} = \begin{bmatrix} 1 \\ f_1 \\ f_2 \\ \cdot \\ \cdot \\ f_n \end{bmatrix} = \mathbf{F}(\mathbf{y})$$

is autonomous.

7.1.1. Example

Write the second-order equation

$$y'' - 2ty' + y = 0$$

as an autonomous system of first-order equations.

Solution

We set $x_1 = t$, $x_2 = y$, $x_3 = y'$. Then $x_1' = 1$, $x_2' = y' = x_3$ and $x_3' = y'' = 2ty' - y = 2x_1x_3 - x_2$. The required autonomous system is, therefore,

$$\begin{aligned} x_1' &= 1 \\ x_2' &= x_3 \\ x_3' &= 2x_1x_3 - x_2 \end{aligned}$$

Exercise 7.1

1. Which of the following are linear? autonomous?

   (a) $x' = \sin tx$ (b) $x' = \dfrac{1}{x}$

   (c) $x'' - x = t$ (d) $x_1' = t \quad x_2' = x_1 + x_2$

   (e) $x_1' = x_1 \quad x_2' = x_2$ (f) $x' = t^2x + \dfrac{1}{t}$

   (g) $x_1' = \dfrac{x_1}{t} \quad x_2' = x_1$

2. Write each nonautonomous system of Problem 1 as an autonomous system.
3. Write

   $$\mathbf{x}' = \mathbf{A}\mathbf{x} + \mathbf{f}(t)$$

   as an autonomous system when A is 2×2.
4. Write

   $$y''' + ay'' + by' + cy = \alpha t$$

   as an autonomous system.

7.2. SOME PREDATORS AND THEIR PREY

Two species of microorganisms live in an environment in which the first has an abundant source of food and the second lives by devouring the first. The prey, Species 1, reproduces so that its rate of population growth is proportional to its population size. They die at a rate proportional to the product of both populations sizes. The predators have no enemies in this fanciful microcosm and reproduce so that their population size grows at a rate proportional to the product of the sizes of the two populations. What mathematical model describes this interplay? To answer this we denote by

x_1, the population size of Species 1, the prey

and by

x_2, the population size of Species 2, the predators

Now, the birth rate minus the death rate is the rate of population change. For Species 1, this means $x_1' = \alpha x_1 - \beta x_1 x_2$ and for Species 2, $x_2' = \gamma x_1 x_2$, since the predators are immortal. If we assume that at some fixed time, $t = 0$, there are s_1 members of Species 1 and s_2 members of Species 2. We can collect all these facts in the following nonlinear initial-value problem:

$$\begin{aligned} x_1' &= \alpha x_1 - \beta x_1 x_2 \\ x_2' &= \gamma x_1 x_2 \\ x_1(0) &= s_1 \\ x_2(0) &= s_2 \end{aligned} \tag{7.2.1}$$

The extent to which the solutions $x_1(t)$ and $x_2(t)$ actually yield the populations depends on how critically our simplifying assumptions alter the biological phenomena. We must remember that in order to obtain the model represented by the system (7.2.1) we assumed x_1 and x_2 varied continuously and were even differentiable! Of course populations vary in discrete integral units. (For large populations this type of simplification is usually without serious consequences.) Without looking very hard, the reader can find many other faults in this model. Nevertheless, such models are extremely helpful in certain areas of biology.† We must also observe that as crude an approximation as (7.2.1) may be, it is a good deal better than any linear model of the form $\mathbf{z}' = \mathbf{A}\mathbf{z}$. For if we assert

$$x_1' = ax_1 + bx_2$$

we find that $b \neq 0$ implies that prey can somehow give birth to predator when $x_1 = 0$ (or create a negative predator!). On the other hand, if $b = 0$ then we are forced to conclude that the predators have lost their appetites. It appears that this simple ecological system is best described by a nonlinear model.

Although the mathematical analysis of the system (7.2.1) is not difficult—a fortuitous circumstance for nonlinear equations—it does not serve our purpose to explore this special equation any further. (Some of the analysis is described in Exercise 7.2.) The reader might wish to try guessing at trial solutions before turning to the problems. This is an excellent lesson in futility and illustrates that our remarkable success with this method on linear equations cannot be expected to continue into nonlinear theory.

Exercise 7.2

1. Suppose $\begin{bmatrix} x(t) \\ y(t) \end{bmatrix}$ is a solution of

$$\begin{bmatrix} x \\ y \end{bmatrix}' = \begin{bmatrix} xy \\ 3xy \end{bmatrix} \qquad \begin{bmatrix} x(0) \\ y(0) \end{bmatrix} = \begin{bmatrix} 1 \\ 1 \end{bmatrix}$$

Suppose there exists f such that

$$y(t) = f(x(t))$$

†See A. J. Lotka, *Elements of Mathematical Biology*, Dover, New York, 1956.

(a) Show that $f'(v) = 3$ for all v.

(b) Show that $y(t) = 3x(t) - 2$ for all t.

(c) Show that

$$\frac{x'(t)}{x(t)[3x(t)-2]} \equiv \frac{1}{2}x'(t)\left[\frac{3}{3x(t)-2} - \frac{1}{x(t)}\right] = 1 \qquad \text{for all } t$$

(d) Show that

$$\frac{1}{2}\int_{x(0)}^{x(t)} \left\{\frac{3}{3u-2} - \frac{1}{u}\right\} du = \frac{1}{2}\ln\left[\frac{3x(t)-2}{x(t)}\right] = t$$

(e) Show that

$$x(t) = \frac{2}{3-e^{2t}} \qquad y(t) = \frac{6}{3-e^{2t}} - 2$$

are the components of a solution.

2. Consider the nonlinear system

$$\begin{bmatrix} x \\ y \end{bmatrix}' = \begin{bmatrix} xy \\ y - xy \end{bmatrix} \qquad \begin{bmatrix} x(0) \\ y(0) \end{bmatrix} = \begin{bmatrix} 1 \\ 1 \end{bmatrix}$$

As in Problem 1 assume the existence of f such that $y(t) = f(x(t))$.

(a) Show that $f'(v) = \frac{1}{v} - 1$ for $v \neq 0$

(b) Show that $y(t) = \ln x(t) - x(t) + 2$

(c) Show that $\dfrac{x'(t)}{x(t)(\ln x(t) - x(t) + 2)} = 1$

7.3. NONLINEAR EQUATIONS IN ONE DIMENSION

The difficulty with nonlinear equations is not due solely to the scarcity of solution techniques but also to the failure of the theorems of linear equations to carry over. We present some examples to illustrate how the uniqueness and superposition principles break down in even simple nonlinear equations. In the course of doing this, we shall incidentally present a number of special techniques for solving particular nonlinear equations.

7.3.1. Example

Show that the initial-value problem

$$y' = \tfrac{3}{2}y^{1/3} \qquad y(0) = 0 \tag{7.3.1}$$

has infinitely many solutions.

Solution

Besides the constant solution $y_1(t) \equiv 0$ we find others by observing

$$\frac{2}{3}\frac{y'}{y^{1/3}} = \frac{d}{dt}(y^{2/3}) = 1$$

and hence $y^{2/3}(t) = t + c$. Choosing $c = 0$ leads to the solution, $y_2(t) = t^{3/2}$.

The functions

$$\begin{aligned} y_b^{(t)} &= 0 && \text{if} \quad t \leqslant b \\ y_b^{(t)} &= (t-b)^{3/2} && \text{if} \quad t > b \end{aligned} \tag{7.3.2}$$

($b > 0$) are also easily verified as solutions. (Take care in computing $y_b'(b)$.)

7.3.2. Example

Show that superposition is violated for the family of solutions of

$$y' = \frac{t}{2} y^{-1} \tag{7.3.3}$$

Solution

Writing (7.3.3) in the form

$$2yy' = t$$

leads to

$$(y^2)' = \left(\frac{t^2}{2} + c\right)'$$

or

$$y^2(t) = \tfrac{1}{2}t^2 + c$$

We verify by substitution that $y(t) = [(t^2/2) + c]^{1/2}$ is a solution of (7.3.3) regardless of the choice of c. (Of course, $[(t^2/2) + c] \geqslant 0$ is required for $[(t^2/2) + c]^{1/2}$ to be defined.) Now it is easy to check that

$$\left(\frac{t^2}{2} + c_0\right)^{1/2} + \left(\frac{t^2}{2} + c_1\right)^{1/2}$$

is never a solution even though both radicals are solutions! Problem 3, Part 2, following this section is devoted to showing that the validity of the superposition principle would imply linearity.

Remarks

(1) Both examples are illustrations of first-order equations which can be solved by the technique known as *separation of variables*. In general terms, suppose $y(t)$ is a solution of

$$y' = f(t, y) \qquad y(t_0) = y_0$$

where

$$f(t, y) = \frac{g(t)}{h(y)} \quad \text{and} \quad h(y_0) \neq 0$$

Thus,

$$h(y(t))y'(t) = g(t)$$

so that

$$\int_{t_0}^{t} h(y(t))y'(t)\,dt = \int_{t_0}^{t} g(t)\,dt$$

By employing the substitution rule for integrals we get

$$\int_{t_0}^{t} h(y(t))y'(t)\,dt = \int_{y_0}^{y(t)} h(u)\,du \equiv \mathrm{H}(y(t))$$
$$= \int_{t_0}^{t} g(t)\,dt \equiv \mathrm{G}(t)$$

which relates (implicitly) y as a function of t. For further illustrations see problems in Exercise 7.3 at the end of this section.

(2) Besides illustrating the failure of superposition, Example 7.3.2 also shows that solutions of nonlinear equations may exist for some restricted set of t's. In this case, $t \geqslant -c$.

The notion of a general solution for nonlinear equations is also a concept that must, at the very least, be carefully scrutinized. Consider the next example.

7.3.3. Example

For each choice of constant c the function $y_c(t) = ct + c^2/2$ is a solution of

$$y' = -t + (2y + t^2)^{1/2} \tag{7.3.4}$$

Clearly $y_c'(t) = c = -t + (2ct + c^2 + t^2)^{1/2} = -t + (t + c),\ t \geqslant -c$.
However, it is a peculiarity of this family of solutions that it misses the solution

$$y(t) = -\frac{t^2}{2}$$

Contrast this with linear equations! Not every solution of (7.3.4) is obtained by appropriate choice of c.

Exercise 7.3

Part 1

1. Show that $y' = -t + (2y + t^2)^{1/2}$, $y(1) = -\frac{1}{2}$ is solved by $y(t) = -t^2/2$ and $y_1(t) = -t + \frac{1}{2}$.
2. Separate variables to find solutions of the following equations
   (a) $ty' = y$
   (b) $y' = e^{t+y}$
   (c) $y' = f(t)y = 0$

Part 2

1. (Bernoulli's Equation) Make the substitution

$$z = y^{-n+1} \qquad (n \neq 1)$$

in the equation $y' + \mathrm{P}(t)y = \mathrm{Q}(t)y^n$ and then solve. (Hint: The resulting equation in z is linear!)

2. Apply the method of Problem 1 to the equations
   (a) $y' = y - te^{-t}y^3$
   (b) $2t^3y' = y(y^2 + 3t^2)$
3.* Let $\mathbf{f}(t, \mathbf{x})$ be a function such that
   (i) Every initial value problem

$$\mathbf{x}' = \mathbf{f}(t, \mathbf{x}) \qquad \mathbf{x}(t_0) = \mathbf{x}_0$$

has a solution.

(ii) The superposition principle holds, that is, if for $i = 1, 2$, $\mathbf{x}_i(t)$ satisfy

$$\mathbf{x}' = \mathbf{f}(t, \mathbf{x}) \qquad \mathbf{x}_i(t_0) = \mathbf{x}_i^0$$

then for any constants a and b, $\mathbf{z} = a\mathbf{x}_1 + b\mathbf{x}_2$ satisfies

$$\mathbf{z}' = \mathbf{f}(t, \mathbf{z}) \qquad \mathbf{z}(t_0) = a\mathbf{x}_1^0 + b\mathbf{x}_2^0$$

(a) Show that $\mathbf{f}(t, \mathbf{z})$ is a linear function of $\mathbf{z}$, that is,

$$\mathbf{f}(t, a\mathbf{x} + b\mathbf{y}) = a\mathbf{f}(t, \mathbf{x}) + b\mathbf{f}(t, \mathbf{y})$$

(b) Show that $\mathbf{f}(t, \mathbf{z}) = \mathrm{A}(t)\mathbf{z}$ where

$$\mathrm{A}(t) = [\mathbf{f}(t, \mathbf{e}_1), \ldots, \mathbf{f}(t, \mathbf{e}_n)]$$

7.4. THE EULER METHOD IN ONE DIMENSION: AN EXAMPLE

The theoretical and practical difficulties with nonlinear equations and systems were evident very early in the study of differential equations. To circumvent these difficulties attention began to be focused on approximate techniques which had wide applicability. Stimulated by the need for solutions to very specific problems in physics, a great many methods of this type were developed. One of the earliest of the numerical methods† was discovered by Euler and, although now outmoded, its underlying ideas are common to a wide class of numerical techniques. For this reason and because of its inherent simplicity, we illustrate the theory of numerical solutions of differential equations by studying what is known as the "Euler Method." Suppose for the sake of this discussion that the solution $x(t) = -t^{-1}$ of the initial-value problem

$$x' = x^2 \qquad x(-1) = 1 \qquad -1 \leq t < 0 \tag{7.4.1}$$

is unknown. By substituting the initial condition into the differential equation, we learn that $x'(-1) = 1$. Hence the graph of the solution (which we call the *solution curve*) passes through the point $(-1, 1)$ in the t, x-plane with slope, 1. In Figure 7.4.1 the solution curve and its tangent at $(-1, 1)$ are drawn. Over "small" intervals the tangent line to a curve and the curve itself differ only slightly. We make this assumption to obtain an approximate

† Power series methods can be viewed as an approximation technique. Numerical methods are qualitatively different.

Figure 7.4.1

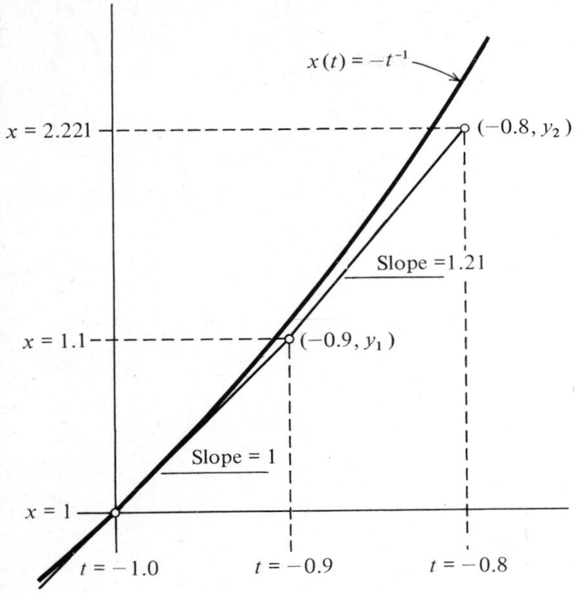

value to x at $t = -0.9$. Call this approximate value y_1; then

$$y_1 = x(-1) + (0.1)\, x'(-1)$$
$$= 1.1$$

(See Figure 7.4.1.)

We apply this same idea again using the approximate value of x at -0.9 to estimate the value of x at -0.8. We accomplish this by considering the new initial-value problem

$$x' = x^2 \qquad x(-0.9) = y_1 = 1.1 \tag{7.4.2}$$

Hence the slope of the solution curve is $(1.1)^2 = 1.21$ and

$$y_2 = y_1 + (0.1)\, y_1^2$$
$$= 1.1 + 0.121 = 1.221$$

is an approximation† to x at $t = -0.8$. This process can be repeated indefinitely, although one suspects a steady accumulation of errors will soon make the closeness of the approximation suspect. Table 7.4.1 presents the results of this computation carried out a total of 10 times. Observe the rapid growth of the error.

† In reality, y_2 is an approximation to the solution of (7.4.2) at $t = -0.8$, namely, $x(t) = [-1/t - (1/110)]$. But $x(t) - \tilde{x}(t)$ is small so that y_2 is also a reasonably good approximation to $x(-0.8)$.

Table 7.4.1

k	t	y_k	$x(t)$	E = error = $\lvert y_k - x(t)\rvert$
1	−0.9	1.100	1.111	$0.01 < E < 0.012$
2	−0.8	1.221	1.250	$0.039 = E$
3	−0.7	1.370	1.429	$0.058 < E < 0.059$
4	−0.6	1.558	1.667	$0.10 < E < 0.11$
5	−0.5	1.801	2.000	$0.19 < E < 0.2$
6	−0.4	2.125	2.500	$0.37 < E < 0.38$
7	−0.3	2.577	3.333	$0.75 < E < 0.76$
8	−0.2	3.241	5.000	$1.7 < E$
9	−0.1	4.291	10.000	$5 < E$
10	0	6.132		

Exercise 7.4

1. Complete the computations that verify Table 7.4.1.
2. For the given initial-value problem

$$x' = x^2 \qquad x(-1) = 1$$

use the Euler Method to compute a table analogous to Table 7.4.1 for $t=-1$, $t=-0.95$, $t=-0.85$, $t=-0.8$, $t=-0.75$, $t=-0.7$, $t=-0.65$, $t=-0.6$, $t=-0.55$, and $t=-0.5$. Compare results with Table 7.4.1 at the corresponding points.
3. Same as Problem 2 except $t=-1$, $t=-0.99$, $t=-0.98$, $t=-0.97$, $t=-0.96, \ldots, t=-0.9$. Again compare with Table 7.4.1 at $t=-0.9$ and with the answers in Problem 2 at $t=-0.95$ and $t=-0.9$.
4. (a) What is the equation of the tangent line to the solution curve at $t=-1$?
(b) What is the equation of the tangent line to the solution curve of Equations (7.4.1) at $t=-0.9$?
(c) Write the equations of the tangent lines to the solution curves of

$$x' = x^2 \qquad x(t_k) = y_k$$

at t_k where $t_k = -1 + k/10$, $k = 2, 3, \ldots, 9$.

THE EULER METHOD

7.5. THE EULER METHOD

We now apply the ideas of the preceding section to the general initial-value problem

$$\mathbf{x}' = \mathbf{f}(t, \mathbf{x}) \qquad \mathbf{x}(a) = \mathbf{x}_0 \tag{7.5.1}$$

We suppose that t_0 is the starting point of our algorithm. Our approximations will be made at the equispaced points, $t_1 = t_0 + h, t_2 = t_1 + h, \ldots, t_N = t_{N-1} + h$. We refer to $h > 0$ as the *step-size* of the method and label $y_k = y(t_k)$ as the approximation to the supposedly unique solution, $x(t)$, of the system (7.5.1).

Our method, therefore, supplies a table as the approximate solution and this table takes the form as shown in (7.5.2).

No. of steps	t_k	$\mathbf{y}_k$	$\mathbf{f}_k$
0	t_0	$\mathbf{y}_0 = \mathbf{x}_0$	$\mathbf{f}_0 = \mathbf{f}(t_0, \mathbf{x}_0)$
1	t_1	$\mathbf{y}_1$	$\mathbf{f}_1 = \mathbf{f}(t_1, \mathbf{y}_1)$
.	.	.	.
.	.	.	.
.	.	.	.
k	t_k	$\mathbf{y}_k$	$\mathbf{f}_k = \mathbf{f}(t_k, \mathbf{y}_k)$
.	.	.	.
.	.	.	.
.	.	.	.
N	t_{N}	$\mathbf{y}_{\mathrm{N}}$	$\mathbf{f}_{\mathrm{N}} = \mathbf{f}(t_{\mathrm{N}}, \mathbf{y}_{\mathrm{N}})$

(7.5.2)

The Euler Algorithm is the method described by Equations (7.5.3), which completes line k of the table from the information in line $k-1$.

$$\begin{aligned} \mathbf{y}_0 &= \mathbf{x}(t_0) \\ \mathbf{y}_{k+1} &= \mathbf{y}_k + h\mathbf{f}_k \qquad k = 0, 1, \ldots, \mathrm{N}-1 \end{aligned} \tag{7.5.3}$$

where

$$\begin{aligned} \mathbf{f}_k &= \mathbf{f}(t_k, \mathbf{y}_k) \\ &= \mathbf{f}(t_0 + kh, \mathbf{y}_k) \end{aligned}$$

We are often asked to solve

$$\mathbf{x}' = \mathbf{f}(t, \mathbf{x}) \qquad \mathbf{x}(a) = \mathbf{x}_0$$

on the interval $a \leqslant t \leqslant b$. For this formulation, we set $t_0 = a$, $t_{\mathrm{N}} = b$ and divide $[a, b]$ into N equal parts by selecting $h = (b-a)/\mathrm{N}$. Of course, in the event that the given initial-value problem is in one dimension, we simply interpret each equation in (7.5.2) as a scalar equation.

7.5.1. Example

Use the Euler Method to find an approximate solution of

$$\mathbf{x}' = \begin{bmatrix} 0 & -1 \\ 1 & 0 \end{bmatrix} \mathbf{x} \qquad \mathbf{x}(0) = \begin{bmatrix} 1 \\ 0 \end{bmatrix}$$

on [0, 1].

Solution

We must first decide on our step size. Suppose we select $h = 0.25$ so that $\mathrm{N} = 4$. Then,

$$\mathbf{y}_0 = \mathbf{x}(0) = \begin{bmatrix} 1 \\ 0 \end{bmatrix} \qquad \mathbf{f}_0 = \begin{bmatrix} 0 & -1 \\ 1 & 0 \end{bmatrix} \begin{bmatrix} 1 \\ 0 \end{bmatrix} = \begin{bmatrix} 0 \\ 1 \end{bmatrix}$$

Hence,

$$\mathbf{y}(0.25) = \mathbf{y}_1 = \begin{bmatrix} 1 \\ 0 \end{bmatrix} + (0.25)\begin{bmatrix} 0 \\ 1 \end{bmatrix} = \begin{bmatrix} 1 \\ 0.25 \end{bmatrix}$$

and

$$\mathbf{f}_1 = \begin{bmatrix} 0 & -1 \\ 1 & 0 \end{bmatrix}\begin{bmatrix} 1 \\ 0.25 \end{bmatrix} = \begin{bmatrix} -0.25 \\ 1 \end{bmatrix}$$

Hence,

$$\mathbf{y}(0.5) = \mathbf{y}_2 = \begin{bmatrix} 1 \\ 0.25 \end{bmatrix} + (0.25)\begin{bmatrix} -0.25 \\ 1 \end{bmatrix} = \begin{bmatrix} 0.9375 \\ 0.5000 \end{bmatrix}$$

$$\mathbf{f}_2 = \begin{bmatrix} 0 & -1 \\ 1 & 0 \end{bmatrix}\begin{bmatrix} 0.9375 \\ 0.5 \end{bmatrix} = \begin{bmatrix} -0.5 \\ 0.94 \end{bmatrix}$$

Hence,

$$\mathbf{y}_3(0.75) = \mathbf{y}_3 = \begin{bmatrix} 0.97 \\ 0.50 \end{bmatrix} + (0.25)\begin{bmatrix} -0.5 \\ 0.94 \end{bmatrix} = \begin{bmatrix} 0.81 \\ 0.74 \end{bmatrix}$$

$$\mathbf{f}_4 = \begin{bmatrix} -0.74 \\ 0.81 \end{bmatrix}$$

Hence,

$$\mathbf{y}_4(1) = \mathbf{y}_4 = \begin{bmatrix} 0.81 \\ 0.74 \end{bmatrix} + (0.25)\begin{bmatrix} -0.74 \\ 0.81 \end{bmatrix} = \begin{bmatrix} 0.63 \\ 0.54 \end{bmatrix}$$

The tabular form is given in Table 7.5.1.

Table 7.5.1

No. of steps	t_k	$\mathbf{y}_k$	$\mathbf{f}_k$	$\mathbf{x}(t_k)$	$E = \lVert\mathbf{x}(t_k) - \mathbf{y}_k\rVert$
0	0.00	$\begin{bmatrix} 1.00 \\ 0.00 \end{bmatrix}$	$\begin{bmatrix} 0.00 \\ 1.00 \end{bmatrix}$	$\begin{bmatrix} 1.00 \\ 0.00 \end{bmatrix}$	0.00
1	0.25	$\begin{bmatrix} 1.00 \\ 0.25 \end{bmatrix}$	$\begin{bmatrix} -0.25 \\ 1.00 \end{bmatrix}$	$\begin{bmatrix} 0.97 \\ 0.25 \end{bmatrix}$	0.03
2	0.50	$\begin{bmatrix} 0.94 \\ 0.50 \end{bmatrix}$	$\begin{bmatrix} -0.50 \\ 0.94 \end{bmatrix}$	$\begin{bmatrix} 0.88 \\ 0.48 \end{bmatrix}$	0.06
3	0.75	$\begin{bmatrix} 0.84 \\ 0.74 \end{bmatrix}$	$\begin{bmatrix} -0.74 \\ 0.81 \end{bmatrix}$	$\begin{bmatrix} 0.73 \\ 0.68 \end{bmatrix}$	0.13
4	1.0	$\begin{bmatrix} 0.66 \\ 0.95 \end{bmatrix}$		$\begin{bmatrix} 0.54 \\ 0.84 \end{bmatrix}$	0.17

The actual solution is

$$\mathbf{x}(t) = \begin{bmatrix} \cos t \\ \sin\ t \end{bmatrix}$$

and hence

$$\mathbf{x}(0.25) = \begin{bmatrix} 0.97 \\ 0.25 \end{bmatrix} \qquad \mathbf{x}(0.5) = \begin{bmatrix} 0.88 \\ 0.48 \end{bmatrix}$$

$$\mathbf{x}(0.75) = \begin{bmatrix} 0.73 \\ 0.68 \end{bmatrix} \qquad \mathbf{x}(1) = \begin{bmatrix} 0.54 \\ 0.84 \end{bmatrix}$$

7.5.2. Example

Obtain an approximate solution of

$$x' = \frac{2x}{t} \qquad x(1) = 1 \qquad 1 \leqslant t \leqslant 2$$

Solution

Here we select $h = 0.2$. Then

$$y_0 = x_0 = 1 \qquad f_0 = \frac{2(1)}{1} = 2.0000$$

Hence,

$$y_1 = 1 + (0.2)2 = 1.4000 \qquad f_1 = \frac{2(1.4)}{1.2} = 2.3333$$

Hence,

$$y_2 = 1.4 + (0.2)(2.3333) = 1.8667 \qquad f_2 = \frac{2(1.8667)}{1.4} = 2.6667$$

Hence,

$$y_3 = 1.87 + (0.2)(2.6667) = 2.4000 \qquad f_3 = \frac{2(2.4)}{1.6} = 3.0000$$

Hence,

$$y_4 = 2.4 + (0.2)(3) = 3 \qquad f_4 = \frac{2.3}{1.8} = 3.3333$$

Hence,

$$y_5 = 3 + (0.2)(3.3) = 3.6667$$

Since the unique solution of the given problem is $x(t) = t^2$, we find $x(1.2) = 1.41$, $x(1.4) = 1.96$, $x(1.6) = 2.56$, $x(1.8) = 3.24$, and $x(2) = 4$. (See Table 7.5.2.)

Table 7.5.2

No. of steps	t_k	y_k	f_n	$x(t_n)$	$E = \lvert x(t_k) - y_k \rvert$
0	1.0	1.0000	2.0000	1.0000	0.0000
1	1.2	1.4000	2.3333	1.4100	0.0100
2	1.4	1.8667	2.6667	1.9600	0.0933
3	1.6	2.4000	3.0000	2.5600	0.1600
4	1.8	3.0000	3.3333	3.2400	0.2400
5	2.0	3.6667		4.0000	0.3333

Exercise 7.5

In the following five problems use Euler's Method to construct approximation solutions. In each case complete a table such as illustrated in Table 7.4.1.

1. $x' = -x, \quad x(0) = 1; \qquad 0 \leqslant t \leqslant 0.5; \qquad h = 0.1$
2. $x'' + x = 0, \quad x(0) = 1, \quad x'(0) = 0; \qquad 0 \leqslant t \leqslant 0.8, \quad h = 0.2$
   Hint: Convert to the companion system.
3. $\mathbf{x}' = \begin{bmatrix} 0 & -1 \\ 1 & 0 \end{bmatrix} \mathbf{x} \qquad \mathbf{x}(0) = \begin{bmatrix} 1 \\ 0 \end{bmatrix}$
   (a) $h = 0.3 \qquad 0 \leqslant t \leqslant 0.9$
   (b) $h = 0.5 \qquad 0 \leqslant t \leqslant 2$
   (c) $h = 0.1 \qquad 0 \leqslant t \leqslant 0.5$
4. $x' = -x + 1, \quad x(0) = 1; \qquad h = 0.1, \quad 0 \leqslant t \leqslant 1$
5. $x_1' = x_1 - x_1 x_2 \qquad x_1(0) = 10$
   $x_2' = x_1 x_2 \qquad x_2(0) = 2$
   $h = 1 \qquad 0 \leqslant t \leqslant 10$

7.6.* EXISTENCE AND UNIQUENESS

In order to approximate a solution by Euler's Method, we must first be assured of the existence of a solution.† Such assurance comes from the so-called "existence" theorems, appropriately enough. We discuss some such theorems in this section.

Basically, we aim to restrict the behavior of $\mathbf{f}$ by sufficient hypotheses to guarantee at least one solution of the initial-value problem

$$\mathbf{x}' = \mathbf{f}(t, \mathbf{x}) \qquad \mathbf{x}(t_0) = \mathbf{x}_0 \tag{7.6.1}$$

for all t in some interval containing t_0. To understand the nature of these hypotheses, it is necessary to view $\mathbf{f}(t, \mathbf{x})$ as a function of the $n+1$ variables t and the n components of $\mathbf{x}$; $x_1, x_2, \ldots, x_n$. Our analysis is assumed to apply throughout the bounded domains defined by $|t - t_0| \leqslant \mathrm{T}$ and $\|\mathbf{x} - \mathbf{x}_0\| \leqslant \mathrm{R}$.

†Some proofs of existence reverse this order and use an approximation method as a basic tool. Given the appropriate hypotheses, the approximation method is shown to converge to a function and then this function is demonstrated to be a solution. The names of Picard and Cauchy are prominent in this type of proof.

We note the inequality $\|\mathbf{x}-\mathbf{x}_0\| \leqslant \mathrm{R}$ implies the n inequalities,

$$|x_i - x_{0i}| \leqslant \|\mathbf{x}-\mathbf{x}_0\| \leqslant \mathrm{R} \qquad i = 1, \ldots, n$$

and *is implied by* the n inequalities

$$|x_i - x_{0i}| \leqslant n^{-1/2}\mathrm{R}$$

We turn to some examples of these ideas.

7.6.1. Example

The scalar equation

$$y' = y^2 + t^2 \qquad y(0) = 0$$

defines $f(t, y) = y^2 + t^2$ in every pair of intervals, $|t| \leqslant \mathrm{T}$ and $|y| \leqslant \mathrm{R}$. This f is continuous because it is a polynomial in t and y.

7.6.2. Example

The second-order equation

$$y'' + ty' = y^2 + t \qquad y(0) = 0, \quad y'(0) = 0$$

is studied by reference to its companion system. Set, as usual, $y = x_1$ and $y' = x_2$. Then $x_1' = y' = x_2$ and $x_2' = y'' = -ty' + y^2 + t = -tx_2 + x_1^2 + t$. Therefore,

$$\mathbf{x}' = \begin{bmatrix} x_1 \\ x_2 \end{bmatrix}' = \mathbf{f}(t, \mathbf{x}) = \begin{bmatrix} x_2 \\ x_1^2 - tx_2 + t \end{bmatrix}$$

f is continuous if its two components, $f_1(t, \mathbf{x}) = x_2$ and $f_2(t, \mathbf{x}) = x_1^2 - tx_2 + t$ are continuous. These components are polynomials in t, x_1, and x_2 and are thus continuous in every triple of intervals, $|t| \leqslant \mathrm{T}$, $|x_1| \leqslant (\mathrm{R}/\sqrt{2})$, $|x_2| \leqslant (\mathrm{R}/\sqrt{2})$, and thus in $|t| \leqslant \mathrm{T}$ and $\|\mathbf{x}\| = (x_1^2 + x_2^2)^{1/2} \leqslant \mathrm{R}$.

7.6.3. Example

The linear system

$$\mathbf{x}' = \mathrm{A}(t)\mathbf{x} \qquad \mathbf{x}(0) = \mathbf{x}_0$$

defines $\mathbf{f}(t, \mathbf{x}) = \mathrm{A}(t)\mathbf{x}$. Suppose $\mathrm{A}(t)$ is continuous on $|t - t_0| \leqslant \mathrm{T}$. The vector **x** is continuous as a function of **x** on every "interval," $\|\mathbf{x}-\mathbf{x}_0\| \leqslant \mathrm{R}$. The product of continuous functions is surely continuous and hence $\mathrm{A}(t)\mathbf{x}$ is continuous for all t in $|t - t_0| \leqslant \mathrm{T}$ and every **x** in $\|\mathbf{x}-\mathbf{x}_0\| \leqslant \mathrm{R}$, R arbitrary.

7.6.4. Example

The nonlinear system

$$\begin{bmatrix} x_1 \\ x_2 \end{bmatrix}' = \begin{bmatrix} (t-5)^{1/2} \sec x_2 \\ (x_1 - 2)^{1/2} \end{bmatrix} \qquad \mathbf{x}(7) = \begin{bmatrix} 3 \\ 0 \end{bmatrix}$$

has $\mathbf{f}(t, \mathbf{x}) = \begin{bmatrix} (t-5)^{1/2} \sec x_2 \\ (x_1-2)^{1/2} \end{bmatrix}$ defined when

$$|t-7| \leqslant 2$$
$$|x_1-3| \leqslant 1$$

and

$$|x_2-0| < \frac{\pi}{2}$$

We may thus take T = 2 and R = 1.

It is a theorem in real analysis that a continuous function **f**, in $|t-t_0| \leqslant \mathrm{T}$, $\|\mathbf{x}-\mathbf{x}_0\| \leqslant \mathrm{R}$ has its norm bounded by a constant independent of t and **x**. Let us define M as any constant for which

$$\|\mathbf{f}\| \leqslant \mathrm{M} \tag{7.6.2}$$

holds. The continuity of **f** is sufficient to insure the existence of a solution (7.6.1) but not, however, on all of the interval $|t-t_0| \leqslant \mathrm{T}$. Define the constant δ by

$$\delta = \text{minimum}\,[\mathrm{T}, \mathrm{R}/\mathrm{M}] \tag{7.6.3}$$

7.6.1. THEOREM (Existence). *If* **f** *is continuous on* $|t-t_0| \leqslant \mathrm{T}$, $\|\mathbf{x}-\mathbf{x}_0\| \leqslant \mathrm{R}$ *then*

$$\mathbf{x}' = \mathbf{f}(t, \mathbf{x}) \qquad \mathbf{x}(t_0) = \mathbf{x}_0$$

has at least one solution valid on $|t-t_0| < \delta$.

PROOF For a proof and a further discussion of this theorem see the work by Hurewicz (see footnote on p. 192).

The hypothesis of continuity is, unfortunately, not sufficient to guarantee the uniqueness of the solution. In Example 7.3.2 it was shown that

$$y' = \tfrac{3}{2}y^{1/3} \qquad \text{and} \qquad y(0) = 0$$

has infinitely many solutions. This is therefore a counterexample to the (incorrect) conjecture that continuity of **f** implies uniqueness. Another hypothesis is required. Study the following inequality for $|t-t_0| \leqslant \mathrm{T}$, $\|\mathbf{u}-\mathbf{x}_0\| \leqslant \mathrm{R}$ and $\|\mathbf{v}-\mathbf{x}_0\| \leqslant \mathrm{R}$

$$\|\mathbf{f}(t, \mathbf{u}) - \mathbf{f}(t, \mathbf{v})\| \leqslant \mathrm{K}\|\mathbf{u}-\mathbf{v}\| \tag{7.6.4}$$

We mean by this inequality that the constant K exists independent of t in $|t-t_0| \leqslant \mathrm{T}$ and **u** and **v** in $\|\mathbf{x}-\mathbf{x}_0\| \leqslant \mathrm{R}$. Not every **f** will satisfy such an inequality. Inequality (7.6.4) is known as a *Lipschitz Condition* on **f**.

7.6.2. THEOREM *If, in addition to continuity,* **f** *satisfies a Lipschitz Condition* (7.6.4), *then*

$$\mathbf{x}' = \mathbf{f}(t, \mathbf{x}) \qquad \mathbf{x}(t_0) = \mathbf{x}_0$$

has one and only one solution valid in $|t-t_0| \leqslant \delta =$ *minimum* [T, R/M].

PROOF We refer to Hurewicz once again (see footnote, p. 192). Note all the constants, T, R, and M are defined as before.

As a consequence of Theorem 7.6.2, we shall prove that linear systems have unique solutions.

7.6.3. THEOREM *If* $A(t)$ *is continuous in* $|t-t_0| \leq T$, *then*

$$\mathbf{x}' = A(t)\mathbf{x} \qquad \mathbf{x}(t_0) = \mathbf{x}_0$$

has a unique solution valid in $|t-t_0| < \delta$.

PROOF We have already remarked that continuity of A on $|t-t_0| \leq T$ implies continuity of $A(t)\mathbf{x}$ for $|t-t_0| \leq T$ and $\|\mathbf{x}-\mathbf{x}_0\| \leq R$, for every R. Since $A(t)$ is continuous in $|t-t_0| \leq T$, $\|A(t)\| \leq M$ in $|t-t_0| \leq T$. Thus,

$$\begin{aligned}\|A(t)\mathbf{u}-A(t)\mathbf{v}\| &= \|A(t)(\mathbf{u}-\mathbf{v})\| \\ &\leq \|A(t)\|\,\|\mathbf{u}-\mathbf{v}\| \\ &\leq M\|\mathbf{u}-\mathbf{v}\|\end{aligned}$$

Hence, $\mathbf{f}(t,\mathbf{x}) = A(t)\mathbf{x}$ satisfies a Lipschitz Condition with $K=M$. From Theorem 7.6.2 there exists a unique solution in $|t-t_0| < \delta$, and this completes the proof.

This is not the best result possible, but it illustrates the point we wish to make.

Exercise 7.6

1. Show that there is no constant K for which

$$|f(t,y_1)-f(t,y_2)| \leq K|y_1-y_2|$$

where

$$f(t,y) = \tfrac{3}{2}y^{1/3} \quad \text{and} \quad |t| < T$$

Hint:

$$\frac{y_1^{1/3}-y_2^{1/3}}{y_1-y_2} = \left[\frac{y_1-y_2}{y_1^{1/3}-y_2^{1/3}}\right]^{-1} = [y_1^{2/3}+y_1^{1/3}y_2^{1/3}+y_2^{2/3}]^{-1}$$

which is unbounded for y_1 and y_2 near zero.

2. Consider $y' = y^2+t^2$, $y(0) = 0$. Suppose $|t| \leq T = 1$ and $|y| \leq R$. Show that $M = R^2+1$ and hence prove that $\delta \leq \frac{1}{2}$. For what choice of R is δ maximized?
3. Recall the mean-value theorem: If f is continuous and differentiable on $a \leq t \leq b$ then there exists $\zeta \in (a,b)$ such that

$$f(b)-f(a) = f'(\zeta)(b-a)$$

Suppose $f(t,y)$ is continuous on $|t| \leq T$ and $|y| \leq R$. Suppose $\partial f/\partial y$ exists and is continuous on $|t| \leq T$, $|y| \leq R$. Show via the mean-value theorem that that

$$|f(t,y_1)-f(t,y_2)| \leq K|y_1-y_2|$$

for t in $|t| \leq T$ and y_1, y_2 in $|y| \leq R$.

4.* Following the technique of Problem 3, show that the continuity on $\|\mathbf{x}-\mathbf{x}_0\| \leqslant R$ of all the first partial derivatives of $\mathbf{f}$ with respect to $x_1, x_2, \ldots, x_n$, respectively, implies a Lipschitz condition of $\mathbf{f}$. Hint: Work with each component of $\mathbf{f}$ separately.

7.7.* THE ACCURACY OF THE EULER METHOD

One consequence of the existence and uniqueness theorems of the previous section is that with the addition of one more hypothesis we can estimate the error in using the Euler Method. We begin with a review of some terminology. Recall the Euler Method consists of computing

$$\begin{aligned} \mathbf{y}_0 &= \mathbf{x}(t_0) \\ \mathbf{y}_{k+1} &= \mathbf{y}_k + h\mathbf{f}_k \qquad k = 0, 1, \ldots, N-1 \end{aligned} \tag{7.7.1}$$

where the following abbreviations have been used:

$$\begin{aligned} t_k &= t_0 + kh \\ \mathbf{y}_k &= \mathbf{y}(t_0 + kh) = \mathbf{y}(t_k) \\ \mathbf{f}_k &= \mathbf{f}(t_k, \mathbf{y}_k) = \mathbf{f}(t_0 + kh, \mathbf{y}(t_0 + kh)) \end{aligned} \tag{7.7.2}$$

Thus we are supplied with a table of values; $\mathbf{y}_0, \mathbf{y}_1, \ldots, \mathbf{y}_N$ as approximations to $\mathbf{x}(t)$, at $t_0, t_1, \ldots, t_N$. It is convenient to construct from these values a continuous, "piecewise" linear function $\mathbf{y}(t)$, defined for all t in $t_0 \leqslant t \leqslant t_N$, such that $\mathbf{y}(t_k) = \mathbf{y}_k$, $k = 0, 1, \ldots, N$. The following is such a function: Let

$$\mathbf{y}(t) = \mathbf{y}_k + (t - t_k)\mathbf{f}_k \qquad t_k \leqslant t \leqslant t_{k+1} \tag{7.7.3}$$

On each interval $[t_k, t_{k+1}]$, $k = 0, 1, \ldots, N-1$, $\mathbf{y}(t)$ is linear and clearly $\mathbf{y}$ is continuous. In the scalar case, the graph of $y(t)$ defined by (7.7.3) is a continuous polygonal line whose "corners" occur at $t_1, t_2, \ldots, t_N$. (See Figure 7.7.1.)

The approximating function, $\mathbf{y}(t)$, is differentiable throughout $[t_0, t_N]$ except at the corners, $\mathbf{y}_0, \mathbf{y}_1, \ldots, \mathbf{y}_N$. In fact, $\mathbf{y}'(t) = \mathbf{f}_k$ for $t_k < t < t_{k+1}$. Let us extend the domain of definition to include the corners by defining

$$\mathbf{y}'(t) = \mathbf{f}_k \quad \text{for} \quad t_k \leqslant t < t_{k+1} \tag{7.7.4}$$

Figure 7.7.1

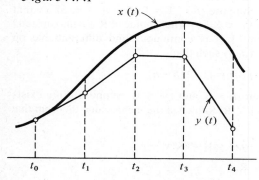

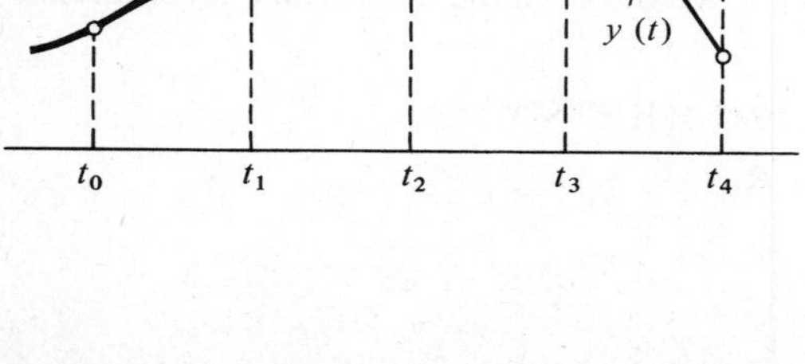

Equation (7.7.4) is not to be interpreted as implying the existence of a derivative of $\mathbf{y}$ at t_k; rather, we have extended $\mathbf{y}'$ so that for all $t \neq t_k$, $k = 0, 1, \ldots, N-1$, $\mathbf{y}'$ is the derivative of $\mathbf{y}$ and at t_k it is defined as $\mathbf{f}_k$.

The first problem we encounter is: for a given step size h, how long can we continue computing $\mathbf{y}_k$? We assume that $\mathbf{f}$ satisfies the hypothesis of Theorem 7.6.2, that is,

(1) $\mathbf{f}$ is continuous in $|t-t_0| \leq T$, $\|\mathbf{x}-\mathbf{x}_0\| \leq R$

(2) $\|\mathbf{f}\| \leq M$ in $|t-t_o| \leq T$, $\|\mathbf{x}-\mathbf{x}_0\| \leq R$

(3) $\|\mathbf{f}(t, \mathbf{u}) - \mathbf{f}(t, \mathbf{v})\| \leq K\|\mathbf{u}-\mathbf{v}\|$, for all t in $|t-t_0| \leq T$ and any $\mathbf{u}$ and $\mathbf{v}$ in $\|\mathbf{x}-\mathbf{x}_0\| \leq R$

In order to compute $\mathbf{y}_{k+1} = \mathbf{y}_k + h\mathbf{f}_k$ we need $kh < T$ to insure $|t-t_0| < T$ and $\|\mathbf{y}_k - \mathbf{y}_0\| \leq R$, $k = 0, 1, \ldots, N$. Suppose $Nh < T$ and $\|\mathbf{y}-\mathbf{y}_0\| \leq R$ but $\|\mathbf{y}_{N+1} - \mathbf{y}_0\| > R$. Then

$$\begin{aligned} R < \|\mathbf{y}_{N+1} - \mathbf{y}_0\| &= \|\mathbf{y}_{N+1} - \mathbf{y}_N + \mathbf{y}_N + \mathbf{y}_{N-1} - \mathbf{y}_{N-1} + \cdots - \mathbf{y}_0\| \\ &\leq \sum_{k=1}^{N+1} \|\mathbf{y}_k - \mathbf{y}_{k-1}\| = \sum_{k=1}^{N+1} h\|\mathbf{f}_k\| \\ &\leq hM(N+1) \end{aligned}$$

Therefore, $h(N+1) > R/M$. Thus our computation procedure may be carried out at least N steps so long as

$$hN \leq T \qquad \text{and} \qquad hN \leq R/M$$

To put it another way, as long as

$$hN \leq \text{minimum}\left\{T, \frac{R}{M}\right\} = \delta$$

Thus, in $|t-t_0| < \delta$ we are guaranteed both a unique solution *and* the computability of an approximate solution.

We now supplement the three conditions (1), (2), and (3) with

(4) $\|\mathbf{f}(t^*, \mathbf{x}) - \mathbf{f}(t^{**}, \mathbf{x})\| \leq F|t^* - t^{**}|$ for any $\mathbf{x}$ in $\|\mathbf{x}-\mathbf{x}_0\| \leq R$ and any choice of t^*, t^{**} belonging to $|t-t_0| \leq T$.

This is another Lipschitz Condition.

Let us define $E(t)$ by

$$E(t) = \|\mathbf{f}(t, \mathbf{y}(t)) - \mathbf{y}'(t)\| \tag{7.7.5}$$

Thus $E(t)$ is a measure of the amount $\mathbf{y}'$ fails to satisfy the differential equation. We prove the following lemma.

7.7.1. LEMMA *Subject to conditions* (1) *to* (4) *above,*

$$E(t) \leq (F + MK)h \tag{7.7.6}$$

for $t_0 \leq t \leq t_N \leq T$.

PROOF First restrict t to the interval, $t_k \leqslant t \leqslant t_{k+1}$. Then, $|t-t_k| < h = |t_{k+1}-t_k|$ and from Equations (7.7.3) and inequality (2),

$$\begin{aligned}\|\mathbf{y}(t)-\mathbf{y}_k\| &= |t-t_k|\,\|\mathbf{f}_k\| \\ &\leqslant h\mathrm{M}\end{aligned} \tag{7.7.7}$$

We have, therefore,

$$\begin{aligned}\mathrm{E}(t) &= \|\mathbf{f}(t,\mathbf{y}(t))-\mathbf{y}'(t)\| \\ &= \|\mathbf{f}(t,\mathbf{y}(t))-\mathbf{f}(t_k,\mathbf{y}(t))+\mathbf{f}(t_k,\mathbf{y}(t))-\mathbf{y}'(t)\| \\ &\leqslant \|\mathbf{f}(t,\mathbf{y}(t))-\mathbf{f}(t_k,\mathbf{y}(t)\|+\|\mathbf{f}(t_k,\mathbf{y}(t))-\mathbf{y}'(t)\|\end{aligned}$$

Applying (7.7.4) and inequality (4) to this latter inequality, we find, first,

$$\mathrm{E}(t) \leqslant \mathrm{F}|t-t_k|+\|\mathbf{f}(t_k,\mathbf{y}(t))-f(t_k,\mathbf{y}_k)\|$$

and from (3),

$$\mathrm{E}(t) \leqslant \mathrm{F}h+\mathrm{K}\|\mathbf{y}(t)-\mathbf{y}_k\|$$

and from (7.7.7),

$$\mathrm{E}(t) \leqslant \mathrm{F}h+\mathrm{K}h\mathrm{M} = (F+\mathrm{KM})h$$

This inequality is true in each of the intervals $[t_k, t_{k+1}]$ $k=0, 1, \ldots, \mathrm{N}-1$ and is independent of t. Therefore it is true for all t in $[t_0, t_\mathrm{N}]$.

With the lemma proved we can now turn to the central question: How does $\|\mathbf{x}(t)-\mathbf{y}(t)\|$ depend upon h? We note that

$$\begin{aligned}\mathbf{x}(t) &= \mathbf{x}(t_0)+\int_{t_0}^{t} \mathbf{x}'(s)\,dx \\ &= \mathbf{x}(t_0)+\int_{t_0}^{t} \mathbf{f}(s,\mathbf{x}(s))\,ds\end{aligned}$$

and

$$\mathbf{y}(t) = \mathbf{y}(t_0)+\int_{t_0}^{t} \mathbf{y}'(s)\,ds \qquad \text{(Why?)}$$

Since $\mathbf{y}(t_0) = \mathbf{x}_0 = \mathbf{x}(t_0)$, we have

$$\begin{aligned}\|\mathbf{x}(t)-\mathbf{y}(t)\| &= \left\|\int_{t_0}^{t} \{\mathbf{f}(s,\mathbf{x}(s))-\mathbf{y}'(s)\}\,ds\right\| \\ &\leqslant \int_{t_0}^{t} \|\mathbf{f}(s,\mathbf{x}(s))-\mathbf{y}'(s)\|\,ds\end{aligned}$$

Therefore, after addition and subtraction of $\mathbf{f}(s,\mathbf{y}(s))$ and an application of the triangle inequality:

$$\begin{aligned}\|\mathbf{x}(t)-\mathbf{y}(t)\| \leqslant &\int_{t_0}^{t} \|\mathbf{f}(s,\mathbf{x}(s))-\mathbf{f}(s,\mathbf{y}(s))\|\,ds \\ &+\int_{t_0}^{t} \|\mathbf{f}(s,\mathbf{y}(s))-\mathbf{y}'(s)\|\,ds\end{aligned}$$

We use inequality (3) in the first integral and the lemma in the second to deduce

$$\begin{aligned}\|\mathbf{x}(t)-\mathbf{y}(t)\| &\leq \int_{t_0}^{t} \mathrm{K}\,\|\mathbf{x}(s)-\mathbf{y}(s)\|\,ds \\ &\quad+\int_{t_0}^{t} \mathrm{E}(s)\,ds \\ &\leq \int_{t_0}^{t} \mathrm{K}\|\mathbf{x}(s)-\mathbf{y}(s)\|ds+(\mathrm{F}+\mathrm{MK})h(t-t_0)\end{aligned}$$

But $|t-t_0| < \delta$ and hence

$$\|\mathbf{x}(t)-\mathbf{y}(t)\| \leq \int_{t_0}^{t} \mathrm{K}\|\mathbf{x}(s)-\mathbf{y}(s)\|\,ds+(\mathrm{F}+\mathrm{MK})h\,\delta$$

We now apply Gronwall's Inequality, Theorem 4.6.2, which states that

$$f(t) \leq \hat{\mathrm{K}}+\int_a^t f(s)g(s)\,ds$$

implies

$$f(t) \leq \hat{\mathrm{K}}\exp\left[\int_a^t g(s)\,ds\right]$$

Setting $\hat{\mathrm{K}}=(\mathrm{F}+\mathrm{MK})h\,\delta$, $g(s) \equiv \mathrm{K}$, and $f(t)=\|\mathbf{x}(t)-\mathbf{y}(t)\|$, we obtain

$$\begin{aligned}\|\mathbf{x}(t)-\mathbf{y}(t)\| &\leq (\mathrm{F}+\mathrm{MK})\,\delta h\exp\left[\int_{t_0}^{t} \mathrm{K}\,ds\right] \\ &= (\mathrm{F}+\mathrm{MK})\,\delta h\exp\,[\mathrm{K}(t-t_0)] \\ &\leq (\mathrm{F}+\mathrm{MK})\,\delta\exp\,(\mathrm{K}\,\delta)\,h\end{aligned}$$

7.7.2. THEOREM *Subject to the constraints listed as* (1), (2), (3), (4),

$$\|\mathbf{x}(t)-\mathbf{y}(t)\| \leq (\mathrm{F}+\mathrm{MK})\,\delta e^{\mathrm{K}\delta}\,h$$

for all t, $|t-t_0| \leq t_{\mathrm{N}} < \delta$

We have assumed throughout this discussion that all the arithmetic computations and each evaluation of **f** is carried out with perfect accuracy—that is, no round-off errors. This is unrealistic in practice. However, the inclusion of this additional source of errors will take us too deeply into an area better reserved for a separate study. The interested reader is referred to any of a large number of works or numerical analysis.†

We conclude with an example.

7.7.1. Example

Consider the scalar equation

$$y'=t^2+y^2 \qquad y(0)=0$$

† The text, D. Moursand and C. Duris, *Elementary Theory and Application of Numerical Analysis*, McGraw-Hill, New York, 1967, is one such.

We restrict $f(t, y) \equiv t^2 + y^2$ to $|y| \leq 2$ and $|t| \leq 1$. Thus $R = 2$ and M may be selected as

$$M = \max |f(t, y)| = 5$$

Hence $\delta =$ minimum $[1, \frac{2}{5}] = \frac{2}{5}$ and we can apply the Euler Method only for $0 \leq t \leq \frac{2}{5}$. We choose K and F so that

$$K = \max \left| \frac{f(t, u) - f(t, v)}{u - v} \right| = \max \left| \frac{u^2 - v^2}{u - v} \right| = 4$$

$$F = \max \left| \frac{f(t^*, x) - f(t^{**}, x)}{t^{**} - t} \right|$$

$$= \max |t^* + t^{**}| = \tfrac{4}{5}$$

According to Theorem 7.7.2,

$$|x(t) - y(t)| \leq (\tfrac{4}{5} + 5 \cdot 4)\tfrac{2}{5}e^{4/5}h$$

$$< 20h$$

To guarantee an error smaller than 0.02 for $0 \leq t \leq \frac{2}{5}$, we would require $h = 0.001$ and hence $N = 200$ steps. An error of 0.04 would require $h = 0.002$ and $N = 100$.

This illustrates the fact that the error in Euler's Method is proportional to the step size, h. One means of judging an approximation scheme is to compare the error change as a function of h. It is not difficult to alter the Euler Method and arrive at a numerical method whose error is proportional to h^2. We do not include this analysis.

Exercise 7.7

1. Consider the problem $y' = y$, $y(0) = 1$. According to our estimate, what size step is necessary to compute $y(0.5)$ to within 0.001 of the true value? Use $R = 1$ to show $h \geq 0.001/1.3$.
2. In the problem $y' = t^2 + y^2$, $y(0) = 1$, restrict the domain of $f(t, y) = t^2 + y^2$ to $|t| \leq 0.246$ and $|y - 1| \leq 1.03$. Show that $\delta = 0.247$, approximately.
3. Same as Problem 2 except $|t| \leq 1$ and $|y - 1| \leq 1$. Show that $\delta = 2$ in this case.

Chapter 8
TWO-DIMENSIONAL AUTONOMOUS SYSTEMS: AN INTRODUCTION TO THE QUALITATIVE THEORY OF DIFFERENTIAL EQUATIONS

8.1. INTRODUCTION: THE PHASE PLANE

We have seen a few of the many techniques for dealing with linear differential equations having variable as well as constant coefficients. In some cases these techniques give us explicit solutions, in other cases they give us a description of the nature of solutions. On the other hand, our treatment in Chapter 7 of nonlinear equations has been limited to obtaining approximate solutions by numerical methods and is satisfactory for many purposes. However, there are certain "qualitative" properties of solutions that are not revealed by such analysis. Some of these properties; such as "equilibrium," "periodicity," and "stability," will be treated in this chapter. The study of the qualitative theory of differential equations is quite difficult. We therefore shall limit our discussion to the two-dimensional autonomous case, which is accessible on an elementary level and for which there exists a good deal of definitive information. Throughout this chapter we shall treat the autonomous system

$$\mathbf{x}' = \mathbf{f}(\mathbf{x})$$

where $\mathbf{f}$ is a two-dimensional vector-valued function of a two-dimensional vector variable. We assume that $\mathbf{f}$ is such that every initial-value problem $\mathbf{x}' = \mathbf{f}(\mathbf{x})$, $\mathbf{x}(t_0) = \mathbf{x}_0$ has a unique solution valid for all t. This will be the case, for example, if $\mathbf{f}$ has continuous first partial derivatives with respect to the components of $\mathbf{x}$, and for some constant K, $\|\mathbf{f}(\mathbf{x})\| \leq K\|\mathbf{x}\|$. (See Hurewicz reference in footnote on p. 192.)

The unique solution of $\mathbf{x}' = \mathbf{f}(\mathbf{x}),\ \mathbf{x}(t_0) = \mathbf{x}_0$ is a function, say

$$\mathbf{x}(t) = \begin{bmatrix} x_1(t) \\ x_2(t) \end{bmatrix}$$

whose components may be viewed as the parametric representation of some curve in the x_1, x_2-plane. Equivalently, $\mathbf{x}(t)$ may be viewed as a position vector emanating from (0, 0) and terminating at (x_1, x_2). As t varies, the "tip" of $\mathbf{x}(t)$ sweeps out a locus in the x_1,x_2-plane. This locus is known as a *solution curve* and the x_1, x_2-plane in which it is drawn is called the *phase plane.* The derivative of the position vector is a velocity vector and, if nonzero, is tangent to the solution curve. (Why?) Thus $\mathbf{x}'(t) = \mathbf{f}(\mathbf{x})$ may be interpreted as the velocity a particle has as it traverses the solution curve. This assigns a direction to the solution curve which we shall portray by an arrow. To make these ideas concrete consider the following examples and figures.

8.1.1. Example

Draw the solution curve for

$$\mathbf{x}' = \begin{bmatrix} 0 & -1 \\ 1 & 0 \end{bmatrix} \mathbf{x} \qquad \mathbf{x}(0) = \begin{bmatrix} 1 \\ 0 \end{bmatrix} \tag{8.1.1}$$

and indicate its orientation.

Solution

We know that

$$\mathbf{x}(t) = \begin{bmatrix} \cos t \\ \sin t \end{bmatrix}$$

is the unique solution to the given initial-value problem. Hence $x_2(t) = \sin t$ and $x_1(t) = \cos t$ is a parametric representation of the solution curve, the circle $x_1^2 + x_2^2 = 1$. A particle moving according to $\mathbf{x}(t)$ traverses the circle in a counterclockwise direction. We indicate this in Figure 8.1.1.

8.1.2. Example

Redo Example 8.1.1 with the initial condition now reading

$$\mathbf{x}(s) = \begin{bmatrix} 1 \\ 0 \end{bmatrix}$$

Solution

Since $x_2(t) = \sin (t - s)$ and $x_1(t) = \cos (t - s)$, the circle $x_1^2 + x_2^2 = 1$ is the solution curve. Although the solutions of

$$\mathbf{x}' = \begin{bmatrix} 0 & -1 \\ 1 & 0 \end{bmatrix} \mathbf{x} \qquad \mathbf{x}(s) = \begin{bmatrix} 1 \\ 0 \end{bmatrix} \tag{8.1.2}$$

Figure 8.1.1. The phase plane.

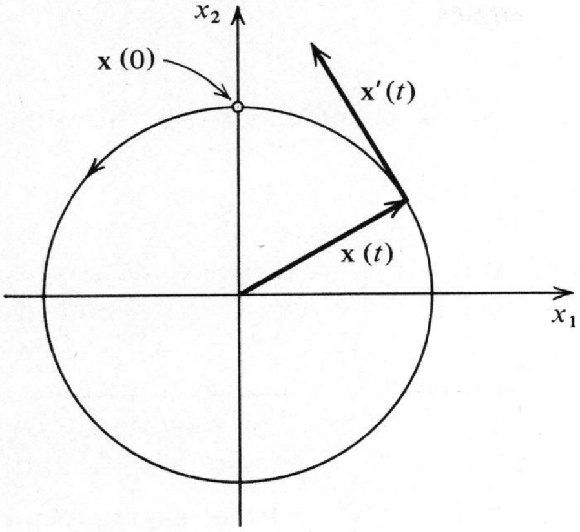

depend on s the solution curves do not! This latter result is a feature of autonomous systems not usually found in the nonautonomous case. Witness,

$$\mathbf{x}' = \begin{bmatrix} 1 & 0 \\ 0 & \frac{1}{t} \end{bmatrix} \mathbf{x} \qquad \mathbf{x}(s) = \begin{bmatrix} 1 \\ 1 \end{bmatrix} \tag{8.1.3}$$

whose solution

$$\mathbf{x}(t) = \begin{bmatrix} e^{t-s} \\ \frac{t}{s} \end{bmatrix}$$

leads to $x_1 = e^{s(x_2-1)}$. The choices $s = 0.5$ and $s = 1$ are illustrated in Figure 8.1.2.

The study of $\mathbf{x}' = \mathbf{f}(\mathbf{x})$ through analysis of the phase plane is often called

Figure 8.1.2. Phase plane.

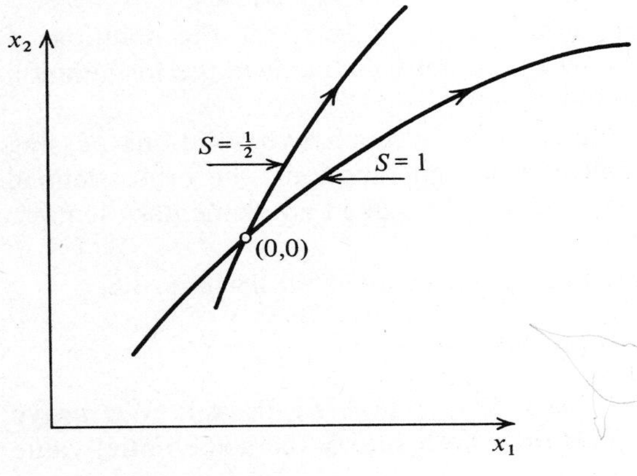

"qualitative" analysis as contrasted to the "quantitative" analysis typified by the numerical techniques of Chapter 7.

Exercise 8.1

1. In Figure 8.1.1, the orientation of $x_1^2+x_2^2=1$ is counterclockwise. Prove this fact.
2. Draw the solution curves of

$$\mathbf{x}' = \begin{bmatrix} 0 & 1 \\ -1 & 0 \end{bmatrix} \mathbf{x} \qquad \mathbf{x}(s) = \begin{bmatrix} 1 \\ 0 \end{bmatrix}$$

   in the phase plane.
3. Same as Problem 2, except

$$\mathbf{x}(s) = \begin{bmatrix} \alpha \\ \beta \end{bmatrix}$$

4. Explain why $\mathbf{x}'$ can be interpreted as a velocity. If $\mathbf{x}'(t) \neq \mathbf{0}$ the velocity vector is tangent to the solution curve. Why?

8.2. ORBITS AND THEIR PROPERTIES: I

In the previous section we introduced the phase plane and portrayed graphically the set of values taken on by a solution, $\mathbf{x}(t)$. We indicated with an arrow the order in which these values were assumed as t increases. We shall generally omit from such drawings the time at which a given value is assumed. In this section we make precise a number of ideas hinted at in the introduction and thus begin a study of the geometry of the solutions of two-dimensional autonomous systems.

8.2.1. DEFINITION *Let* $\mathbf{x}(t, t_0; \mathbf{x}_0)$ *be the unique solution of* $\mathbf{x}' = \mathbf{f}(\mathbf{x})$, $\mathbf{x}(t_0) = \mathbf{x}_0$. *Then the set*

$$O(t_0, \mathbf{x}_0) = \{\mathbf{x}(t, t_0; \mathbf{x}_0) \,|\, -\infty < t < \infty\}$$

is called the orbit or trajectory through $\mathbf{x}_0$.

(The use of the "expanded" notation, $\mathbf{x}(t, t_0; \mathbf{x}_0)$, for the solution of $\mathbf{x}' = \mathbf{f}(\mathbf{x})$, $\mathbf{x}(t_0) = \mathbf{x}_0$ is a temporary expedient which aids in the formulation of many of the results in this chapter.)

Our first theorem says essentially that although two solutions passing through $\mathbf{x}_0$ at different times may be different functions, the orbits defined by these solutions are the same sets. First we need an elementary lemma.

8.2.2. LEMMA *Let* $\mathbf{x}_0$ *be a fixed vector and* t_0 *an arbitrary time. Then*

$$\mathbf{x}(t, t_0; \mathbf{x}_0) = \mathbf{x}(t - t_0, 0; \mathbf{x}_0)$$

PROOF Let $\boldsymbol{\alpha}(t) = \mathbf{x}(t, t_0; \mathbf{x}_0)$ and $\boldsymbol{\beta}(t) = \mathbf{x}(t - t_0, 0; \mathbf{x}_0)$. We prove $\boldsymbol{\alpha}(t) \equiv \boldsymbol{\beta}(t)$ by demonstrating that they both satisfy the same initial-value problem, $\mathbf{x}' = \mathbf{f}(\mathbf{x}), \mathbf{x}(t_0) = \mathbf{x}_0$. We see immediately that

$$\boldsymbol{\alpha}(t_0) = \mathbf{x}(t_0, t_0; \mathbf{x}_0) = \mathbf{x}_0$$

and

$$\boldsymbol{\beta}(t_0) = \mathbf{x}(0, 0; \mathbf{x}_0) = \mathbf{x}_0$$

for by definition, $\mathbf{x}(s, s; \mathbf{x}_0) = \mathbf{x}_0$ for any s. Next,

$$\begin{aligned}\boldsymbol{\alpha}'(t_0) &= \mathbf{x}'(t, t_0; \mathbf{x}_0) = \mathbf{f}(\mathbf{x}(t, t_0; \mathbf{x}_0)) \\ &= \mathbf{f}(\boldsymbol{\alpha}(t))\end{aligned}$$

and

$$\begin{aligned}\boldsymbol{\beta}'(t) &= \frac{d}{dt}\mathbf{x}(t-t_0, 0; \mathbf{x}_0) \\ &= \mathbf{x}'(t-t_0, 0; \mathbf{x}_0)\frac{d}{dt}(t-t_0) \\ &= \mathbf{f}(\mathbf{x}(t-t_0, 0; \mathbf{x}_0)) \\ &= \mathbf{f}(\boldsymbol{\beta}(t))\end{aligned}$$

The uniqueness theorem, assumed to hold throughout this chapter, thus implies $\boldsymbol{\alpha}(t) \equiv \boldsymbol{\beta}(t)$.

8.2.3. THEOREM *For each* $\mathbf{x}_0$ *and* t_0,

$$\mathrm{O}(t_0, \mathbf{x}_0) = \mathrm{O}(0, \mathbf{x}_0)$$

PROOF We have, using Lemma 8.2.2,

$$\begin{aligned}\mathrm{O}(t_0, \mathbf{x}_0) &= \{\mathbf{x}(t, t_0; \mathbf{x}_0) | -\infty < t < \infty\} \\ &= \{\mathbf{x}(t-t_0, 0; \mathbf{x}_0) | -\infty < t < \infty\} \\ &= \{\mathbf{x}(s, 0; \mathbf{x}_0) | -\infty < s + t_0 < \infty\} \\ &= \{\mathbf{x}(s, 0; \mathbf{x}_0) | -\infty < s < \infty\} \\ &= \mathrm{O}(0, \mathbf{x}_0)\end{aligned}$$

(The reader should apply a reason for each equality in the above argument.) Since orbits through $\mathbf{x}_0$ do not depend on t_0 we may as well suppress the reference to t_0 in $\mathrm{O}(t_0; \mathbf{x}_0)$ and write $\mathrm{O}(\mathbf{x}_0)$ instead. Also, in referring to solutions of $\mathbf{x}' = f(\mathbf{x})$ whose solution curve passes through $\mathbf{x}_0$ at some time t_0, we may always select a solution for which $t_0 = 0$, by Lemma 8.2.2. Let us write, for simplicity, $\mathbf{x}(t; \mathbf{x}_0)$ to mean this solution. That is, $\mathbf{x}(t, 0; \mathbf{x}_0) = \mathbf{x}(t; \mathbf{x}_0)$. In slightly different terms, through each $\mathbf{x}_0$ passes an orbit and we may assume that a particle moving on this orbit arrives at $\mathbf{x}_0$ at $t = 0$.

A very important result, often referred to as the "group property" is our next theorem.

8.2.4. THEOREM *For each fixed* $\mathbf{x}_0$ *and every* t *and* s

$$\mathbf{x}(t+s; \mathbf{x}_0) = \mathbf{x}(t; \mathbf{x}(s; \mathbf{x}_0)) \tag{8.2.1}$$

PROOF Our principal weapon will again be the uniqueness theorem. Fix s and set

$$\mathbf{x}_1 = \mathbf{x}(s; \mathbf{x}_0)$$
$$\boldsymbol{\alpha}(t) = \mathbf{x}(t; \mathbf{x}_1)$$
$$\boldsymbol{\beta}(t) = \mathbf{x}(t+s; \mathbf{x}_0)$$

Then,

$$\boldsymbol{\alpha}(0) = \mathbf{x}(0; \mathbf{x}_1) = \mathbf{x}_1 \qquad \text{(Why?)} \tag{8.2.2}$$

and

$$\boldsymbol{\beta}(0) = \mathbf{x}(s; \mathbf{x}_0) = \mathbf{x}_1$$

Also,

$$\begin{aligned}\boldsymbol{\alpha}'(t) = \mathbf{x}'(t; \mathbf{x}_1) &= \mathbf{f}(\mathbf{x}'(t; \mathbf{x}_1)) \\ &= \mathbf{f}(\boldsymbol{\alpha}(t))\end{aligned}$$

and

$$\begin{aligned}\boldsymbol{\beta}'(t) &= \frac{d}{dt}\mathbf{x}(t+s; \mathbf{x}_0) \\ &= \mathbf{x}'(t+s; \mathbf{x}_0)\frac{d}{dt}(t+s) \\ &= \mathbf{f}(\mathbf{x}(t+s; \mathbf{x}_0)) \\ &= \mathbf{f}(\boldsymbol{\beta}(t))\end{aligned}$$

Hence $\boldsymbol{\alpha}(t)$ and $\boldsymbol{\beta}(t)$ solve $\mathbf{x}' = \mathbf{f}(\mathbf{x})$, $\mathbf{x}(0) = \mathbf{x}_1$ and are therefore identical.

8.2.1. Example

Verify Lemma 8.2.2 and Theorem 8.2.4 for the linear system

$$\mathbf{x}' = A\mathbf{x} \qquad \mathbf{x}(t_0) = \mathbf{x}_0$$

Solution

Let Φ be the fundamental matrix for $\mathbf{x}' = A\mathbf{x}$; normalized at 0. Then

$$\mathbf{x}(t, t_0; \mathbf{x}_0) = \Phi(t)\Phi^{-1}(t_0)\mathbf{x}_0$$

and

$$\begin{aligned}\mathbf{x}(t-t_0, 0; \mathbf{x}_0) &= \Phi(t-t_0)\mathbf{x}_0 \\ &= \Phi(t)\Phi^{-1}(t_0)\mathbf{x}_0\end{aligned}$$

So much for Lemma 8.2.2. To verify Theorem 8.2.4 consider,

$$\begin{aligned}\mathbf{x}(t+s; \mathbf{x}_0) &= \Phi(t+s)\mathbf{x}_0 \\ &= \Phi(t)\Phi(s)\mathbf{x}_0 \qquad \text{(Why?)} && (8.2.3) \\ &= \Phi(t)\,\mathbf{x}(s; \mathbf{x}_0) \qquad \text{(Why?)} && (8.2.4) \\ &= \Phi(t)\mathbf{x}_1 \\ &= \mathbf{x}(t; \mathbf{x}_1) \\ &= \mathbf{x}(t; \mathbf{x}(s; \mathbf{x}_0))\end{aligned}$$

We may interpret Theorem 8.2.4 dynamically. The left-hand side of (8.2.1) may be viewed as asserting that a particle subject to $\mathbf{x}' = \mathbf{f}(\mathbf{x})$ and starting at $\mathbf{x}_0$ when $t = 0$ will arrive at $\mathbf{x}(t+s; \mathbf{x}_0)$ in time $t+s$. Along the way, this particle will arrive at $\mathbf{x}_1 = \mathbf{x}'(s; \mathbf{x}_0)$ in time s. A second particle subject to $\mathbf{x}' = \mathbf{f}(\mathbf{x})$ but beginning its journey from $\mathbf{x}_1$ at time 0 will arrive at $\mathbf{x}(t+s; \mathbf{x}_0)$ in time $t = t+s-s$. If the system were nonautonomous we could not be sure that the first particle when it reaches $\mathbf{x}_1$ would then traverse the same path that the second particle travels since the two particles are at $\mathbf{x}_1$ at different times!

One special orbit of considerable importance occurs when $\mathbf{x}'$ vanishes.

8.2.5. DEFINITION *The point* $\mathbf{x}^*$ *is called a critical point of* $\mathbf{x}' = \mathbf{f}(\mathbf{x})$ *if*

$$\mathbf{f}(\mathbf{x}^*) = \mathbf{0}$$

Such points are also called equilibrium points.

8.2.6. THEOREM *The point* $\mathbf{x}^*$ *is a critical point of* $\mathbf{x}' = \mathbf{f}(\mathbf{x})$ *if and only if the constant function* $\mathbf{x}(t) \equiv \mathbf{x}^*$ *is a solution of* $\mathbf{x}' = \mathbf{f}(\mathbf{x})$.

PROOF If $\mathbf{x}(t) \equiv \mathbf{x}^*$ is a solution of $\mathbf{x}' = \mathbf{f}(\mathbf{x})$ then

$$\frac{d\mathbf{x}^*}{dt} = \mathbf{0} = \mathbf{f}(\mathbf{x}^*)$$

which implies that $\mathbf{x}^*$ is a critical point of $\mathbf{x}' = \mathbf{f}(\mathbf{x})$. Conversely, if $\mathbf{f}(\mathbf{x}^*) = \mathbf{0}$, then since $d\mathbf{x}^*/dt = \mathbf{0}$, $\mathbf{x}(t) = \mathbf{x}^*$ is a solution of $\mathbf{x}' = \mathbf{f}(\mathbf{x})$.

8.2.2. Example

What are the critical points of $\mathbf{x}' = A\mathbf{x}$?

Solution

Since $\mathbf{f}(\mathbf{x}) = A\mathbf{x}$ in this example, $A\mathbf{x}^* = \mathbf{0}$ is necessary and sufficient for $\mathbf{x}^*$ to be a critical point. Hence, if A is invertible (i.e., if det $A \neq 0$) $\mathbf{x}^* = \mathbf{0}$ is the sole critical point. If det $A = 0$ then there are infinitely many critical points. The case $A = 0$ shows, in fact, that every point may be a critical point. (See Exercise 8.2, Part 2, Problems 2,3, and 4.)

Exercise 8.2

Part 1

Find all the critical points for the following systems

1. $\mathbf{x}' = \begin{bmatrix} 1 & 2 \\ -2 & -4 \end{bmatrix} \mathbf{x}$

2. $\mathbf{x}' = \begin{bmatrix} 1 & -1 \\ 1 & 1 \end{bmatrix} \mathbf{x}$

3. $\mathbf{x}' = \begin{bmatrix} x_1 - x_1^2 \\ x_2 \end{bmatrix}$
4. What is the difference between the expressions $(d/dt)\ \mathbf{x}(t-t_0, 0; \mathbf{x}_0)$ and $\mathbf{x}'(t-t_0, 0; \mathbf{x}_0)$? Does your explanation clarify why $(d/dt)f(2t) = 2f'(2t)$?
5. Explain why

$$\{\mathbf{x}(t-t_0, 0; \mathbf{x}_0) \mid -\infty < t < \infty\} = \{\mathbf{x}(s, 0, \mathbf{x}_0) \mid -\infty < s+t_0 < \infty\}$$
$$= \{\mathbf{x}(s, 0; \mathbf{x}_0) \mid -\infty < s < \infty\}$$

Part 2

1. Supply reasons for the equalities named as (8.2.3) and (8.2.4).
2. If A has only nonzero eigenvalues, prove that $\mathbf{x}^* = \mathbf{0}$ is the only critical point of $\mathbf{x}' = \mathrm{A}\mathbf{x}$.
3. If A has a zero eigenvalue prove that there is a line of critical points through $(0, 0)$ for $\mathbf{x}' = \mathrm{A}\mathbf{x}$.
4. Prove: Every point is a critical point of $\mathbf{x}' = \mathrm{A}\mathbf{x}$ if and only if $\mathrm{A} = \mathrm{O}$.

8.3. ORBITS AND THEIR PROPERTIES: II

Orbits are point sets in the phase plane and the sketch of suitably many typical ones constitute what we call the *phase portrait* of $\mathbf{x}' = \mathbf{f}(\mathbf{x})$. In the next section a number of phase portraits are given for the linear equation $\mathbf{x}' = \mathrm{A}\mathbf{x}$. In this section we continue our study by examining the kinds of point sets which are allowable orbits.

We know that single points are orbits of constant solutions of $\mathbf{x}' = \mathbf{f}(\mathbf{x})$. We call these orbits *critical orbits* to distinguish them from the remaining orbits which constitute smooth continuous arcs† in the phase plane.

That orbits are defined via solutions of $\mathbf{x}' = \mathbf{f}(\mathbf{x})$ places great restrictions on their geometry. We list a number of these prohibitions in the following theorem. We phrase these properties in heuristic, nontechnical terms and then restate them in somewhat more precise language. The proofs are left to the exercises.

8.3.1 THEOREM

(1) *Noncritical orbits do not pass through critical points.*
(2) *Different orbits cannot intersect.*
(3) *Orbits cannot terminate at noncritical points.*

In more precise language:

(1′) *If* $\mathbf{x}(t;\ \mathbf{x}_0)$ *defines the noncritical orbit* $\mathrm{O}(\mathbf{x}_0)$, *then* $\mathbf{x}(t;\ \mathbf{x}_0)$ *cannot assume the value* $\mathbf{x}^*$, *a critical point of* $\mathbf{x}' = \mathbf{f}(\mathbf{x})$.

See Figure 8.3.1(a) and Exercise 8.3, Part 2, Problem 1.

† By a smooth continuous arc we mean a curve parametrized by a continuously differentiable function, $\boldsymbol{\phi}'(t)$, such that $\boldsymbol{\phi}'(t)$ never vanishes.

Figure 8.3.1. Situations which do not occur in autonomous systems with uniqueness property.

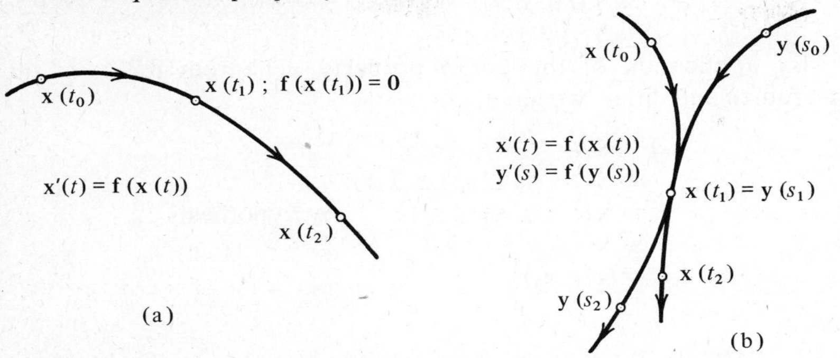

(2') *Suppose* $O_1(\mathbf{x}_1)$ *and* $O_2(\mathbf{x}_2)$ *are two orbits of* $\mathbf{x}' = \mathbf{f}(\mathbf{x})$ *with at least one point in common. Then* $O_1(\mathbf{x}_1) = O_2(\mathbf{x}_2)$.

See Figure 8.3.1(b) and Exercise 8.3, Part 2, Problem 2.

(3') *If* $\lim_{t\to+\infty} \mathbf{x}(t) = \mathbf{x}^*$ $(\lim_{t\to-\infty} \mathbf{x}(t) = \mathbf{y}^*)$ *then* $\mathbf{x}^*(\mathbf{y}^*)$ *is a critical point.*

See Figure 8.3.2 and Exercise 8.3, Part 2, Problems 3, 4, and 5.

Figure 8.3.2. An orbit terminating at critical points.

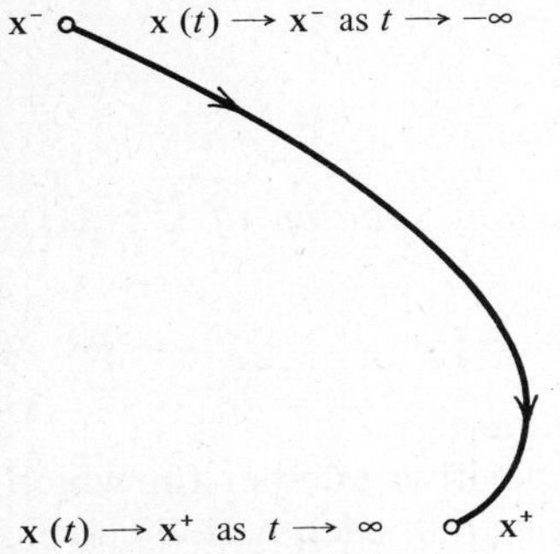

Although two orbits cannot intersect, a single orbit may intersect itself. Physical intuition suggests that a particle which arrives at the same points at two different times must repeat its motion periodically *for autonomous systems.*† Before we make this concept precise, we consider the following important result.

8.3.2. THEOREM *Let s be a real number such that*

$$\mathbf{x}(t_0 + s; \mathbf{x}_0) = \mathbf{x}(t_0; \mathbf{x}_0) \tag{8.3.1}$$

† For nonautonomous systems the future position of a particle depends not only on its present position but the time at which it is assumed to be at this present position.

Then for all t,

$$\mathbf{x}(t+s; \mathbf{x}_0) = \mathbf{x}(t; \mathbf{x}_0) \tag{8.3.2}$$

PROOF By application of the group property, Theorem 8.2.4, in the second and fourth equalities, we have,

$$\begin{aligned}\mathbf{x}(t+s; \mathbf{x}_0) &= \mathbf{x}(t-t_0+t_0+s; \mathbf{x}_0)\\ &= \mathbf{x}(t-t_0; \mathbf{x}(t_0+s; \mathbf{x}_0))\\ &= \mathbf{x}(t-t_0; \mathbf{x}(t_0; \mathbf{x}_0)) \qquad \text{by hypothesis}\\ &= \mathbf{x}(t-t_0+t_0; \mathbf{x}_0)\\ &= \mathbf{x}(t; \mathbf{x}_0)\end{aligned}$$

as required.

Figure 8.3.3 exhibits three curves which cannot be orbits by this theorem.

Figure 8.3.3. Curves which cannot be orbits.

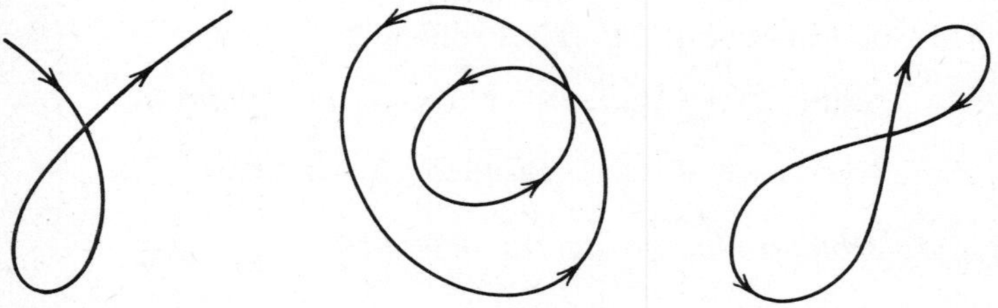

This motivates our definition of periodic orbits.

8.3.3. DEFINITION *If* $\mathbf{x}(t; \mathbf{x}_0)$ *is a nonconstant solution of* $\mathbf{x}' = \mathbf{f}(\mathbf{x})$ *for which* $\mathbf{x}(t_1; \mathbf{x}_0) = \mathbf{x}(t_2; \mathbf{x}_0)$ *with* $t_1 \neq t_2$, *then*

$$O(\mathbf{x}_0) = \{\mathbf{x}(t; \mathbf{x}_0) \mid -\infty < t < \infty\}$$

is called a periodic orbit and $\mathbf{x}(t; \mathbf{x}_0)$ *a periodic solution.*

We call $s = \rho$, the *period* if it is the smallest positive number for which (8.3.2) holds. It is reasonable but not at all obvious that each periodic orbit has a period in this sense.† In fact it does, although the proof is not trivial.

8.3.1. Example

Show that all noncritical orbits of

$$\mathbf{x}' = \begin{bmatrix} 0 & 1 \\ -4 & 0 \end{bmatrix} \mathbf{x}$$

are periodic and determine their periods.

† It is conceivable, but false, that there is a sequence of positive numbers $s_1, s_2, \ldots$ tending to zero for which (8.3.2) holds. If there were such sequences the solution in question would necessarily be constant.

Solution

Since

$$\det\begin{bmatrix} 0 & 1 \\ -4 & 0 \end{bmatrix} \neq 0$$

the only critical point of this system is $\mathbf{x} = \mathbf{0}$. (Why?) A general solution can be verified as

$$\begin{aligned} x_1(t) &= \quad r\cos(2t-\alpha) \\ x_2(t) &= -2r\sin(2t-\alpha) \end{aligned}$$

from which the periodicity of the orbits is obvious. Clearly $\rho = \pi$ is the period. Eliminating t from these parametric equations yields

$$\frac{x_1^2}{r^2} + \frac{x_2^2}{4r^2} = 1$$

a family of ellipses, centered at (0, 0).

We conclude this section with two theorems that delineate the set of numbers for which (8.3.2) holds. Basically we prove that all such numbers are integral multiples of the period.

8.3.4. THEOREM *Under the hypothesis of Theorem 8.3.2,*

$$\mathbf{x}(t+ks; \mathbf{x}_0) = \mathbf{x}(t; \mathbf{x}_0) \tag{8.3.3}$$

for all t and any integer k.

PROOF The proof is accomplished by induction. We outline the essential feature of the argument.

$$\begin{aligned} \mathbf{x}(t+ks; \mathbf{x}_0) &= \mathbf{x}(t+(k-1)s+s; \mathbf{x}_0) \\ &= \mathbf{x}(t+(k-1)s; \mathbf{x}_0) \end{aligned}$$

by hypothesis. If k is positive, repeated application yields (8.3.3). If k is negative,

$$\begin{aligned} \mathbf{x}(t+ks; \mathbf{x}_0) &= \mathbf{x}(t+ks+s; \mathbf{x}_0) \\ &= \mathbf{x}(t+(k+1)s; \mathbf{x}_0) \end{aligned}$$

and again we arrive at (8.3.3) after k steps.

A more significant result is the following theorem.

8.3.5. THEOREM *Let ρ be the period of $\mathbf{x}(t; \mathbf{x}_0)$ and s be any real number for which*

$$\mathbf{x}(t+s; \mathbf{x}_0) = \mathbf{x}(t; \mathbf{x}_0) \tag{8.3.4}$$

for all t. Then $s = k\rho$ for some integer k.

PROOF Define k as the unique integer for which $s \geqslant k\rho$ but $s < (k+1)\rho$. Then $s = k\rho + r$ where $0 \leqslant r < \rho$. Then

$$\begin{aligned} \mathbf{x}(t; \mathbf{x}_0) &= \mathbf{x}(t+s; \mathbf{x}_0) \\ &= \mathbf{x}(t+k\rho+r; \mathbf{x}_0) \\ &= \mathbf{x}(t+r; \mathbf{x}_0) \end{aligned}$$

from Theorem 8.3.3. But $r < \rho$ and hence ρ would not be the period. Thus $r = 0$ and this implies $k\rho = s$, as required.

Exercise 8.3

Part 1

Find the critical points and any periodic orbits in the following four problems.

1. $x_1' = x_1 + x_2^2 \qquad x_2' = x_2$
2. $\mathbf{x}' = \begin{bmatrix} 1 & 2 \\ -2 & -4 \end{bmatrix}\mathbf{x}$
3. $\mathbf{x}' = \begin{bmatrix} 0 & \nu \\ -\nu & 0 \end{bmatrix}\mathbf{x} \qquad \nu \neq 0$
4. $\mathbf{x}' = \begin{bmatrix} 2 & -4 \\ 2 & -2 \end{bmatrix}\mathbf{x}$
5. Explain why a particle moving subject to the "law" $\mathbf{x}' = \mathbf{f}(\mathbf{x})$ never has a zero velocity unless it is always at rest.
6. Can a particle moving subject to the "law" $\mathbf{x}' = \mathbf{f}(\mathbf{x})$ retrace a portion of its path? Explain!
7. Can a particle moving subject to the "law" $\mathbf{x}' = \mathbf{f}(\mathbf{x})$ oscillate between (0, 0) and (0, 1) along the straight line joining these points? Explain!
8. Prove Theorem 8.3.4 by induction on k.
9. Explain why

$$\det\begin{bmatrix} 0 & 1 \\ -4 & 0 \end{bmatrix} \neq 0$$

implies that

$$\mathbf{x}' = \begin{bmatrix} 0 & 1 \\ -4 & 0 \end{bmatrix}\mathbf{x}$$

has $\mathbf{x} = \mathbf{0}$ as the only critical point.

Part 2

1. Prove Theorem 8.3.1 (1). Hint: Show that $\mathbf{x}(t_1; \mathbf{x}_0) = \mathbf{x}^*$ implies that $\mathbf{x}(t; \mathbf{x}_0) \equiv x^*$ by the uniqueness theorem.
2. Prove Theorem 8.3.1 (2). Hint: Find two solutions of $\mathbf{x}' = \mathbf{f}(\mathbf{x})$ which, at some common time, say t_3, take a common value of $\mathrm{O}_1(\mathbf{x}_1)$ and $\mathrm{O}_2(\mathbf{x}_2)$. Use the uniqueness theorem to prove $\mathrm{O}_1(\mathbf{x}_1) \equiv \mathrm{O}_2(\mathbf{x}_2)$.
3.* Show that $|x_1'(t)| \geqslant \mathrm{K}$ for all $t > \mathrm{T}$ implies that $|x_1(t_2) - x_1(t_1)| \geqslant \mathrm{K}|t_2 - t_1|$ if $t_1, t_2 > \mathrm{T}$.

4.* Use the conclusion of Problem 3 to show that $\lim_{t\to+\infty} x_1(t) = a_1$ and $|x_1'(t)| \geqslant$ K > 0 for all $t >$ T are contradictory. Hint:

$$\begin{aligned} |x_1(t_2) - x_1(t_1)| &= |x_1(t_2) - a_1 + a_1 - x_1(t_1)| \\ &\leqslant |x_1(t_2) - a_1| + |x_1(t_1) - a_1| \\ &\leqslant \epsilon \end{aligned}$$

for all $t >$ T.

5.* Prove Theorem 8.3.1 (3). Hint: If $\lim_{t\to+\infty} \mathbf{x}(t) = \mathbf{a}$ and $\mathbf{f}(\mathbf{a}) \neq \mathbf{0}$ we may assume $\lim_{t\to+\infty} |x'(t)| = \mathrm{K} > 0$. Explain! Now use Problems 3 and 4.

8.4. THE GEOMETRY OF CRITICAL POINTS OF $\mathbf{x}' = \mathbf{A}\mathbf{x}$

In this section we shall investigate the geometric or qualitative behavior of solutions of $\mathbf{x}' = \mathrm{A}\mathbf{x}$ near the critical point $\mathbf{x}_0 = \mathbf{0}$; where A is a 2×2 matrix with real coefficients. We shall consider the so-called nondegenerate cases where A is nonsingular. This is equivalent to requiring that no eigenvalue of A is 0 (Exercise 8.4, Part 2, Problem 3). Thus, according to Definition 8.2.5, $\mathbf{x}_0 = \mathbf{0}$ is the only critical point since $\mathbf{f}(\mathbf{x}) = \mathrm{A}\mathbf{x} = \mathbf{0}$ if and only if $\mathbf{x} = \mathbf{0}$. We shall study six cases determined by the nature of the eigenvalues and eigenvectors of A.

Case 1 A *has one eigenvalue*, λ, *with two linearly independent eigenvectors*, $\mathbf{u}_1$, $\mathbf{u}_2$.

In this case, since $\bar{\lambda}$ is also an eigenvalue, $\lambda = \bar{\lambda}$ is real. Moreover since there are two linearly independent eigenvectors, and we are dealing with two-dimensional vectors, the eigenvectors form a basis and every vector, $\mathbf{v}$, can be written as $\mathbf{v} = \alpha\mathbf{u}_1 + \beta\mathbf{u}_2$. Thus $\mathrm{A}\mathbf{v} = \lambda\mathbf{v}$ for all $\mathbf{v}$. This implies that $\mathrm{A} = \lambda\mathrm{I}$, and every nonzero vector is an eigenvector. We find, therefore, that every solution is of the form $\mathbf{x}(t) = e^{\lambda t}\mathbf{x}_0$.

Figure 8.4.1

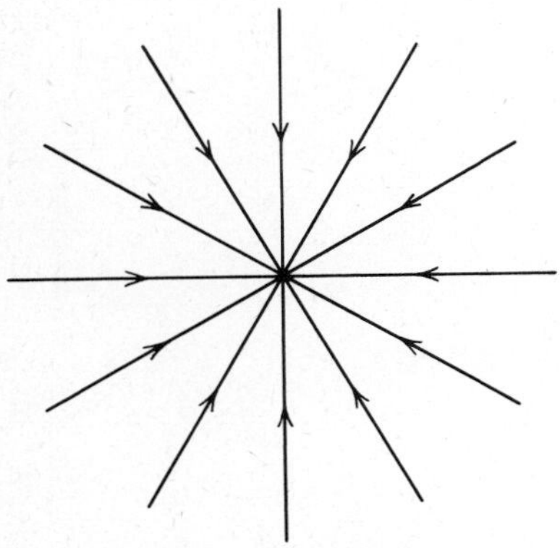

Figure 8.4.2

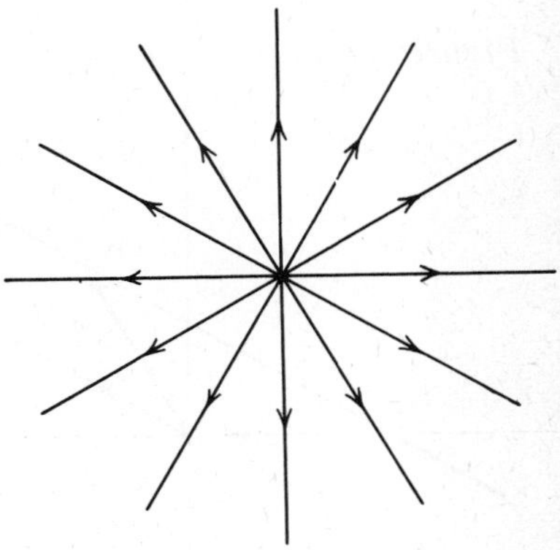

The orbit of every nonzero solution is just the ray $\{e^{\lambda t}\mathbf{x}_0 | -\infty < t < \infty\} = \{s\mathbf{x}_0 | 0 < s < \infty\}$. The phase portrait of $\mathbf{x}' = \mathrm{A}\mathbf{x}$ is Figure 8.4.1 if $\lambda < 0$ and Figure 8.4.2 if $\lambda > 0$.

Case 2 *A has one eigenvalue*, λ, *one eigenvector*, $\mathbf{u}_1$, *and a corresponding second-order root vector* $\mathbf{u}_2$.

Again λ must be real. Every vector $\mathbf{v} = \alpha\mathbf{u}_1 + \beta\mathbf{u}_2$. We have $(\mathrm{A} - \lambda\mathrm{I})\mathbf{v} = \beta(\mathrm{A} - \lambda\mathrm{I})\mathbf{u}_2 = \beta\mathbf{u}_1$ and $(\mathrm{A} - \lambda\mathrm{I})^2\mathbf{v} = \mathbf{0}$. So that every vector is a root vector, eigenvector, or $\mathbf{0}$. Thus all solutions of $\mathbf{x}' = \mathrm{A}\mathbf{x}$ are of the form

$$\mathbf{x}(t) = \alpha e^{\lambda t}\mathbf{u}_1$$

or

$$\mathbf{x}(t) = e^{\lambda t}[\alpha\mathbf{u}_2 + \beta t\mathbf{u}_1]$$

Let us consider the example where $\mathbf{u}_1$ lies as in Figure 8.4.3. Orbits of solutions with initial values of the form $\mathbf{x}_0 = \alpha\mathbf{u}_1$, $\alpha \neq 0$, will correspond to the rays R_1^+ or R_1^- in Figure 8.4.4, if α is positive or negative (respectively), assuming $\lambda > 0$; Figure 8.4.5 describes the case, $\lambda < 0$.

Figure 8.4.3

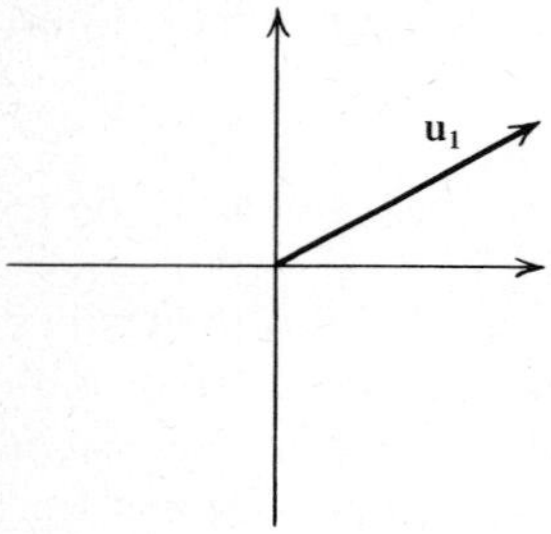

Figure 8.4.4

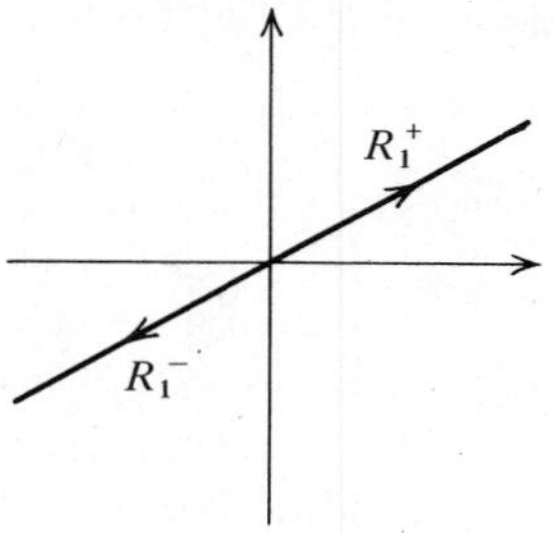

Figure 8.4.5

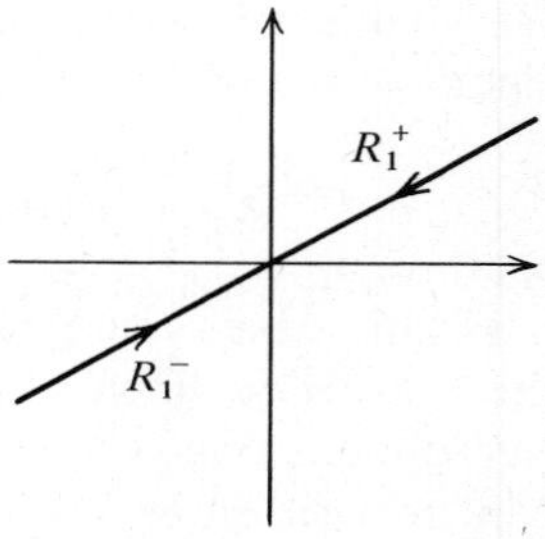

Now let us consider an initial value $\mathbf{x}_0 \neq \beta\mathbf{u}_1$ for any β. Then $\mathbf{x}_0$ will be a root vector of order 2. The solution $\mathbf{x}(t)$ with $\mathbf{x}(0) = \mathbf{x}_0$ will be $\mathbf{x}(t) = e^{\lambda t}\{\mathbf{x}_0 + t(\mathrm{A} - \lambda\mathrm{I})\mathbf{x}_0\} = e^{\lambda t}\{\mathbf{x}_0 + t\alpha\mathbf{u}_1\}$, for some real number α. Let us *assume* that $\mathbf{x}_0$ lies as in Figure 8.4.6 *and* $\alpha > 0$. The set $\{\mathbf{x}_0 + t\alpha\mathbf{u}_1\}$ is depicted in

Figure 8.4.6

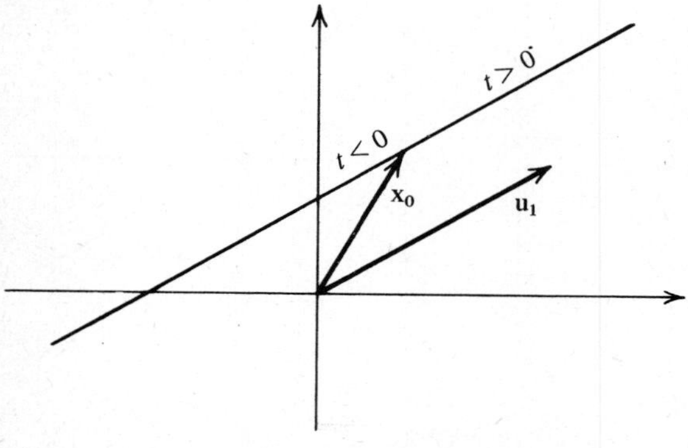

Figure 8.4.6. Thus $\{e^{\lambda t}[\mathbf{x}_0 + t\alpha\mathbf{u}_1]\}$ will be depicted by the heavy line in Figure 8.4.7, since for $t > 1$, assuming $\lambda < 0$,

$$\|\mathbf{x}(t)\| \leqq e^{\lambda t}t\|\mathbf{x}_0 + \alpha\mathbf{u}_1\| \to 0$$

In general the phase portrait will resemble Figure 8.4.8 if $\lambda < 0$.

Figure 8.4.7

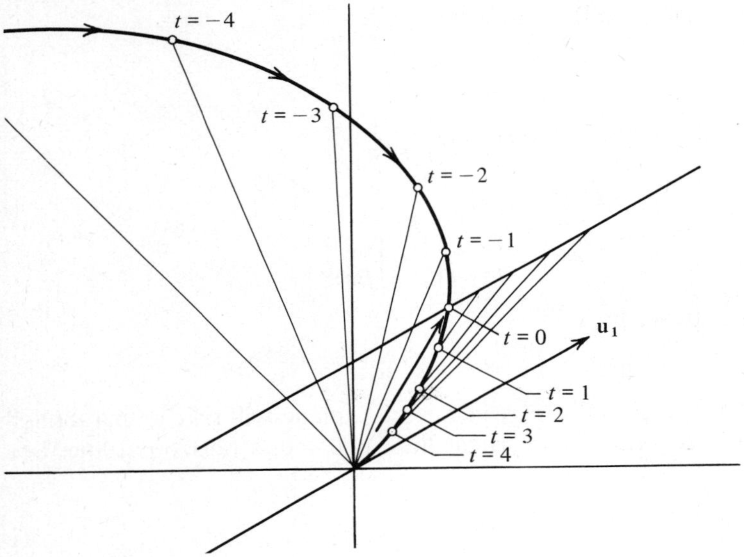

Figure 8.4.8

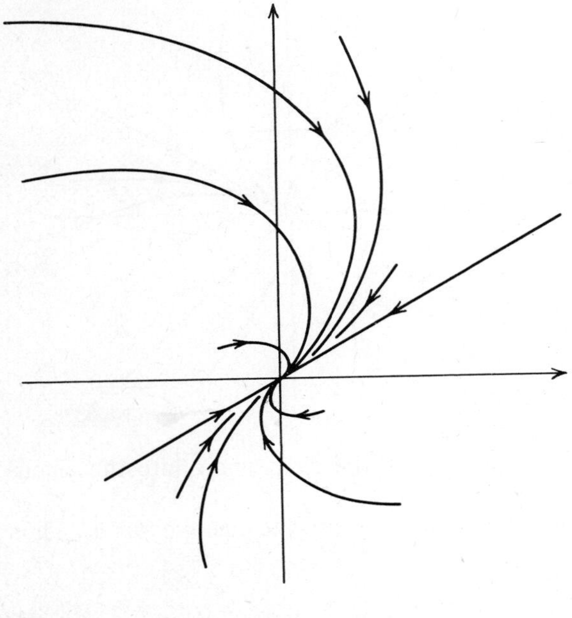

Case 3 *A has distinct real eigenvalues of the same sign.*

Assume $\lambda_2 > \lambda_1 > 0$ with corresponding eigenvectors $\mathbf{u}_1$, $\mathbf{u}_2$. Solutions will have the form

$$\mathbf{x}(t) = \alpha e^{\lambda_1 t}\,\mathbf{u}_1 + \beta e^{\lambda_2 t}\,\mathbf{u}_2$$

Let

$$\mathrm{B} = [\mathbf{u}_1, \mathbf{u}_2] = \begin{bmatrix} u_{11} & u_{12} \\ u_{21} & u_{22} \end{bmatrix}$$

and observe that

$$\mathbf{x}(t) = \mathrm{B}\begin{bmatrix} \alpha e^{\lambda_1 t} \\ \beta e^{\lambda_2 t} \end{bmatrix}$$

Set

$$\mathbf{z}(t) = \begin{bmatrix} z_1(t) \\ z_2(t) \end{bmatrix} = \begin{bmatrix} \alpha e^{\lambda_1 t} \\ \beta e^{\lambda_2 t} \end{bmatrix}$$

and assume $\alpha \neq 0$. We have

$$z_2(t) = \beta e^{\lambda_2 t} = \beta|\alpha|^{-\lambda_2/\lambda_1}|\alpha e^{\lambda_1 t}|^{\lambda_2/\lambda_1} = \gamma|z_1(t)|^{\lambda_2/\lambda_1}$$

Since $\lambda_2/\lambda_1 > 1$, the graph of $\mathbf{z}(t)$ in the z_1, z_2-plane will take a "parabolic" form. If $\alpha = 0$, the orbit is a vertical line; if $\beta = 0$, a horizontal line. See Figure 8.4.9.

Figure 8.4.9

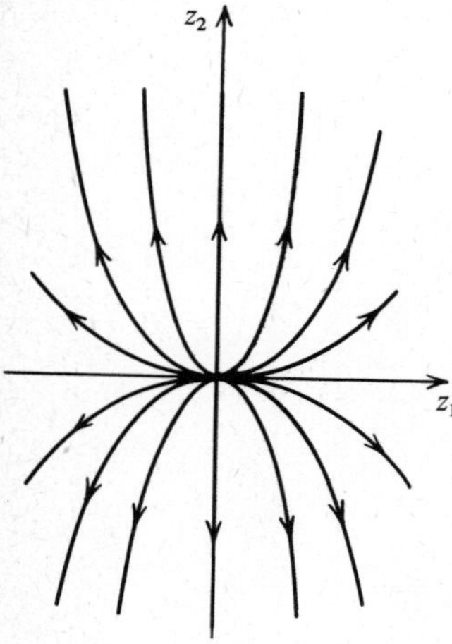

Figure 8.4.10

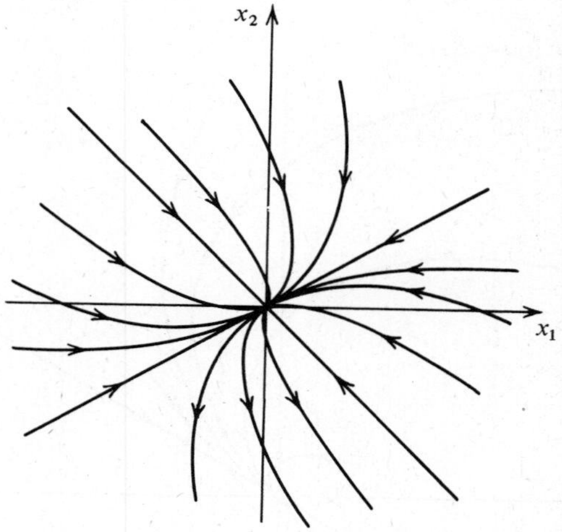

The map, $\mathbf{x} = \mathrm{B}\mathbf{z}$, carries the vector $\begin{bmatrix} 1 \\ 0 \end{bmatrix}$ in the z_1, z_2-plane into the eigenvector $\mathbf{u}_1$ in the x_1, x_2-plane and the vector $\begin{bmatrix} 0 \\ 1 \end{bmatrix}$ into the eigenvector $\mathbf{u}_2$. Thus in the x_1, x_2-plane the family of curves, $\mathbf{x}(t)$, will appear as in Figure 8.4.10. If $\lambda_2 < \lambda_1 < 0$, we get the same phase portrait as in Figure 8.4.10 with the arrows reversed.

Case 4 *A has real eigenvalues*, $\lambda_1 < 0 < \lambda_2$, *of opposite sign.*

As in the previous case, suppose $\mathbf{u}_1$ and $\mathbf{u}_2$ are corresponding eigenvectors and that

$$B = [\mathbf{u}_1, \mathbf{u}_2] = \begin{bmatrix} u_{11} & u_{21} \\ u_{21} & u_{22} \end{bmatrix}$$

$$\mathbf{z}(t) = \begin{bmatrix} z_1(t) \\ z_2(t) \end{bmatrix} = \begin{bmatrix} \alpha e^{\lambda_1 t} \\ \beta e^{\lambda_2 t} \end{bmatrix}$$

Then

$$\mathbf{x}(t) = B\mathbf{z}(t)$$

But

$$\begin{aligned} z_2(t)^{|\lambda_1|} z_1(t)^{|\lambda_2|} &= \alpha^{|\lambda_2||\lambda_1|} \exp\{|\lambda_2|\lambda_1 + |\lambda_1|\lambda_2\}\, t \\ &= \alpha^{|\lambda_2||\lambda_1|} = \gamma \end{aligned}$$

γ a constant, because

$$|\lambda_2|\lambda_1 + |\lambda_1|\lambda_2 = \lambda_2\lambda_1 - \lambda_1\lambda_2 = 0$$

when $\lambda_1 < 0 < \lambda_2$.

Thus the family of curves, $\mathbf{z}(t)$, will be "hyperbolic" and appear as in Figure 8.4.11. The curves $\mathbf{x}(t)$ will appear as in Figure 8.4.12.

We now consider complex eigenvalues.

Figure 8.4.11

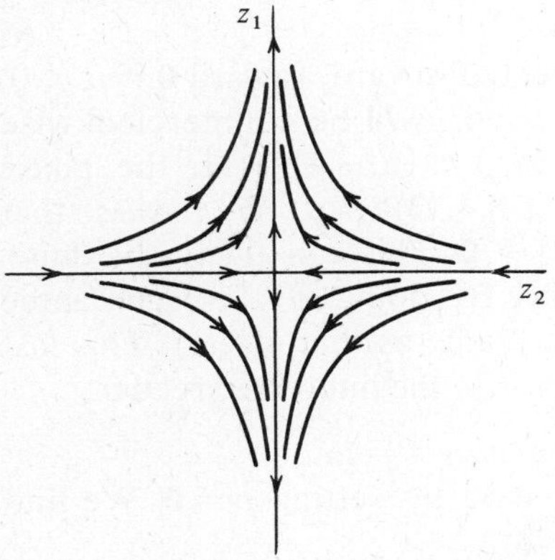

Figure 8.4.12

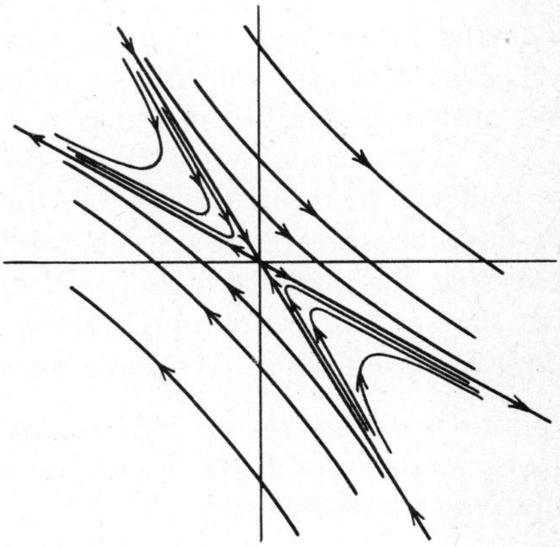

Case 5 *A has the complex eigenvalue* $\rho + \sigma i$ $(\rho \neq 0)$.

The corresponding eigenvector is $\mathbf{u} + i\mathbf{v}$. We recall that $\mathbf{u} \neq \mathbf{0}$ and $\mathbf{v} \neq \mathbf{0}$

since $\mathbf{u}-i\mathbf{v}$ is also an eigenvector and $\mathbf{u}-i\mathbf{v}$, $\mathbf{u}+i\mathbf{v}$ must be linearly independent. We obtain two linearly independent solutions by taking the real and imaginary parts of

$$e^{(\rho+i\sigma)t}\,[u+i\mathbf{v}] = e^{\rho t}\,\{(\cos\sigma t\mathbf{u}-\sin\sigma t\mathbf{v})+i(\cos\sigma t\mathbf{v}+\sin\sigma t\mathbf{u})\}$$

which may be conveniently written

$$\mathbf{x}_1(t) = e^{\rho t}\begin{bmatrix} u_1 & -v_1 \\ u_2 & -v_2 \end{bmatrix}\begin{bmatrix} \cos\sigma t \\ \sin\sigma t \end{bmatrix} \tag{8.4.1}$$

$$\mathbf{x}_2(t) = e^{\rho t}\begin{bmatrix} u_1 & -v_1 \\ u_2 & -v_2 \end{bmatrix}\begin{bmatrix} \sin\sigma t \\ -\cos\sigma t \end{bmatrix} \tag{8.4.2}$$

where

$$\begin{bmatrix} u_1 & -v_1 \\ u_2 & -v_2 \end{bmatrix} = [\mathbf{u}, -\mathbf{v}]$$

In general the solutions take the form

$$\mathbf{x}(t) = \begin{bmatrix} u_1 & -v_1 \\ u_2 & -v_2 \end{bmatrix}\mathbf{z}(t) \tag{8.4.3}$$

where

$$\mathbf{z}(t) = e^{\rho t}\,\mathbf{w}(t) \tag{8.4.4}$$

and

$$\begin{aligned} \mathbf{w}(t) &= \alpha\begin{bmatrix} \cos\sigma t \\ \sin\sigma t \end{bmatrix} + \beta\begin{bmatrix} \sin\sigma t \\ -\cos\sigma t \end{bmatrix} \\ &= \alpha\mathbf{a}(t)+\beta\mathbf{b}(t) \end{aligned} \tag{8.4.5}$$

Since the summands of $\mathbf{w}(t)$ are perpendicular, one easily computes

$$\|\mathbf{w}(t)\|^2 = \mathbf{w}(t)\cdot\mathbf{w}(t) = \alpha^2+\beta^2$$

Thus the path of $\mathbf{w}(t)$ is a circle in each case.

The path of $\mathbf{z}(t)$ will thus be a spiral directed inward, toward $\mathbf{0}$ if $\rho < 0$, and outward, away from $\mathbf{0}$ if $\rho > 0$. The motion will be counterclockwise if $\sigma > 0$ and clockwise if $\sigma < 0$. Since $\mathbf{x}(t) = [\mathbf{u}, -\mathbf{v}]\ \mathbf{z}(t)$, the phase portrait will be typified by one of the Figures 8.4.13(a)–(d). We *assume* that we have chosen $\mathbf{u}$ and $\mathbf{v}$ so that $\det[\mathbf{u}, -\mathbf{v}] > 0$. This can always be done, reversing the signs of $\mathbf{v}$ and σ if necessary. By doing this, we guarantee that the sense of rotation of $\mathbf{x}(t)$ will be the same as that of $\mathbf{z}(t)$. This last remark is *not* obvious. We leave its justification to the interested reader.

Case 6 A has the purely imaginary eigenvalue $\lambda = i\sigma$.

The analysis of Case 5 may be applied here by setting $\rho = 0$. We find solutions are of the form

$$\mathbf{x}(t) = [\mathbf{u}, -\mathbf{v}]\,[\alpha\mathbf{a}(t)+\beta\mathbf{b}(t)]$$

so that the phase portrait is typified by Figure 8.4.14.

Figure 8.4.13

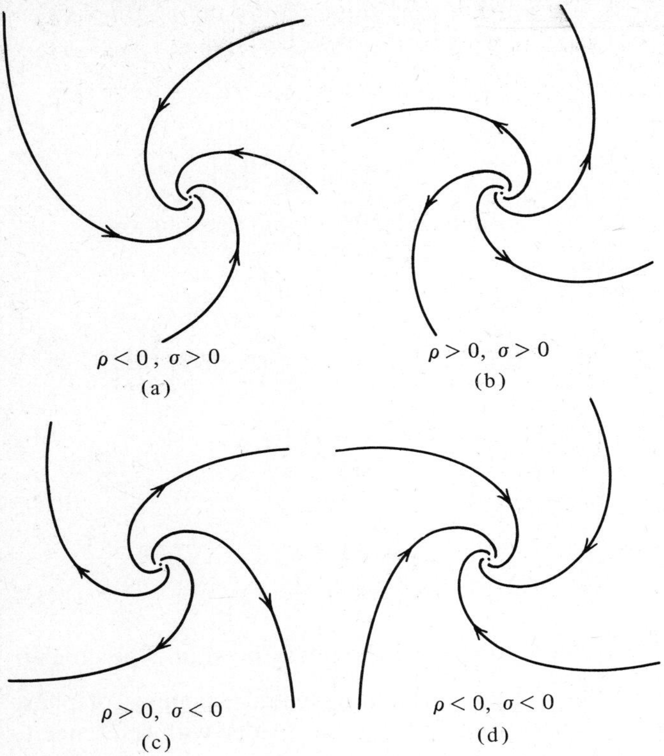

Figure 8.4.14

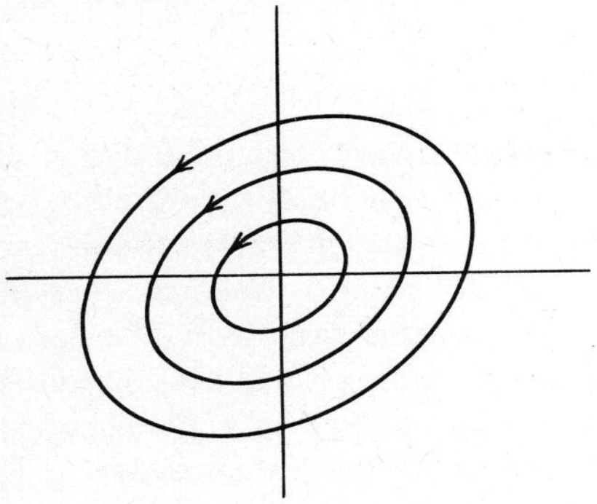

8.4.1. Example

Consider the differential equation

$$\mathbf{x}' = \begin{bmatrix} 4 & -5 \\ 2 & -2 \end{bmatrix} \mathbf{x}$$

We find that the eigenvalues and corresponding eigenvectors are

$$\lambda_1 = (1+i) \qquad \mathbf{w}_1 = \begin{bmatrix} 3+i \\ 2 \end{bmatrix} = \begin{bmatrix} 3 \\ 2 \end{bmatrix} + i \begin{bmatrix} 1 \\ 0 \end{bmatrix}$$

and

$$\lambda_2 = (1-i) \qquad \mathbf{w}_2 = \begin{bmatrix} 3-i \\ 2 \end{bmatrix} = \begin{bmatrix} 3 \\ 2 \end{bmatrix} - i \begin{bmatrix} 1 \\ 0 \end{bmatrix}$$

so that the discussion of Case 5 is applicable. We obtain the solutions

$$\mathbf{x}(t) = e^{(1+i)t}\left\{\begin{bmatrix} 3 \\ 2 \end{bmatrix} + i \begin{bmatrix} 1 \\ 0 \end{bmatrix}\right\}$$

$$= e^t\left\{\left(\cos t \begin{bmatrix} 3 \\ 2 \end{bmatrix} - \sin t \begin{bmatrix} 1 \\ 0 \end{bmatrix}\right) + i\left(\cos t \begin{bmatrix} 1 \\ 0 \end{bmatrix} + \sin t \begin{bmatrix} 3 \\ 2 \end{bmatrix}\right)\right\}$$

$$\mathbf{x}_1(t) = e^t \begin{bmatrix} 3 & -1 \\ 2 & 0 \end{bmatrix} \begin{bmatrix} \cos t \\ \sin t \end{bmatrix}$$

and

$$\mathbf{x}_2(t) = e^t \begin{bmatrix} 3 & -1 \\ 2 & 0 \end{bmatrix} \begin{bmatrix} \sin t \\ -\cos t \end{bmatrix}$$

Since $\det \begin{bmatrix} 3 & -1 \\ 2 & 0 \end{bmatrix} > 0$, $\rho = 1$ and $\sigma = 1$ the phase portrait is given as in Figure 8.4.13(b). In Figure 8.4.15 we have collected examples of phase portraits for $\mathbf{x}' = \mathrm{A}\mathbf{x}$. These portraits provide the reader with reference to the various possibilities inherent in this equation.

Exercise 8.4

Part 1

1. Sketch the phase portraits of the following systems

(a) $\mathbf{x}' = -\mathbf{x}$ (b) $\mathbf{x}' = \begin{bmatrix} 1 & -1 \\ 1 & 1 \end{bmatrix}\mathbf{x}$

(c) $\mathbf{x}' = \begin{bmatrix} 1 & -2 \\ 2 & -1 \end{bmatrix}\mathbf{x}$ (d) $\mathbf{x}' = \begin{bmatrix} 3 & 4 \\ 4 & -3 \end{bmatrix}\mathbf{x}$

(e) $\mathbf{x}' = \begin{bmatrix} 0 & 1 \\ -2 & 3 \end{bmatrix}\mathbf{x}$ (f) $\mathbf{x}' = \begin{bmatrix} 0 & 1 \\ -1 & 2 \end{bmatrix}\mathbf{x}$

2. Verify by substitution that the functions $\mathbf{x}_1(t)$ and $\mathbf{x}_2(t)$ given by Equations (8.4.1) and (8.4.2) solve $\mathbf{x}' = \mathrm{A}\mathbf{x}$ when A has complex eigenvalues $\lambda = \rho + i\sigma$, $\rho \neq 0$ and eigenvectors $\mathbf{u} + i\mathbf{v}$, $\mathbf{u} - i\mathbf{v}$.
3. Use the result of Problem 2 to show that $\mathbf{x}(t)$ given by Equation (8.4.3) is a solution of $\mathbf{x}' = \mathrm{A}\mathbf{x}$.

Figure 8.4.15. A phase portrait gallery.

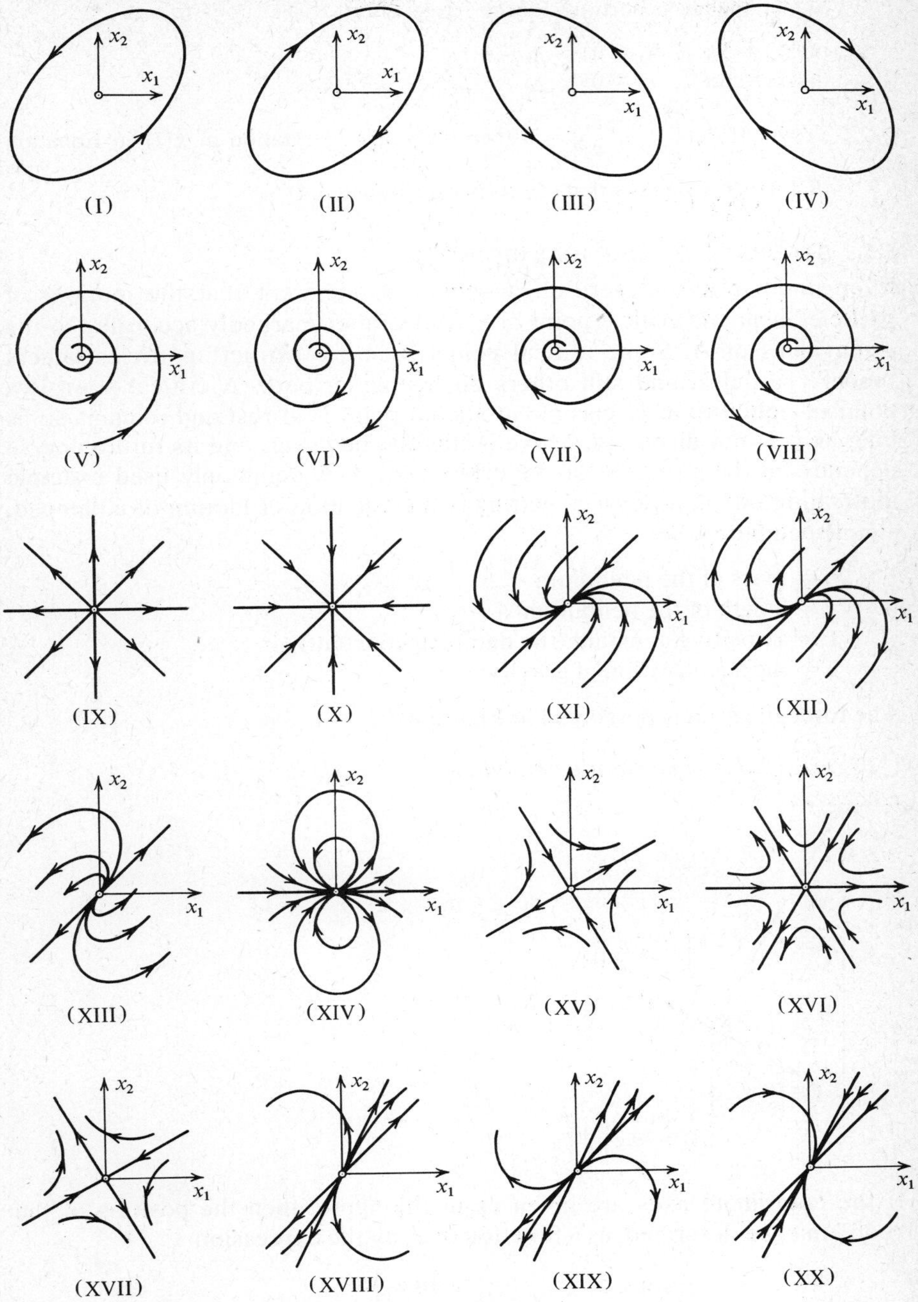

Part 2

1. Sketch the phase portraits for the "degenerate" cases:

   (a) $\mathbf{x}' = \mathbf{A}\mathbf{x} \quad \lambda_1 = 0 \quad \lambda_2 = 1$
   (b) $\mathbf{x}' = \mathbf{A}\mathbf{x} \quad \lambda_1 = 0 \quad \lambda_2 = -1$

2.* Prove: If $\det \begin{bmatrix} u_1 & -v_1 \\ u_2 & -v_2 \end{bmatrix} > 0$ then the sense of rotation of $\mathbf{x}(t)$ in Equation (8.4.3) is the same as that of $\mathbf{z}(t)$ in Equation (8.4.4).

8.5. STABILITY OF CRITICAL POINTS

Even under a casual perusal, it should be apparent that the motions of particles near the critical point of $\mathbf{x}' = \mathbf{A}\mathbf{x}$ differ markedly according to the eigenvalues of A. Some critical points seem to "attract" particles, others "repel" particles, and still others do neither or both. A critical point is a point of equilibrium; a particle at such a point is at rest and remains so as long as it is not displaced. Move it slightly, however, and its future course depends on the structure of the orbits near **0**. A commonly used example illustrating this in a physical setting is the equation of motion of a damped, simple pendulum. Set

m = mass of the pendulum
l = length of the pendulum
k = a positive constant (the damping constant)
g = the acceleration of gravity

The force diagram is portrayed in Figure 8.5.1.

Figure 8.5.1. The simple pendulum.

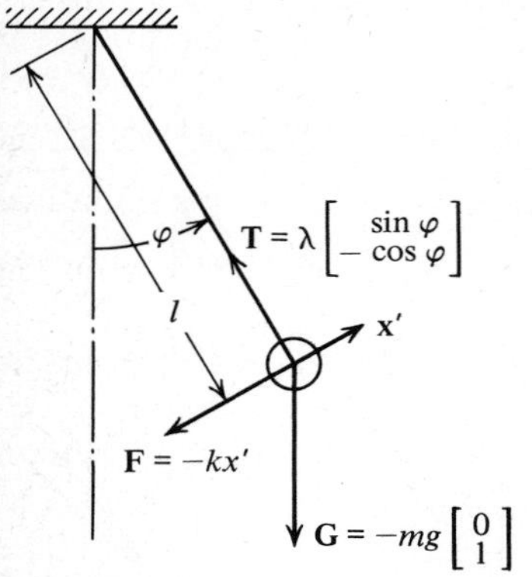

If the coordinate axes are given as in the figure, then the position of the "bob" may be described, as a function of φ, by the expression

$$\mathbf{z} = l \begin{bmatrix} \sin\varphi \\ -\cos\varphi \end{bmatrix}$$

Two differentiations lead to

$$\mathbf{z}' = l\varphi' \begin{bmatrix} \cos\varphi \\ \sin\varphi \end{bmatrix}$$

$$\mathbf{z}'' = l(\varphi')^2 \begin{bmatrix} -\sin\varphi \\ \cos\varphi \end{bmatrix} + l\varphi'' \begin{bmatrix} \cos\varphi \\ \sin\varphi \end{bmatrix}$$

From Newton's Second Law, we obtain

$$m\mathbf{z}'' = -k\mathbf{z}' - mg \begin{bmatrix} 0 \\ 1 \end{bmatrix} + \lambda \begin{bmatrix} \sin\varphi \\ -\cos\varphi \end{bmatrix}$$

We obtain a scalar equation by taking the inner product of each term in this latter expression with $\begin{bmatrix} \cos\varphi \\ \sin\varphi \end{bmatrix}$ using the expressions derived earlier for $\mathbf{z}'$ and $\mathbf{z}''$. That is

$$\begin{aligned} m\mathbf{z}'' \begin{bmatrix} \cos\varphi \\ \sin\varphi \end{bmatrix} &= ml\varphi'' \\ &= -kl\varphi' - mg\sin\varphi \end{aligned}$$

or,

$$\varphi'' + \frac{k}{m}\varphi' + \frac{g}{l}\sin\varphi = 0 \tag{8.5.1}$$

Setting $x_1 = \varphi$ and $x_2 = \varphi'$ leads to the companion system,

$$\mathbf{x}' = \begin{bmatrix} x_1' \\ x_2' \end{bmatrix} = \begin{bmatrix} x_2 \\ \dfrac{-g}{l}\sin x_1 - \dfrac{k}{m}x_2 \end{bmatrix} \tag{8.5.2}$$

Clearly $\mathbf{x} = \mathbf{0}$ and $\mathbf{x} = \begin{bmatrix} \pi \\ 0 \end{bmatrix}$ are critical points and these correspond to the two vertical positions with zero velocity. If the pendulum is displaced from $\mathbf{x} = \mathbf{0}$ it tends to return ultimately to its "stable" resting position at $\mathbf{x} = \mathbf{0}$. No matter how small the displacement from the "unstable" equilibrium position, $\mathbf{x} = \begin{bmatrix} \pi \\ 0 \end{bmatrix}$, however, the pendulum swings through $\mathbf{x} = \mathbf{0}$. The mathematical analysis which leads to these physically obvious conclusions is presented in Example 8.6.4.

We present precise definitions of stable and unstable below. First let us assume that the critical point has been placed at $\mathbf{x} = \mathbf{0}$ by a change in coordinate system if necessary. Also, we assume that only "isolated" critical points are under discussion. (By an isolated critical point we mean one about which a small circle may be drawn excluding all other critical points.) The system $\mathbf{x} = \mathrm{A}\mathbf{x}$ has an isolated critical point at $\mathbf{x} = \mathbf{0}$ if and only if $\det \mathrm{A} \neq 0$.

8.5.1. DEFINITION *A critical point is called stable if given $\epsilon > 0$ there exists $\delta > 0$ depending on ϵ, such that*

(1) $\|\mathbf{x}_0\| < \delta$ *implies* $\|\mathbf{x}(t, \mathbf{x}_0)\| < \epsilon$ *for all* $t > 0$.

To understand this definition first note that $\|\mathbf{x}_0\| < \delta$ means $\mathbf{x}_0$ is closer to $\mathbf{0}$ than δ; it is within a circle centered at $\mathbf{0}$, radius δ. The expression, $\mathbf{x}(t, \mathbf{x}_0)$, denotes the unique solution of $\mathbf{x}' = \mathbf{f}(\mathbf{x})$, $\mathbf{x}(0) = \mathbf{x}_0$. Hence (1) may be interpreted as asserting that a particle subject to $\mathbf{x}' = \mathbf{f}(\mathbf{x})$ and beginning its motion near $\mathbf{0}$ (that is, within δ of $\mathbf{0}$) at $t = 0$ will forever after remain near $\mathbf{0}$ (that is, within ϵ of $\mathbf{0}$).

8.5.1. Example

Examine Figure 8.5.2. Consider a 2×2 matrix A which has complex conjugate eigenvalues $\rho \pm i\sigma$ with $\rho < 0$, and corresponding eigenvectors $\mathbf{u} \pm i\mathbf{v}$. Following the analysis of Case 5, Section 8.4 we may construct $\mathbf{x}(t; \mathbf{x}_0)$ for $\mathbf{x}' = \mathrm{A}\mathbf{x}$ as follows: Let

$$\mathrm{B} = [\mathbf{u}, -\mathbf{v}]$$
$$\mathbf{x}_0 = \mathrm{B}(\mathrm{B}^{-1}\mathbf{x}_0) = \mathrm{B}\mathbf{y}_0$$

Figure 8.5.2

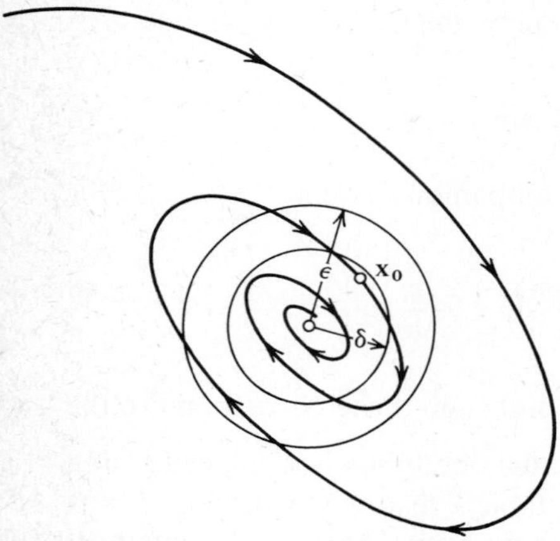

We assume that $\mathbf{x}_0 \neq \mathbf{0}$ hence $\mathbf{y}_0 \neq \mathbf{0}$, and write

$$\mathbf{y}_0 = \|\mathbf{y}_0\| \begin{bmatrix} \cos \varphi_0 \\ \sin \varphi_0 \end{bmatrix}$$

for suitably chosen φ_0.

In studying Case 5 we found that

$$\mathbf{x}(t) = \|\mathbf{y}_0\| e^{\rho t} \mathrm{B} \begin{bmatrix} \cos\,(t + \varphi_0) \\ \sin\,(t + \varphi_0) \end{bmatrix}$$

is a solution of $\mathbf{x}' = \mathrm{A}\mathbf{x}$. Furthermore we may verify that $\mathbf{x}(0) = \mathbf{x}_0$. Hence

$$\mathbf{x}(t) = \mathbf{x}(t; \mathbf{x}_0)$$

and thus

$$\|\mathbf{x}(t; \mathbf{x}_0)\| \leqslant e^{\rho t}\|\mathbf{y}_0\|\,\|\mathrm{B}\| \leqslant e^{\rho t}\|\mathrm{B}^{-1}\|\,\|\mathrm{B}\|\,\|\mathbf{x}_0\|$$

From this computation we see that for any given $\epsilon > 0$, we may choose

$$\delta = \frac{\epsilon}{\|B^{-1}\| \|B\|}$$

for which

$$\|\mathbf{x}_0\| < \delta$$

implies

$$\|\mathbf{x}(t, \mathbf{x}_0)\| < \epsilon \qquad \text{for} \quad t \geqslant 0$$

We note that the critical point is stable if $\rho < 0$ or even if $\rho = 0$. But in the former case we also have $\lim_{t\to\infty} \mathbf{x}(t; \mathbf{x}_0) = \mathbf{0}$.

The reader should note that according to the definition of stability, a critical point may be unstable even though *some* orbits starting nearby, remain nearby. For example, the critical point, $\mathbf{0}$, of $x_1' = -x_1$, $x_2' = x_2$ is *not* stable.

The critical points of $\mathbf{x}' = \mathrm{A}\mathbf{x}$ in the case where the eigenvalues of A are complex with negative real parts are stable. The critical points of $\mathbf{x}' = \mathrm{A}\mathbf{x}$ when the eigenvalues of A are purely imaginary are also stable. Yet it is apparent that the motions of particles near $\mathbf{0}$ are quite different in these two cases. We are thus led to the following definition which distinguishes these cases.

8.5.2. DEFINITION *A stable critical point is called asymptotically stable if there exists* $\delta > 0$ *such that*

(2) $\|\mathbf{x}_0\| < \delta_0$ *implies* $\lim_{t\to+\infty} \mathbf{x}(t; \mathbf{x}_0) = \mathbf{0}$.

Unstable critical points are never asymptotically stable even if (2) were to hold. Surprisingly, condition (2) does not imply condition (1) so we must include the word "stable" as part of the definition of asymptotically stable. See Exercise 8.5, Part 2, Problem 1.

Condition (2) in essence requires particles near to $\mathbf{0}$ not only to remain near $\mathbf{0}$ but what is more, approach $\mathbf{0}$ as t increases. Table 8.5.1 lists the stability properties for the various cases arising in the study of $\mathbf{x}' = \mathrm{A}\mathbf{x}$. The results of this table may be put in the form given in Theorem 8.5.3.

8.5.3. THEOREM *If* $\mathbf{x} = \mathbf{0}$ *is an isolated critical point of* $\mathbf{x}' = \mathrm{A}\mathbf{x}$ *then*

(1) $\mathbf{x} = \mathbf{0}$ *is asymptotically stable if the eigenvalues of* A *have negative real parts.*†

(2) $\mathbf{x} = \mathbf{0}$ *is stable but not asymptotically stable if the eigenvalues of* A *are purely imaginary.*

(3) *If at least one eigenvalue has positive real part*† $\mathbf{x} = \mathbf{0}$ *is unstable.*

† A real number, w, may be considered complex with zero imaginary part. Hence a negative (positive) real number has negative (positive) real part.

Table 8.5.1

Nature of the eigenvalues	Critical point	Figure reference
Real and positive	Unstable	8.4.2
Real and negative	Asymptotically stable	8.4.1, 8.4.8
Real and of opposite sign	Unstable	8.4.11, 8.4.12
Complex with real part positive	Unstable	8.4.14(b), (d)
Complex with real part zero	Stable but not asymptotically stable	8.4.15
Complex with real part negative	Asymptotically stable	8.4.14(a), (c)

Exercise 8.5

Part 1

1. The change $t \to -t$ in $\mathbf{x}' = \mathbf{f}(\mathbf{x})$ reverses the orientation of the phase portrait. (Why?). Hence explain why $t \to -t$ converts an asymptotically stable critical point to an unstable critical point.
2. Give an example illustrating that $t \to -t$ may retain a stable critical point.
3. Show that if $t \to -t$, a point which is not an asymptotically stable point goes into stable critical point which may or may not be asymptotically stable.
4. Review Table 8.5.1 and show by examining the cases that Theorem 8.5.3 is valid.
5. Decide which of the critical points of Exercise 8.4, Part 1, Problem 1 are stable, asymptotically stable, unstable.

Part 2

1. Consider the phase portrait given in Figure 8.5.3. Note that every solution tends to $\mathbf{0}$ as $t \to \infty$. Is $\mathbf{0}$ stable?

8.6. PERTURBATIONS OF LINEAR SYSTEMS

It is common practice among scientists to simplify formidable problems by ignoring or modifying factors whose effects are judged small. Mathematicians attack difficult problems in an analogous way. The numerical methods, of which Euler's Method is one example, may be viewed as the replacement of one problem with a simpler one whose solution is a first approximation to the

Figure 8.5.3

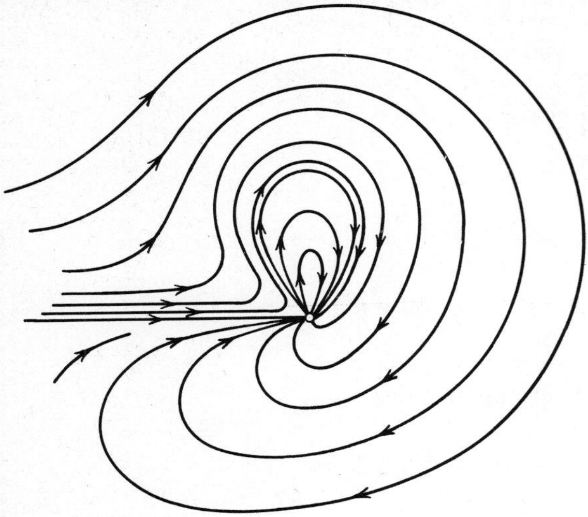

desired solution. It is our intent in this section to introduce a similar procedure for the qualitative theory of differential equation.

We consider the nonlinear system

$$\mathbf{x}' = A\mathbf{x} + \mathbf{h}(\mathbf{x}) \tag{8.6.1}$$

and ask how the nature of the critical point $\mathbf{x} = \mathbf{0}$ of $\mathbf{x}' = A\mathbf{x}$ is affected by the introduction of the nonlinear "perturbation" term $\mathbf{h}(\mathbf{x})$. As usual we insist that every initial-value problem derived from (8.6.1) have a unique solution.† The hypothesis directly related to the question at hand concerns the size of $\|\mathbf{h}(\mathbf{x})\|$ near $\mathbf{x} = \mathbf{0}$. We assume throughout this section:

There exists a constant, $M > 0$ *such that for all* $\mathbf{x}$, $\|\mathbf{x}\| < M$

$$\|\mathbf{h}(\mathbf{x})\| \leq K\|\mathbf{x}\|^2 \tag{8.6.2}$$

for some constant, K.

We call $\mathbf{h}(\mathbf{x})$ a *perturbation* and the system (8.6.1) a *perturbed system.* Sometimes (8.6.1) is called "almost linear" when $\mathbf{h}(\mathbf{x})$ satisfies the above hypothesis or its equivalent.

8.6.1. Example

Show that the companion system for the simple pendulum, Equation (8.5.2), may be put in a form for which (8.6.2) attains.

† We assume the components of $\mathbf{h}(\mathbf{x})$ have continuous first partial derivatives with respect to x_1 and x_2 the components of $\mathbf{x}$ and that (8.6.2) holds.

Solution

We have

$$\mathbf{x}' = \begin{bmatrix} x_2 \\ -\dfrac{g}{l}\sin x_1 - \dfrac{k}{m}x_2 \end{bmatrix} = \begin{bmatrix} 0 & 1 \\ -\dfrac{g}{l} & -\dfrac{k}{m} \end{bmatrix}\mathbf{x} + \frac{g}{l}\begin{bmatrix} 0 \\ x_1 - \sin x_1 \end{bmatrix}$$

Hence

$$\mathbf{h}(\mathbf{x}) = \frac{g}{l}\begin{bmatrix} 0 \\ x_1 - \sin x_1 \end{bmatrix}$$

In Exercise 8.6, Part 2, Problem 1 we show for $|u| < 1$, $|u - \sin u| < u^3/3$. Granting this

$$\begin{aligned} \|\mathbf{h}(\mathbf{x})\| &= \frac{g}{l}|x_1 - \sin x_1| \\ &\leqslant \frac{g}{3l}|x_1|^3 \end{aligned}$$

for $|x_1| < 1$.

Returning to (8.6.2), observe that $\mathbf{h}(\mathbf{0}) = \mathbf{0}$ is a consequence of the assumptions on $\mathbf{h}$, above, and thus (8.6.1) has a critical point at $\mathbf{x} = \mathbf{0}$. We now come to the central result of this section, a theorem due to A. M. Liapunov, a Russian mathematician who contributed much to the study of the stability and perturbation theorem.

8.6.1. THEOREM *Under the assumptions given on* $\mathbf{h}(\mathbf{x})$: *If the critical point of the unperturbed system* $\mathbf{x}' = \mathrm{A}\mathbf{x}$ *is unstable (asymptotically stable) then* $\mathbf{x} = \mathbf{0}$ *is an unstable (asymptotically stable) critical point of* $\mathbf{x}' = \mathrm{A}\mathbf{x} + \mathbf{h}(\mathbf{x})$.

This theorem offers no conclusion if $\mathbf{x} = \mathbf{0}$ is stable but not asymptotically stable and we shall show by example that the perturbed system may then be stable in either sense—asymptotically or nonasymptotically—as the case may be.

One of the ways of studying the critical point, $\mathbf{x} = \mathbf{0}$, of $\mathbf{x}' = \mathrm{A}\mathbf{x} + \mathbf{h}(\mathbf{x})$ is to introduce polar coordinates in the phase plane. As usual, set

$$\begin{aligned} \|\mathbf{x}\|^2 &= r^2 = x_1^2 + x_2^2 \\ \theta &= \arctan\frac{x_2}{x_1} \end{aligned} \tag{8.6.3}$$

We compute $d(r^2)/dt$ and $d\theta/dt$ as follows.

$$\begin{aligned} \frac{d\|\mathbf{x}\|^2}{dt} &= \frac{d}{dt}(\mathbf{x}\cdot\mathbf{x}) \\ &= 2\mathbf{x}\cdot\mathbf{x}' \\ &= 2\mathbf{x}(\mathrm{A}\mathbf{x} + \mathbf{h}(\mathbf{x})) \\ &= 2\mathbf{x}\cdot\mathrm{A}\mathbf{x} + 2\mathbf{x}\cdot\mathbf{h}(\mathbf{x}) \end{aligned}$$

For $\mathbf{x} \neq \mathbf{0}$ define $\mathbf{u}$ via

$$\mathbf{x} = \|\mathbf{x}\|\mathbf{u} \tag{8.6.4}$$

This means $\|\mathbf{u}\| = 1$ and it is reasonable to expect the existence of a differentiable function † $\theta(t)$ such that

$$\mathbf{u}(t) = \begin{bmatrix} \cos\theta(t) \\ \sin\theta(t) \end{bmatrix} \tag{8.6.5}$$

Assuming this we conclude

$$\frac{d\mathbf{u}}{dt} = \frac{d\theta}{dt}\begin{bmatrix} -\sin\theta \\ \cos\theta \end{bmatrix}$$

$$= \frac{d\theta}{dt}\mathrm{R}\mathbf{u}$$

where R is the matrix

$$\mathrm{R} = \begin{bmatrix} 0 & -1 \\ 1 & 0 \end{bmatrix} \tag{8.6.6}$$

The matrix, R, rotates every nonzero vector counterclockwise by a right angle. Therefore,

$\mathbf{v} \cdot \mathrm{R}\mathbf{v} = 0$ for every $\mathbf{v}$ (including, incidentally, $\mathbf{v} = \mathbf{0}$).

We now find, from differentiating (8.6.4),

$$\frac{d\mathbf{x}}{dt} = \frac{d\|\mathbf{x}\|}{dt}\mathbf{u} + \|\mathbf{x}\|\frac{d\mathbf{u}}{dt} = \frac{d\|\mathbf{x}\|}{dt}\mathbf{u} + \frac{d\theta}{dt}\mathrm{R}\,\mathbf{u}\,\|\mathbf{x}\|$$

Since $\mathrm{R}\mathbf{x} = \|\mathbf{x}\|\,\mathrm{R}\mathbf{u}$, we have

$$\frac{d\mathbf{x}}{dt} = \frac{d\|\mathbf{x}\|}{dt}\mathbf{u} + \frac{d\theta}{dt}\mathrm{R}\mathbf{x}$$

Hence,

$$\mathrm{R}\mathbf{x} \cdot \frac{d\mathbf{x}}{dt} = \frac{d\|\mathbf{x}\|}{dt}(\mathrm{R}\mathbf{x} \cdot \mathbf{u}) + \frac{d\theta}{dt}(\mathrm{R}\mathbf{x} \cdot \mathrm{R}\mathbf{x})$$

$$= \frac{d\theta}{dt}\|\mathbf{x}\|^2$$

because $\mathrm{R}\mathbf{x} \cdot \mathbf{u} = (\mathrm{R}\mathbf{x} \cdot \mathbf{x})\,\|\mathbf{x}\|^{-1} = 0$ and $\mathrm{R}\mathbf{x} \cdot \mathrm{R}\mathbf{x} = \|\mathrm{R}\mathbf{x}\|^2 = \|\mathbf{x}\|^2$

From this latter equation,

$$\frac{d\theta}{dt} = \frac{\mathrm{R}\mathbf{x} \cdot (d\mathbf{x}/dt)}{\|\mathbf{x}\|^2} = \frac{(\mathrm{R}\mathbf{x} \cdot \mathrm{A}\mathbf{x}) + (\mathrm{R}\mathbf{x} \cdot \mathbf{h}(\mathbf{x}))}{\|\mathbf{x}\|^2}$$

The next theorem summarizes the foregoing.

† See Exercise 8.6, Part 2, Problem 7.

8.6.2. THEOREM *If* $\mathbf{x}(t)$ *is a solution of* $\mathbf{x}' = \mathrm{A}\mathbf{x}+\mathbf{h}(\mathbf{x})$ *then*

$$\frac{d\|\mathbf{x}(t)\|^2}{dt} = 2\mathbf{x}(t)\cdot \mathrm{A}\mathbf{x}(t) + 2\mathbf{x}(t)\cdot \mathbf{h}(\mathbf{x}(t)) \tag{8.6.7}$$

and

$$\frac{d\theta(t)}{dt} = \frac{\mathrm{R}\mathbf{x}(t)\cdot \mathrm{A}\mathbf{x}(t) + \mathrm{R}\mathbf{x}(t)\cdot \mathbf{h}(\mathbf{x}(t))}{\|\mathbf{x}(t)\|^2} \tag{8.6.8}$$

8.6.2. Example

Study the behavior of the orbits of

$$\mathbf{x}' = \begin{bmatrix} -1 & 0 \\ 0 & -1 \end{bmatrix}\mathbf{x} - \|\mathbf{x}\|\,\mathrm{R}\mathbf{x}$$

Solution

Since the eigenvalues of $\mathrm{A} = -\mathrm{I}$ are negative, Theorem 8.5.1 asserts that $\mathbf{x} = \mathbf{0}$ is an asymptotically stable critical point of the unperturbed system, $\mathbf{x}' = -\mathrm{I}\mathbf{x}$. Theorem 8.6.1 then leads to the conclusion that $\mathbf{x} = \mathbf{0}$ is an asymptotically stable critical point of $\mathbf{x}' = -\mathrm{I}\mathbf{x} - \|\mathbf{x}\|\mathrm{R}\mathbf{x}$ as well. Let us verify this directly by using the tools provided in Theorem 8.6.2. From the simple structure of A and R and from (8.6.7) and (8.6.8) we obtain,

$$\begin{aligned}\frac{d\|\mathbf{x}\|^2}{dt} &= -2\mathbf{x}\cdot\mathbf{x} - 2\|\mathbf{x}\|\mathbf{x}\cdot\mathrm{R}\mathbf{x} \\ &= -2\|\mathbf{x}\|^2\end{aligned}$$

Also,

$$\begin{aligned}\frac{d\theta}{dt} &= \frac{-\mathrm{R}\mathbf{x}\cdot\mathbf{x} - \|\mathbf{x}\|\mathrm{R}\mathbf{x}\cdot\mathrm{R}\mathbf{x}}{\|\mathbf{x}\|^2} \\ &= -\frac{\|\mathbf{x}\|^3}{\|\mathbf{x}\|^2} = -\|\mathbf{x}\|\end{aligned}$$

But

$$\|\mathbf{x}(t)\|^2 = c_1 e^{-2t}$$

and hence

$$\theta(t) = \frac{c_1}{2}e^{-2t} + c_2$$

We see therefore, that $\mathbf{x}(t)$ tends to $\mathbf{0}$ in such a way that $\lim_{t\to+\infty} \theta(t) = c_2$. Since c_2 is arbitrary, every angle is approached. One such approach is indicated in Figure 8.6.1(a). The unperturbed system has a phase portrait as in Figure 8.6.1(b). Here every ray is an orbit and $\|\mathbf{x}\|$ "decays" exponentially for

$$\frac{d\|\mathbf{x}\|^2}{dt} = 2\mathbf{x}\cdot(-\mathrm{I}\mathbf{x}) = -2\|\mathbf{x}\|^2$$

Figure 8.6.1

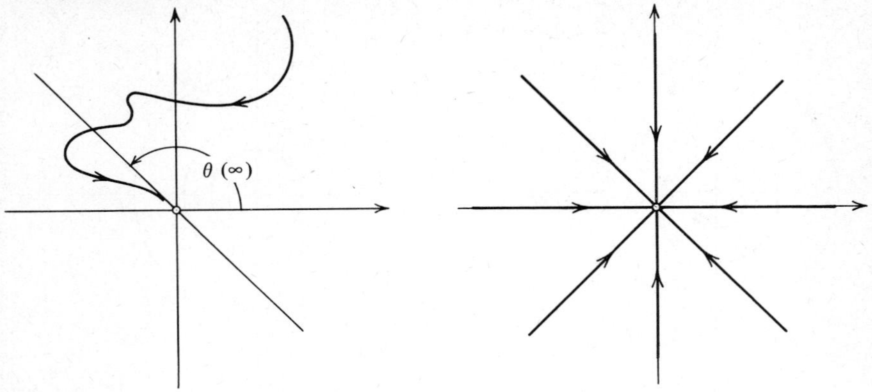

(a) (b)

as in the perturbed system. In a small neighborhood of $\mathbf{x} = \mathbf{0}$ the phase portraits of the perturbed and unperturbed systems are very similar.

8.6.3. Example

Study the behavior of the orbits of

$$\mathbf{x}' = \begin{bmatrix} 0 & -1 \\ 1 & 0 \end{bmatrix}\mathbf{x} - \|\mathbf{x}\|\mathbf{x}$$

Solution

Since the eigenvalues of A = R are $\lambda = \pm i$, the origin is now a stable but not asymptotically stable critical point of $\mathbf{x}' = A\mathbf{x}$. Theorem 8.6.1 does not apply but our methods yield

$$\frac{d}{dt}\|\mathbf{x}(t)\|^2 = -2\|\mathbf{x}(t)\|^3$$

with solution

$$\|\mathbf{x}(t)\| = \frac{1}{(t+c_1)^2}$$

Also, since A = R in this problem and $R\mathbf{x}\cdot\mathbf{x} = 0$, we have

$$\frac{d\theta}{dt} = \frac{R\mathbf{x}(t) \cdot R\mathbf{x}(t)}{\|\mathbf{x}(t)\|^2} = 1$$

Hence $\theta(t) = t + \theta_0$, where $\theta(0) = \theta_0$. The phase portrait of the perturbed system is drawn in Figure 8.6.2(a) and the unperturbed system in Figure 8.6.2(b). The perturbed system has an asymptotically stable critical point in contrast to the stable, nonasymptotically stable critical point of $\mathbf{x}' = R\mathbf{x}$.

The next example indicates the alternative possibility that can be realized when $\mathbf{x} = \mathbf{0}$ is a stable but not asymptotically stable critical point of $\mathbf{x}' = A\mathbf{x}$.

Figure 8.6.2

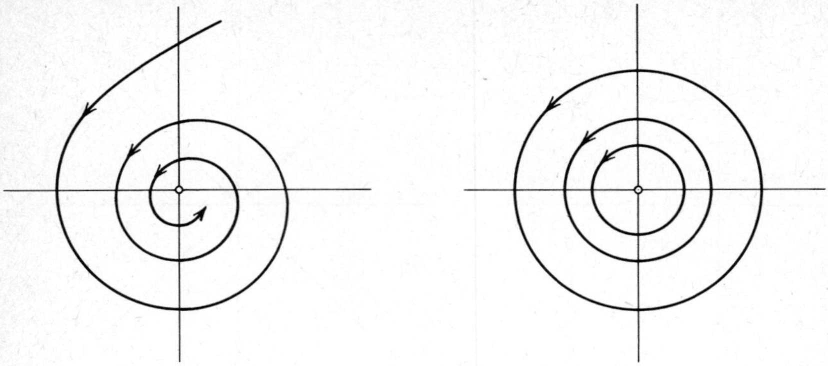

(a) (b)

8.6.4. Example

Study the behavior of the orbits of

$$\mathbf{x}' = \begin{bmatrix} 0 & -1 \\ 1 & 0 \end{bmatrix}\mathbf{x} - \|\mathbf{x}\|\mathbf{R}\mathbf{x}$$

Solution

We learn from (8.6.2) that

$$\frac{d}{dt}\|\mathbf{x}\|^2 = 0 \qquad \frac{d\theta}{dt} = 1 - \|\mathbf{x}\|$$

Hence $\|\mathbf{x}\| = c_1$ and $\theta = (1-c_1)t + c_2$. The orbits of the perturbed system are circle periodic as in the unperturbed system. The periods are more complicated, however, since $(1-c_1)(t_2-t_1) = 2\pi$ implies $\rho = 2\pi/(1-c_1)$ a function of the radius of the circle.

8.6.5. Example

Discuss the critical points $\mathbf{x} = \mathbf{0}$ and $\mathbf{x} = \begin{bmatrix} \pi \\ 0 \end{bmatrix}$ of the system

$$\mathbf{x}' = \begin{bmatrix} x_2 \\ \dfrac{-g}{l}\sin x_1 - \dfrac{k}{m}x_2 \end{bmatrix} \tag{8.6.9}$$

representing the simple pendulum.

Solution

For the critical point $\mathbf{x} = \mathbf{0}$ we put (8.6.9) in the form

$$\mathbf{x}' = \begin{bmatrix} 0 & 1 \\ \dfrac{-g}{l} & \dfrac{-k}{m} \end{bmatrix}\mathbf{x} + \frac{g}{l}\begin{bmatrix} 0 \\ x_1 - \sin x_1 \end{bmatrix}$$

The "unperturbed" system

$$\mathbf{x}' = \begin{bmatrix} 0 & 1 \\ -\dfrac{g}{l} & -\dfrac{k}{m} \end{bmatrix} \mathbf{x}$$

has $\mathbf{x} = \mathbf{0}$ as an asymptotically stable critical point as we can see by examining the eigenvalues,

$$\lambda_1 = -\frac{k}{2m} + \frac{1}{2}\left(\frac{k^2}{m^2} - 4\frac{g}{l}\right)^{1/2}$$

$$\lambda_2 = -\frac{k}{2m} - \frac{1}{2}\left(\frac{k^2}{m^2} - 4\frac{g}{l}\right)^{1/2}$$

If $(k^2/m^2) - 4(g/l) < 0$ then the roots are complex and $\operatorname{Re}\lambda_1 = \operatorname{Re}\lambda_2 = -k/2m < 0$ since k and m are positive. If $(k^2/m^2) - 4(g/l) \geqslant 0$ then, since $k^2/m^2 > (k^2/m^2) - 4(g/l)$, $\lambda_2 \leqslant \lambda_1 < 0$ and in either case Theorem 8.5.3 asserts that $\mathbf{x} = \mathbf{0}$ is asymptotically stable. Theorem 8.6.1, in turn, asserts that $\mathbf{x} = \mathbf{0}$ is an asymptotically stable critical point of (8.6.9).

For $\mathbf{x} = \begin{bmatrix} \pi \\ 0 \end{bmatrix}$, we replace $\mathbf{x}$ by $\mathbf{y} + \begin{bmatrix} \pi \\ 0 \end{bmatrix}$ in (8.6.9) obtaining

$$\mathbf{y}' = \begin{bmatrix} y_2 \\ \dfrac{g}{l}\sin y_1 - \dfrac{k}{m} y_2 \end{bmatrix} \tag{8.6.10}$$

To examine this system near $\mathbf{y} = \mathbf{0}$ (i.e., $\mathbf{x} = \begin{bmatrix} \pi \\ 0 \end{bmatrix}$), we write (8.6.10) in the form

$$\mathbf{y}' = \begin{bmatrix} 0 & 1 \\ \dfrac{g}{l} & -\dfrac{k}{m} \end{bmatrix} \mathbf{y} - \frac{g}{l} \begin{bmatrix} 0 \\ y_1 - \sin y_1 \end{bmatrix}$$

In this case the eigenvalues of the "unperturbed" system are

$$\lambda_1 = -\frac{k}{2m} + \left(\frac{k^2}{m^2} + \frac{4g}{l}\right)^{1/2}$$

$$\lambda_2 = -\frac{k}{2m} - \left(\frac{k^2}{m^2} + \frac{4g}{l}\right)^{1/2}$$

But, clearly, $\lambda_2 < 0 < \lambda_1$.

Using Theorems 8.5.3 and 8.6.1, as above, we find $\mathbf{x} = \begin{bmatrix} \pi \\ 0 \end{bmatrix}$ is an unstable critical point of (8.6.9), with only *two* orbits approaching it asymptotically. Exercise 8.6, Part 2, Problem 11 shows that there are no periodic orbits. See Figure 8.6.3. These results have been anticipated from physical considerations in Section 8.5.

Figure 8.6.3. A rough sketch of the phase portrait of $\mathbf{x}' = \begin{bmatrix} x_2 \\ -\alpha \sin x_1 - \beta x_2 \end{bmatrix}$, $\alpha > 0, \beta > 0$.

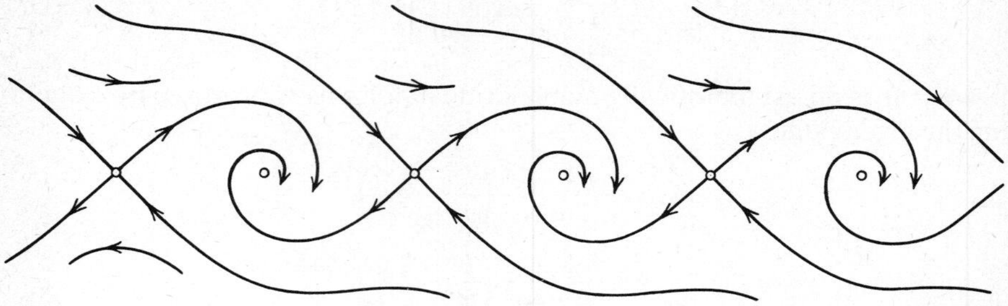

Exercise 8.6

Part 1

1. Which of the following perturbations, $\mathbf{h}(\mathbf{x})$, meet the hypothesis (8.6.2)?

(a) $\begin{bmatrix} x_1^2 \\ x_2^2 \end{bmatrix}$ (b) $\|\mathbf{x}\|^2$ (c) $\begin{bmatrix} x_1 x_2 \\ x_1^2 \end{bmatrix}$

(d) $\begin{bmatrix} 0 \\ r - g(r) \end{bmatrix}$ $g(r) = r + a_2 r^2 + \cdots$ $r = \|\mathbf{x}\|$ $g(r)$ analytic

(e) $\begin{bmatrix} \dfrac{e^{x_1} - e^{-x_1}}{2} \\ \dfrac{e^{x_1} + e^{-x_1}}{2} - 1 \end{bmatrix}$ (f) $\begin{bmatrix} a \\ b \end{bmatrix}$ a and b are constants

(g) $f(r)r^2$, where $r = \|\mathbf{x}\|$ and f is continuous and differentiable in $0 \leqslant r \leqslant M$.

(h) $\dfrac{\mathbf{R}\mathbf{x}}{\log \|\mathbf{x}\|}$ (i) $x_1 \mathbf{R}\mathbf{x}$ (*j*) $\|\mathbf{x}\| \mathbf{A}\mathbf{x}$ A arbitrary

2. Find all the critical points of the systems companion to the following second-order equations.

(a) $y'' - y + y^3 = 0$ (b) $y'' + y - y^3 = 0$
(c) $y'' + y + y^3 = 0$ (d) $y'' - y - y^3 = 0$
(e) $y'' + ay' + by + y^2 = 0$ $a > 0, b > 0$
(f) $y'' + (y')^2 + y^2 + y = 0$

3. Examine each critical point of the equations given in Problem 2 above and using Theorem 8.6.1, determine whether they are stable, asymptotically stable, or unstable.
4. Use Theorem 8.6.2 to examine the critical point $\mathbf{x} = \mathbf{0}$ of

(a) $\mathbf{x}' = -\mathbf{I}\mathbf{x} - \dfrac{\mathbf{R}\mathbf{x}}{\log \|\mathbf{x}\|}$

(b) $\mathbf{x}' = \mathbf{I}\mathbf{x} - (1 + \alpha r)\mathbf{R}\mathbf{x}, r = \|\mathbf{x}\|$

Note $\mathbf{R} = \begin{bmatrix} 0 & -1 \\ 1 & 0 \end{bmatrix}$

Part 2

1. Show that $|u - \sin u| < \frac{1}{3}$ if $|u| < 1$.
   Hint: Use $u - \sin u = (u^3/3!) - (u^5/5!) + \cdots$.
2. Derive Equation (8.6.8) from $\theta = \arctan x_2/x_1$ assuming $\mathbf{x}' = \mathbf{A}\mathbf{x} + \mathbf{h}(\mathbf{x})$.
3. Show that $\mathbf{u}(t) \cdot \mathbf{u}'(t) = 0$ if $\|\mathbf{u}(t)\| = 1$ for all t.
4. Let $\|\mathbf{u}(t_0)\| = 1$ and

$$\mathrm{R} = \begin{bmatrix} 0 & -1 \\ 1 & 0 \end{bmatrix}$$

Show that every vector $\mathbf{a}$ can be written as

$$\mathbf{a} = (\mathbf{a} \cdot \mathbf{u}(t_0))\mathbf{u}(t_0) + (\mathbf{a} \cdot \mathbf{R}\mathbf{u}(t_0))\mathbf{R}\mathbf{u}(t_0)$$

5. Use the result of Problem 4 to conclude:
   If $\|\mathbf{u}(t)\| = 1$ for all t then

$$\mathbf{u}'(t) = (\mathbf{u}'(t) \cdot \mathbf{R}\mathbf{u}(t))\ \mathbf{R}\mathbf{u}(t)$$

6. Let

$$\mathbf{v}(\tau) = \begin{bmatrix} \cos \tau \\ \sin \tau \end{bmatrix}$$

and R be defined as in Problem 4. Suppose τ_0 is picked so that $\mathbf{u}(0) = \mathbf{v}(\tau_0)$, and

$$\tau = \tau_0 + \int_0^t \mathbf{u}'(s) \cdot \mathbf{R}\mathbf{u}(s)\, ds$$

Use the chain rule to show that

$$\frac{d}{dt}\mathbf{v}(\tau) = (\mathbf{u}'(t) \cdot \mathbf{R}\mathbf{u}(t))\ \mathbf{R}\mathbf{u}(t)$$

7.* Prove: If $\mathbf{u}(t)$ is differentiable for all t and $\|\mathbf{u}(t)\| = 1$ then there exists a differentiable function $\theta(t)$ such that

$$\mathbf{u}(t) = \begin{bmatrix} \cos \theta(t) \\ \sin \theta(t) \end{bmatrix}$$

Hint: Define $\theta = \tau$ and use Problems 5 and 6 above to show that $\mathbf{u}(t)$ and $\mathbf{v}(\theta(t))$ satisfy the same initial-value problem.

8. Show that the assumptions on $\mathbf{h}(\mathbf{x})$ imply

$$\lim_{\|\mathbf{x}\| \to 0} \frac{\|\mathbf{h}(\mathbf{x})\|}{\|\mathbf{x}\|} = 0$$

9. (a) Use Theorem 8.6.2 to show: If $\mathbf{x}(t)$ is a solution of

$$\mathbf{x}' = \mathrm{I}\mathbf{x} + f(r)\mathrm{R}\mathbf{x},\ r = \|\mathbf{x}\|, \text{ then } \|\mathbf{x}(t)\| = c_1 e^{-t}$$

and

$$\theta(t) = \int_{t_0}^t f(c_1 e^{-s})\, ds + \theta(t_0)$$

(b) Use the result of (a) to examine the critical point $\mathbf{x} = \mathbf{0}$ of the system:

$$\mathbf{x}' = \mathrm{I}\mathbf{x} + \mathrm{P}(r)\ \mathrm{R}\mathbf{x}$$

where $\mathrm{P}(r)$ is a polynomial in $r = \|\mathbf{x}\|$.

10. Using Example 8.6.5 as a guide study the critical point of the companion system of

$$y''+y'+g(y)=0$$

where $g(y)=g_1y+g_2y^2+\cdots, \quad g_1>0$.

11. Corresponding to Equations (8.5.1) and (8.5.2), we consider the so-called total energy function

$$\mathrm{T}(x_1, x_2)=mgl\,(1-\cos x_1)+\tfrac{1}{2}ml^2x_2^2$$

(a) What has $\mathrm{T}(x_1, x_2)$ go to do with total energy?
(b) For any solution of (8.5.2), $\mathbf{x}(t)$, show that

$$\frac{d}{dt}\mathrm{T}(\mathbf{x}(t))=-kl^2x_2^2\leqslant 0$$

(c) Show that $x_2(t)\equiv 0$ for a solution $\mathbf{x}(t)$ only if $\mathbf{x}(t)\equiv\begin{bmatrix} n\pi \\ 0\end{bmatrix}$ for some integer n.

(d) Show that except for the constant solutions $\mathbf{x}(t)\equiv\begin{bmatrix} n\pi \\ 0\end{bmatrix}$, $\mathrm{T}(\mathbf{x}(t))$ is strictly decreasing.
(e) Show that (8.5.2) has no nonconstant periodic solutions.

8.7.* THE EQUATION OF FIRST VARIATION

The present section involves some concepts of multivariable calculus which we will introduce and discuss briefly. These ideas are then applied to the study of nonlinear equations. The reader who is unfamiliar with the calculus of several variables should attempt to grasp the scope of the discussion even if he cannot master the details. Although the definitions and assertions of this section are intended to be fairly precise, we have omitted many arguments necessary for their proof.

We consider a two-dimensional vector-valued function of a two-dimensional vector variable,

$$\mathbf{f}(\mathbf{x})=\begin{bmatrix} f_1(\mathbf{x}) \\ \\ f_2(\mathbf{x})\end{bmatrix}=\begin{bmatrix} f_1\left(\begin{bmatrix} x_1 \\ x_2\end{bmatrix}\right) \\ f_2\left(\begin{bmatrix} x_1 \\ x_2\end{bmatrix}\right)\end{bmatrix}$$

To study the variation of $\mathbf{f}(\mathbf{x})$ it is useful to introduce the *Jacobian Matrix* of $\mathbf{f}(\mathbf{x})$, namely,

$$\mathbf{Jf}(\mathbf{x})=\begin{bmatrix} \dfrac{\partial f_1}{\partial x_1}(\mathbf{x}) & \dfrac{\partial f_1}{\partial x_2}(\mathbf{x}) \\ \dfrac{\partial f_2}{\partial x_1}(\mathbf{x}) & \dfrac{\partial f_2}{\partial x_2}(\mathbf{x})\end{bmatrix}$$

Note that the Jacobian Matrix depends on both **f** *and* **x**. Its four components representing the variation of each of the components of **f** with respect to each of the components of **x**.

For a given **f**, $\mathbf{Jf}(\mathbf{x})$ is a function of **x**. If each of the partial derivatives $\partial f_i/\partial x_j$, $i = 1, 2$; $j = 1, 2$ varies continuously for values of **x** near some fixed $\mathbf{x}_0$, then the Jacobian Matrix may be used to *approximate* the change in $\mathbf{f}(\mathbf{x})$, $\mathbf{f}(\mathbf{x}) - \mathbf{f}(\mathbf{x}_0)$, as follows:

$$\mathbf{f}(\mathbf{x}) - \mathbf{f}(\mathbf{x}_0) = \mathbf{Jf}(\mathbf{x}_0)(\mathbf{x} - \mathbf{x}_0) + \mathrm{E}(\mathbf{x} - \mathbf{x}_0) \tag{8.7.1}$$

where the "error" term $\mathrm{E}(\mathbf{x} - \mathbf{x}_0)$ is "small" compared to $\mathbf{x} - \mathbf{x}_0$ in the sense that

$$\frac{\|\mathrm{E}(\mathbf{x} - \mathbf{x}_0)\|}{\|\mathbf{x} - \mathbf{x}_0\|} \to 0$$

as $\mathbf{x} \to \mathbf{x}_0$. We express this as

$$\mathbf{f}(\mathbf{x}) - \mathbf{f}(\mathbf{x}_0) \approx \mathbf{Jf}(\mathbf{x}_0)(\mathbf{x} - \mathbf{x}_0)$$

8.7.1 Example

Consider the function

$$\mathbf{f}(\mathbf{x}) = \begin{bmatrix} x_1^2 \\ x_1 x_2^3 \end{bmatrix}.$$

We find

$$\mathbf{Jf}(\mathbf{x}) = \begin{bmatrix} 2x_1 & 0 \\ x_2^3 & 3x_1 x_2^2 \end{bmatrix}$$

For $\mathbf{x}_0 = \begin{bmatrix} 1 \\ 2 \end{bmatrix}$, we have

$$\mathbf{f}(\mathbf{x}) - \mathbf{f}(\mathbf{x}_0) = \mathbf{f}(\mathbf{x}) - \mathbf{f}\left(\begin{bmatrix} 1 \\ 2 \end{bmatrix}\right) \approx \mathbf{Jf}(\mathbf{x}_0)(\mathbf{x} - \mathbf{x}_0)$$

$$= \begin{bmatrix} 2 & 0 \\ 8 & 12 \end{bmatrix} \begin{bmatrix} x_1 - 1 \\ x_2 - 2 \end{bmatrix}$$

Problem 1 at the end of this section is to evaluate the error of this approximation.

It is also possible to show that if $\mathbf{Jf}(\mathbf{x})$ is continuous and $\mathbf{x}(t)$ is differentiable then

$$\frac{d}{dt}\mathbf{f}(\mathbf{x}(t)) = \mathbf{Jf}(\mathbf{x}(t))\mathbf{x}'(t)$$

Note that here we have strict equality, not an approximation.

8.7.2 Example

Let

$$\mathbf{f}(\mathbf{x}) = \begin{bmatrix} x_1^2 \\ x_1 x_2^3 \end{bmatrix}$$

and

$$\mathbf{x}(t) = \begin{bmatrix} \sin t \\ \cos t \end{bmatrix}.$$

Then

$$\mathbf{f}(\mathbf{x}(t)) = \begin{bmatrix} \sin^2 t \\ \sin t \cos^3 t \end{bmatrix}$$

so that

$$\frac{d}{dt}\mathbf{f}(\mathbf{x}(t)) = \begin{bmatrix} 2 \sin t \cos t \\ \cos^4 t - 3 \sin^2 t \cos^2 t \end{bmatrix}$$

On the other hand,

$$\mathbf{Jf}\begin{bmatrix} \sin t \\ \cos t \end{bmatrix} \cdot \begin{bmatrix} \cos t \\ -\sin t \end{bmatrix} = \begin{bmatrix} 2 \sin t & 0 \\ \cos^3 t & 3 \sin t \cos^2 t \end{bmatrix} \begin{bmatrix} \cos t \\ -\sin t \end{bmatrix}$$

$$= \begin{bmatrix} 2 \sin t \cos t \\ \cos^4 t - 3 \sin^2 t \cos^2 t \end{bmatrix}$$

as well.

A rule similar to that above is the following: Let $\mathbf{f}(\mathbf{x})$ and $\mathbf{g}(\mathbf{x})$ be given functions. Form the composite function, $\mathbf{h}$, defined by

$$\mathbf{h}(\mathbf{x}) = \mathbf{f}(\mathbf{g}(\mathbf{x}))$$

In other words set

$$\mathbf{h}(\mathbf{x}) = \mathbf{f}(\mathbf{y})$$

where

$$\mathbf{y} = \mathbf{g}(\mathbf{x})$$

If both $\mathbf{Jf}(\mathbf{x})$ and $\mathbf{Jg}(\mathbf{x})$ are continuous then

$$\mathbf{Jh}(\mathbf{x}) = \mathbf{Jf}(\mathbf{g}(\mathbf{x}))\mathbf{Jg}(\mathbf{x})$$

That is, we form the product of the matrix $\mathbf{Jf}(\mathbf{y})$, where $\mathbf{Jf}(\mathbf{y})$ is evaluated at $\mathbf{y} = \mathbf{g}(\mathbf{x})$, and the matrix $\mathbf{Jg}(\mathbf{x})$ and obtain the matrix $\mathbf{Jh}(\mathbf{x})$.

8.7.3 Example

Let

$$\mathbf{f}(\mathbf{y}) = \begin{bmatrix} y_1^2 \\ y_1 y_2^2 \end{bmatrix}$$

and

$$\mathbf{g}(\mathbf{x}) = \begin{bmatrix} x_1 + x_2 \\ x_1 - x_2 \end{bmatrix}.$$

Then

$$\mathbf{h}(\mathbf{x}) = \mathbf{f}(\mathbf{g}(\mathbf{x})) = \begin{bmatrix} (x_1 + x_2)^2 \\ (x_1 + x_2)(x_1 - x_2)^2 \end{bmatrix}$$

and

$$\mathbf{Jh}(\mathbf{x}) = \begin{bmatrix} 2(x_1+x_2) & 2(x_1+x_2) \\ 3x_1^2 - x_2^2 - 2x_1x_2 & 3x_2^2 - x_1^2 - 2x_1x_2 \end{bmatrix}$$

On the other hand,

$$\begin{aligned} \mathbf{Jf}(\mathbf{g}(\mathbf{x})) &= \begin{bmatrix} 2y_1 & 0 \\ y_2^2 & 2y_1y_2 \end{bmatrix}_{\mathbf{y}=\mathbf{g}(\mathbf{x})} \\ &= \begin{bmatrix} 2(x_1+x_2) & 0 \\ (x_1-x_2)^2 & 2(x_1+x_2)(x_1-x_2) \end{bmatrix} \end{aligned}$$

and

$$\mathbf{Jg}(\mathbf{x}) = \begin{bmatrix} 1 & 1 \\ 1 & -1 \end{bmatrix}$$

so that

$$\begin{aligned} \mathbf{Jf}(\mathbf{g}(\mathbf{x}))\mathbf{Jg}(\mathbf{x}) &= \begin{bmatrix} 2(x_1+x_2) & 0 \\ (x_1-x_2)^2 & 2(x_1+x_2)(x_1-x_2) \end{bmatrix} \begin{bmatrix} 1 & 1 \\ 1 & -1 \end{bmatrix} \\ &= \begin{bmatrix} 2(x_1+x_2) & 2(x_1+x_2) \\ (3x_1+x_2)(x_1-x_2) & (-x_1-3x_2)(x_1-x_2) \end{bmatrix} \\ &= \begin{bmatrix} 2(x_1+x_2) & 2(x_1+x_2) \\ 3x_1^2-2x_1x_2-x_2^2 & 3x_2^2-2x_1x_2-x_1^2 \end{bmatrix} \\ &= \mathbf{Jh}(\mathbf{x}) \end{aligned}$$

The last two examples are cases of the so-called *Chain Rule* which relates the derivative of a composite function to the derivatives of its components. The value of the chain rule lies *not* in computation but rather in the insight it gives about the general behavior of composite functions.

When considering a function $\mathbf{f}(t, \mathbf{x})$ of the real variable t as well as the vector variable $\mathbf{x}$, the Jacobian Matrix will be computed with respect to $\mathbf{x}$, treating t as a constant; that is, we take

$$\mathbf{Jf}(t, \mathbf{x}) = \begin{bmatrix} \dfrac{\partial f_1}{\partial x_1}(t, \mathbf{x}) & \dfrac{\partial f_1}{\partial x_2}(t, \mathbf{x}) \\ \dfrac{\partial f_2}{\partial x_1}(t, \mathbf{x}) & \dfrac{\partial f_2}{\partial x_2}(t, \mathbf{x}) \end{bmatrix}$$

We now apply these concepts to the study of the equation

$$\mathbf{x}' = \mathbf{f}(\mathbf{x}) \tag{8.7.2}$$

where $\mathbf{f}(\mathbf{x})$ is a given two-dimensional vector-valued function of a two-dimensional vector available. We suppose that

$$\mathbf{x}_0(t) = \mathbf{x}(t; \mathbf{x}_0)$$

is a *known* solution of (8.7.2) with initial value $\mathbf{x}_0(\mathbf{0}) = \mathbf{x}_0$. We consider the solution

$$\mathbf{x}_1(t) = \mathbf{x}(t; \mathbf{x}_1)$$

which has the initial value $\mathbf{x}_1(0) = \mathbf{x}_1$ near $\mathbf{x}_0$. We shall attempt to get some approximate information about the difference

$$\boldsymbol{\delta}(t) = \mathbf{x}_1(t) - \mathbf{x}_0(t) = \mathbf{x}(t; \mathbf{x}_1) - \mathbf{x}(t; \mathbf{x}_0)$$

We find

$$\begin{aligned} \boldsymbol{\delta}'(t) &= \mathbf{x}_1'(t) - \mathbf{x}_0'(t) \\ &= \mathbf{f}(\mathbf{x}_1(t)) - \mathbf{f}(\mathbf{x}_0(t)) \\ &= J\mathbf{f}(\mathbf{x}_0(t))(\mathbf{x}_1(t)) - \mathbf{x}_0(t)) + E(\mathbf{x}_1(t) - \mathbf{x}_0(t)) \\ &\approx J\mathbf{f}(\mathbf{x}_0(t))\boldsymbol{\delta}(t) \end{aligned}$$

Thus we find that $\boldsymbol{\delta}(t)$ satisfies a differential equation which is approximated by the equation

$$\mathbf{z}' = \mathrm{A}(t)\mathbf{z} \tag{8.7.3}$$

where

$$\mathrm{A}(t) = \mathbf{Jf}(\mathbf{x}_0(t))$$

is known since $\mathbf{f}(\mathbf{x})$ and $\mathbf{x}_0(t)$ are known. Equation (8.7.3) is called the *equation of first variation*. Since it is linear, there is some hope of solving it or at least determining the character of its solutions.

8.7.4. Example

Consider the equation

$$\mathbf{x}' = \mathbf{f}(\mathbf{x}) = \begin{bmatrix} -x_2 + (1 - x_1^2 - x_2^2)x_1 \\ x_1 + (1 - x_1^2 - x_2^2)x_2 \end{bmatrix}$$

One can *verify* that

$$\mathbf{x}_0(t) = \begin{bmatrix} \cos t \\ \sin t \end{bmatrix}$$

is a solution. (Problem 3 at the end of this section.)

By straightforward computation we find that

$$\mathbf{Jf}(\mathbf{x}) = \begin{bmatrix} 1 - 3x_1^2 - x_2^2 & -1 - 2x_2x_1 \\ 1 - 2x_1x_2 & 1 - x_1^2 - 3x_2^2 \end{bmatrix} \tag{8.7.4}$$

so that

$$\begin{aligned} \mathbf{Jf}(\mathbf{x}_0(t)) &= \mathbf{Jf}\begin{bmatrix} \cos t \\ \sin t \end{bmatrix} \\ &= \begin{bmatrix} -2\cos^2 t & -1 - \sin 2t \\ 1 - \sin 2t & -2\sin^2 t \end{bmatrix} \end{aligned}$$

Thus the equation of first variation corresponding to

$$\mathbf{x}' = \mathbf{f}(\mathbf{x}) \qquad \mathbf{x}_0(t) = \begin{bmatrix} \cos t \\ \sin t \end{bmatrix}$$

is given by

$$\mathbf{z}' = \begin{bmatrix} -2\cos^2 t & -1-\sin 2t \\ 1-\sin 2t & -2\sin^2 t \end{bmatrix} \mathbf{z}$$

Exercise 8.7

1. Let $\mathbf{f}(\mathbf{x}) = \begin{bmatrix} x_1^2 \\ x_1 x_2^3 \end{bmatrix}$ and $\mathbf{x}_0 = \begin{bmatrix} 1 \\ 2 \end{bmatrix}$. Show that

$$\begin{aligned} \mathrm{E}(\mathbf{x}-\mathbf{x}_0) &\equiv \mathbf{f}(\mathbf{x}) - \mathbf{f}(\mathbf{x}_0) - \mathbf{Jf}(\mathbf{x}_0)(\mathbf{x}-\mathbf{x}_0) \\ &= \begin{bmatrix} (x_1-1)^2 \\ (x_1-1)(x_2-2)(x_2^2+2x_2+4) + (x_2-2)^2(x_2+4) \end{bmatrix} \end{aligned}$$

   Hint: To obtain the second component, factor!

2. Show that $\mathrm{E}(\mathbf{x}-\mathbf{x}_0)$, above, satisfies:

$$\frac{\mathrm{E}(\mathbf{x}-\mathbf{x}_0)}{\|(\mathbf{x}-\mathbf{x}_0)\|} \to 0 \qquad \text{as} \quad \mathbf{x} \to \mathbf{x}_0$$

3. In Example 8.5.4, show that $\mathbf{x}_0(t)$ as given is a solution.
4. Let $\mathbf{x}(t; \mathbf{x})$ denote the solution of $\mathbf{x}' = \mathbf{f}(\mathbf{x})$ satisfying $\mathbf{x}(0, \mathbf{y}) = \mathbf{y}$. Note this implies

$$\frac{\partial}{\partial t}\mathbf{x}(t; \mathbf{y}) = \mathbf{f}(\mathbf{x}(t; \mathbf{y}))$$

   Assuming that

$$\mathbf{Jx}(t; \mathbf{y}) \text{ as well as } \frac{\partial}{\partial t}\mathbf{Jx}(t; \mathbf{y}) = \mathbf{J}\frac{\partial}{\partial t}\mathbf{x}(t; \mathbf{y})$$

   exist, show that $\mathbf{Jx}(t; \mathbf{y})$ is a *matrix* solution of $z' = A(t)\mathbf{z}$ where

$$A(t) = \mathbf{Jf}(\mathbf{x}(t; \mathbf{y}))$$

8.8.* STABILITY OF PERIODIC ORBITS

In some systems, periodic solutions may represent desirable performance. Such is the case, for example, with an electronic oscillator. Now it is not enough to know that an oscillator will oscillate if certain precise initial conditions are met. The user of the oscillator must be given some tolerance in the choice of initial states which lead to oscillatory behavior. In order to study this problem, let us introduce some mathematical notions which are useful. We assume the differential equation $\mathbf{x}' = \mathbf{f}(\mathbf{x})$ and the solutions $\mathbf{x}(t; \mathbf{x}_0)$ are given.

8.8.1. DEFINITION *Let* $O(\mathbf{x}_0) = \{\mathbf{x}(t; \mathbf{x}_0) \mid -\infty < t < \infty\}$ *be a periodic orbit, and* $\mathbf{x}$ *a vector. We define the distance of* $\mathbf{x}$ *to* $O(\mathbf{x}_0)$, $\|\mathbf{x}, O(\mathbf{x}_0)\|$, *by*

$$\|\mathbf{x}, O(\mathbf{x}_0)\| = \min \{\|\mathbf{x} - \mathbf{y}\| \,\big|\, \mathbf{y} \text{ in } O(\mathbf{x}_0)\}$$

that is, $\|\mathbf{x}, O(\mathbf{x}_0)\|$ *is the distance from* $\mathbf{x}$ *to the nearest point in* $O(\mathbf{x}_0)$.

8.8.2. DEFINITION *Let* $O(\mathbf{x}_0)$ *be a periodic orbit and* $\mathbf{x}$ *a vector. We say* $O(\mathbf{x})$ *approaches* $O(\mathbf{x}_0)$ *asymptotically as* $t \to \infty$, *if* $\|\mathbf{x}(t; \mathbf{x}), O(\mathbf{x}_0)\| \to 0$ *as* $t \to \infty$.

8.8.3. DEFINITION *Let* $O(\mathbf{x}_0)$ *be a periodic orbit. We say* $O(\mathbf{x}_0)$ *is stable if for any given* $\epsilon > 0$ *there is a* $\delta > 0$ *(depending on* ϵ*) such that whenever* $\|\mathbf{x}, O(\mathbf{x}_0)\| < \delta$ *it follows that* $\|\mathbf{x}(t; \mathbf{x}), O(\mathbf{x}_0)\| < \epsilon$ *for* $0 < t < \infty$.

Roughly speaking, stability implies that solutions starting close to $O(\mathbf{x}_0)$, remain close to $O(\mathbf{x}_0)$, externally. This form of stability is sometimes called *stability in the sense of Liapunov.*

8.8.4. DEFINITION *If* $O(\mathbf{x}_0)$ *is a stable periodic orbit such that for some* $\delta > 0$, *whenever* $\|\mathbf{x}, O(\mathbf{x}_0)\| < \delta$, $O(\mathbf{x})$ *approaches* $O(\mathbf{x}_0)$ *asymptotically, we say that* $O(\mathbf{x}_0)$ *is asymptotically stable.*

Now, if we were designing an oscillator we would like to have its operation represented by a differential equation with an asymptotically stable periodic orbit, $O(\mathbf{x}_0)$, of the desired frequency. Then if we could get the system sufficiently close to $O(\mathbf{x}_0)$, it would tend toward $O(\mathbf{x}_0)$ as a steady oscillatory state.

We note that in the case of two-dimensional systems, if a periodic orbit, $O(\mathbf{x}_0)$, has at least one orbit on each side tending toward it asymptotically, then it must asymptotically be stable. However, this is not the case in higher dimensional systems. Its proof depends on the special geometric properties of the plane.

We end this section with the statement of a theorem which is also limited to two-dimensional systems. Although this theorem is fairly simple to state, it is very difficult to prove. The interested reader may refer to a work of Pontryagin.†

8.8.5. THEOREM *Let* $\mathbf{x}' = \mathbf{f}(\mathbf{x})$ *be a differential equation with a periodic orbit* $O(\mathbf{x}_0)$ *having period* p. *Suppose*

$$\tau = \int_0^p \left\{ \frac{\partial f_1}{\partial x_1} (\mathbf{x}_0(t)) + \frac{\partial f_2}{\partial x_2} (\mathbf{x}_0(t)) \right\} dt < 0 \tag{8.8.1}$$

then $O(\mathbf{x}_0)$ *is asymptotically stable.*

We have ended this final chapter with what is really the beginning of the qualitative study of nonlinear differential equations.

†L. S. Pontryagin, *Ordinary Differential Equations*, Addison-Wesley, Reading, Massachusetts, 1962, Chapter 5.

Exercise 8.8

1. Show that $O\left(\begin{bmatrix}1\\0\end{bmatrix}\right)=\left\{\begin{bmatrix}\cos t\\ \sin t\end{bmatrix}\middle| -\infty < t < \infty\right\}=\{\mathbf{x}\mid \|\mathbf{x}\|=1\}$ is an orbit of $\mathbf{x}'=\begin{bmatrix}0 & -1\\ 1 & 0\end{bmatrix}\mathbf{x}$ and is stable but not asymptotically stable.
2. Let $\mathbf{f}(\mathbf{x})=\begin{bmatrix}0 & -1\\ 1 & 0\end{bmatrix}\mathbf{x}+\mathbf{h}(\mathbf{x})$ where $\mathbf{h}(\mathbf{x})=\lambda(\mathbf{x})\mathbf{x}$, $\lambda(\mathbf{x})>0$ for $\|\mathbf{x}\|<1$, $\lambda(\mathbf{x})=0$ for $\|\mathbf{x}\|=1$, $\lambda(\mathbf{x})<0$ for $\|\mathbf{x}\|>1$. Show that $O\left(\begin{bmatrix}1\\0\end{bmatrix}\right)=\{\mathbf{x}\mid \|\mathbf{x}\|=1\}$ is an orbit of $\mathbf{x}'=\mathbf{f}(\mathbf{x})$ and is asymptotically stable. Hint: $d/dt\,\|\mathbf{x}\|^2=d/dt(\mathbf{x}\cdot\mathbf{x})=2\mathbf{x}\cdot\mathbf{x}'=2\lambda(\mathbf{x})\,\|\mathbf{x}\|^2$.
3. Consider the equation

$$\mathbf{x}'=\begin{bmatrix}-x_2+x_1-(x_1^2+x_2^2)x_1\\ x_1+x_2-(x_1^2+x_2^2)x_2\end{bmatrix}$$

(a) Show that

$$\mathbf{x}_0(t)=\begin{bmatrix}\cos t\\ \sin t\end{bmatrix}$$

is a solution. (See Problem 2.)

(b) What is p?

(c) Compute

$$\tau=\int_0^p\left\{\frac{\partial f_1}{\partial x_1}(\mathbf{x}_0(t))+\frac{\partial f_2}{\partial x_2}(\mathbf{x}_0(t))\right\}dt$$

(d) What does the result of (c) imply?

ANSWERS TO ODD-NUMBERED PROBLEMS

Chapter 1

Section 1.1.

1. $x = 2c_1 + c_2, \quad y = -4c_1 + 2c_2, \quad z = c_1, \quad t = c_2$
3. $x = -3c_1, \quad y = 0, \quad z = c_1, \quad t = 0$
5. $x = -c_1 - \frac{4}{3}, \quad y = c_1 + \frac{5}{3}, \quad z = c_1$

Section 1.2, Part 1

3. Let the unknowns be labeled $x_1, x_2, \ldots, x_i, \ldots, x_n$, and assume the system's matrix is upper triangular. Then, if there is solution, the ith equation determines x_j $(j \geqslant i)$ in terms of $x_{j+1}, x_{j+2}, \ldots, x_n$. That is, if the last k unknowns are determined, the $k-1$ may be computed as a combination of these. A system with no solutions is

$$\begin{aligned} x+y+z &= 0 \\ z &= 0 \\ z &= 1 \end{aligned}$$

The system's matrix is upper triangular.

5. Let a_{ij} be the entry of A in the ith row and jth column. Then ka_{ij} is the entry of kA in ith row, jth column, whereas ka_{ji} is the entry of $k\mathrm{A}^{\mathrm{T}}$ in the ith row jth column. Thus ka_{ji} is also the entry of $(k\mathrm{A})^{\mathrm{T}}$ in the ith row, jth column.

Section 1.2, Part 2

2.
$$\begin{aligned} (\mathrm{C}+\mathrm{A})-\mathrm{B} &= (\mathrm{C}+\mathrm{A})+(\mathrm{O}+(-1)\mathrm{B}) \\ &= (\mathrm{C}+\mathrm{A})+((-1)\mathrm{B}+\mathrm{O}) \\ &= (\mathrm{C}+\mathrm{A}+(-1)\mathrm{B})+\mathrm{O} \\ &= (\mathrm{C}+\mathrm{A})+(-1)\mathrm{B} \\ &= \mathrm{C}+(\mathrm{A}+(-1)\mathrm{B}) \\ &= \mathrm{C}+(\mathrm{A}-\mathrm{B}) \end{aligned}$$

9. $-(B-A) = O-(B-A) = O-(B+(-1)A)$
$= O+(-1)(B+(-1)A)$
$= O+((-1)B+A)$
$= O+(A-B) = A-B$

11. $-(KA) = O-(KA) = O+(-1)(KA)$
$= O+(-K)A = (-K)A$

Section 1.3, Part 1

1. (a) $\begin{bmatrix} 13 & 7 \\ 7 & 13 \end{bmatrix}$, (b) $\begin{bmatrix} 1 & 0 & 3 \\ 0 & 2 & 4 \\ 0 & 0 & 1 \end{bmatrix}$, (c) $\begin{bmatrix} 2x \\ y \\ -2z \end{bmatrix}$,

(d) $[a^2+b^2+c^2]$, (e) $\begin{bmatrix} 1 & 0 \\ 0 & 1 \end{bmatrix}$, (f) $\begin{bmatrix} 1 & 0 \\ 0 & -1 \end{bmatrix}$, (g) $[0,0,0]$

3. $A^2 = \begin{bmatrix} 0 & 0 & 1 \\ 0 & 0 & 0 \\ 0 & 0 & 0 \end{bmatrix}$, $A^3 = O$ 5. (a), (b), (c): $\begin{bmatrix} 0 & -6 \\ 0 & 0 \end{bmatrix}$

7. (a) $[x_1^2+x_2^2-x_3^2-x_1x_2+2x_2x_3]$, (b) $[2x_1^2+x_2^2+x_1x_3+x_2x_3]$

Section 1.3, Part 2

5. $\begin{bmatrix} 1 & 1 \\ 0 & 0 \end{bmatrix}\begin{bmatrix} 1 & -1 \\ -1 & 1 \end{bmatrix} = \begin{bmatrix} 1 & 1 \\ 0 & 0 \end{bmatrix}\begin{bmatrix} 0 & 0 \\ 0 & 0 \end{bmatrix}$

Section 1.4, Part 1

1. $n \times n$. Yes. 3. $n \times 1$

Section 1.4, Part 2

3. Since $B^{-1}A^{-1}$ is the inverse of AB, AB is nonsingular if A and B are both nonsingular. The contrapositive may be written: If AB is singular then either A or B is singular.

5. $\begin{bmatrix} -\frac{1}{3} & \frac{2}{3} & 0 & 0 \\ \frac{2}{3} & -\frac{1}{3} & 0 & 0 \\ 0 & 0 & \frac{3}{7} & -\frac{1}{7} \\ 0 & 0 & -\frac{2}{7} & \frac{3}{7} \end{bmatrix}$

Section 1.5, Part 1

1. (a) -20, (b) $(-1)^{[n(n-1)]/2}a_{1n}a_{2(n-1)}\cdots a_{n1}$, (c) -7, (d) -4
5. $\lambda^2+a_1\lambda+a_0$

Section 1.5, Part 2

1. Since det (AB) = (det A)(det B), we have $AA^{-1} = I$ implying (det A)(det A^{-1}) = det I = 1.
3. The proof is by induction and uses det (AB) = (det A)(det B). The critical observation is det (A^n) = (det A^{n-1}) det A.
5. Suppose the ith and jth rows of A are proportional. Then k times the ith row is the jth row. Let A^1 be constructed from A by subtracting k times the ith row

from the jth row. Thus A^1 has its jth row of zeros and therefore $\det A^1 = 0$. But $\det A = \det A^1$.

Section 1.6, Part 1

1. $\mathbf{c} = [-1, 0, 1]$ 5. (b)

Section 1.6, Part 2

3. All points (x, y, z) lying on a plane through the origin satisfy an equation of the form $ax + by + cz = 0$ (a, b, c real constants) and, conversely, if a point's coordinates satisfy this equation the point is on the plane. It is now trivial to verify that if (x_1, y_1, z_1) and (x_2, y_2, z_2) lie on a plane so does $k_1(x_1, y_1, z_1) + k_2(x_2, y_2, z_2)$ for all real k_1 and k_2.
5. The span of $\mathbf{x}$ is the set $\{k\mathbf{x}\}$ where k is any scalar.

Section 1.7, Part 1

1. None.
3. Yes. If $\mathbf{x}_1, \mathbf{x}_2, \mathbf{x}_3$ were linearly dependent, then $a_1\mathbf{x}_1 + a_2\mathbf{x}_2 + a_3\mathbf{x}_3 = \mathbf{0}$, with not all of the scalars zero. Hence $a_1\mathbf{x}_1 + a_2\mathbf{x}_2 + a_3\mathbf{x}_3 + 0\mathbf{x}_4 = \mathbf{0}$ with not all of the scalars zero. Hence $\mathbf{x}_1, \mathbf{x}_2, \mathbf{x}_3, \mathbf{x}_4$ would be linearly dependent. Generalization: If $\mathbf{x}_1, \mathbf{x}_2, \ldots, \mathbf{x}_n$ is linearly independent, so is every subsequence.
5. (a) $2[1, 0, 1] - [1, 1, -1] + [-1, 1, -3] = [0, 0, 0]$
   (b) $2[-1, 0, 0, 1] + [2, -1, 1, 1] - [0, -1, 1, 3] = [0, 0, 0, 0]$
   (c) $-[-1, 2, 0, 0] - [1, 2, -1, 0] - [1, 1, 0, 1] + [1, 5, -1, 1] = [0, 0, 0, 0]$
   (d) $-[1, 1, 0, 1] + 2[1, 0, 0, 1] - [1, -1, 0, 1] = [0, 0, 0, 0]$

Section 1.7, Part 2

1. Because $k\mathbf{x} = \mathbf{0}$, $\mathbf{x} \neq \mathbf{0}$ implies $k = 0$.
3. There exists scalars $a_1, a_2, \ldots, a_k$ not all zero such that $a_1\mathbf{x}_1 + a_2\mathbf{x}_2 + \cdots + a_k\mathbf{x}_k = \mathbf{0}$ because $\mathbf{x}_1, \mathbf{x}_2, \ldots, \mathbf{x}_k$ is assumed to be linearly dependent. Suppose $a_i \neq 0$. Then

$$\mathbf{x}_i = -\frac{a_1}{a_i}\mathbf{x}_1 - \cdots - \frac{a_{i-1}}{a_i}\mathbf{x}_{i-1} - \frac{a_{i+1}}{a_i}\mathbf{x}_{i+1} - \cdots - \frac{a_k}{a_i}\mathbf{x}_k$$

5. (a) The given row operation takes a matrix whose rows are linearly independent into a matrix whose rows are linearly independent.
   (b) If the rows of a matrix can be reduced to a matrix with a row of zeros then, since the latter matrix has linearly dependent rows, the former matrix cannot have linearly independent rows by (a) of this problem.

Section 1.8, Part 1

1. (a) Basis: $[1, 1, 0, 1]$, $[1, 0, 0, 1]$, $[0, 0, 1, 0]$; dimension $= 3$
   (b) $[-1, 2, 0, 0]$, $[1, 2, -1, 0]$, $[1, 1, 0, 1]$; dimension $= 3$
   (c) $[1, 0, 1]$, $[1, 1, 1]$ $[-1, -1, 3]$; dimension $= 3$
3. (a) No solutions   (b) $x = -1 - c, y = -1 - c, z = c$
5. (a) $x = -c, y = z = c$
   (b) $x = -c - \frac{4}{3}, y = \frac{5}{3} + c, z = c$
   (c) $x = 1, y = z = 0$
7. $\mathbf{x}_3 = [1, 0, 0]$

Section 1.8, Part 2

1. (a) No vector from a basis may be removed without destroying the fact that the basis spans the space.
(b) No vector may be adjoined to a basis and still preserve its linear independence.

Section 1.9, Part 1

1. (a) $|x_1^2|+|x_2^2|+|x_3^2|$, (b) -2, (c) 1

Section 1.10, Part 1

1. $\dfrac{1}{t(2-3t)}\begin{bmatrix} 2 & -t^2 \\ -3 & t \end{bmatrix}$

7. $\begin{bmatrix} \dfrac{t^2}{2} & \dfrac{t^3}{3} \\ 3t & 2t \end{bmatrix}$

Chapter 2

Section 2.1, Part 1

1. $s' = -(k/v)s$. 3. $m' = -km$. The solution is $m(t) = m(0)e^{-kt}$ from which $m(10) = m(0)e^{-k10} = m(0)/2$. Thus $-10k = \ln \frac{1}{2}$ and $k = \frac{1}{10} \ln 2$.

Section 2.2

1. From $\mathbf{x}(t) = k_2\begin{bmatrix}1\\0\\1\end{bmatrix}e^{4t}$ we find $x_1(t) = k_2e^{4t}$, $x_2(t) = 0$, $x_3(t) = k_2e^{4t}$. From

$\mathbf{x}(t) = k_3\begin{bmatrix}1\\1\\1\end{bmatrix}e^{6t}$, $x_1(t) = x_2(t) = x_3(t) = k_3e^{6t}$.

Section 2.3, Part 1

1. (a) $\lambda^2-4\lambda-5$; $\lambda = 5, -1$ (c) $-\lambda^3+5\lambda^2-4\lambda$; $\lambda = 0, 1, 4$
(d) $(1-\lambda)^3$; $\lambda = 1, 1, 1$ (e) $(1-\lambda)^3$, $\lambda = 1, 1, 1$
(f) $(1-\lambda)(2-\lambda)(-1-\lambda)\lambda$; $\lambda = 0, -1, 1, 2$
3. (a) $\lambda^2+a\lambda+b$ (b) $\lambda^3+a\lambda^2+b\lambda+c$
5. (a) $\begin{bmatrix}2\\-1\end{bmatrix}e^{-t}, \begin{bmatrix}1\\1\end{bmatrix}e^{5t}$ (c) $\begin{bmatrix}1\\-1\\1\end{bmatrix}, \begin{bmatrix}-2\\1\\1\end{bmatrix}e^{t}, \begin{bmatrix}1\\1\\1\end{bmatrix}e^{4t}$

(d) $\begin{bmatrix}1\\0\\0\\0\end{bmatrix}e^{t}$ (e) $\begin{bmatrix}1\\0\\0\\0\end{bmatrix}e^{t}$ (f) $\begin{bmatrix}-1\\-1\\2\\2\end{bmatrix}, \begin{bmatrix}1\\0\\0\\0\end{bmatrix}e^{t}, \begin{bmatrix}1\\1\\0\\0\end{bmatrix}e^{2t}, \begin{bmatrix}1\\0\\2\\0\end{bmatrix}e^{-t}$

Section 2.3, Part 2

1. $\det \mathrm{A} = \det(\mathrm{A}-0\mathrm{I}) = 0$ if and only if A is singular.
3. Suppose A^{-1} exists. Then $\det \mathrm{A} \neq 0$ and if λ is an eigenvalue of A, $\lambda \neq 0$. Thus

$0 = \det(A - \lambda I) = \det[-\lambda A(A^{-1} - (1/\lambda)I)] = (-\lambda)^n \det A \det(A^{-1} - (1/\lambda)I)$. Hence $\det(A^{-1} - (1/\lambda)I) = 0$.

5. By subtracting the first row of $J_n - \lambda I$ from each of the other rows we obtain

$$\det(J_n - \lambda I) = \det \begin{bmatrix} 1-\lambda & 1 & 1 \cdots & 1 \\ \lambda & -\lambda & 0 \cdots & 0 \\ \lambda & 0 & -\lambda \cdots & 0 \\ \cdot & \cdot & \cdot & \cdot \\ \cdot & \cdot & \cdot & \cdot \\ \cdot & \cdot & \cdot & \cdot \\ \lambda & 0 & 0 \cdots & 0 \\ \lambda & 0 & 0 \cdots & -\lambda \end{bmatrix}_{n \times n}$$

We factor λ from each row other than the first and then add each of these rows to the first obtaining

$$\det(J_n - \lambda I) = \lambda^{n-1} \det \begin{bmatrix} n-\lambda & 0 & 0 & \cdots & 0 \\ 1 & -1 & 0 & \cdots & 0 \\ 1 & 0 & -1 & \cdots & 0 \\ \cdot & \cdot & \cdot & & \cdot \\ \cdot & \cdot & \cdot & & \cdot \\ \cdot & \cdot & \cdot & & \cdot \\ 1 & 0 & 0 & \cdots & -1 \end{bmatrix}_{n \times n}$$

In this latter determinant, expand via the entries in the first row obtaining

$$\begin{aligned} \det(J_n - \lambda I) &= \lambda^{n-1}(n-\lambda) \det - I_{(n-1)\times(n-1)} \\ &= (-\lambda)^{n-1}(n-\lambda) \end{aligned}$$

Section 2.4, Part 1

1. $\mathbf{x}(t) = k_1 \begin{bmatrix} 3 \\ 1 \end{bmatrix} e^{2t} + k_2 \begin{bmatrix} 1 \\ -1 \end{bmatrix} e^{-2t}$

3. $\mathbf{x}(t) = k_1 \begin{bmatrix} 1+\alpha \\ 1 \end{bmatrix} e^{\alpha t} + k_2 \begin{bmatrix} 1-\alpha \\ 1 \end{bmatrix} e^{-\alpha t}$

5. $\mathbf{x}(t) = k_1 \begin{bmatrix} \cos\theta + 1 \\ \sin\theta \end{bmatrix} e^{t} + k_2 \begin{bmatrix} \cos\theta - 1 \\ \sin\theta \end{bmatrix} e^{-t}$

7. $\mathbf{x}(t) = k_1 \begin{bmatrix} 1 \\ 0 \\ 0 \end{bmatrix} e^{t} + k_2 \begin{bmatrix} 1 \\ 1 \\ 0 \end{bmatrix} e^{2t} + k_3 \begin{bmatrix} 1 \\ 1 \\ 1 \end{bmatrix} e^{3t}$

9. $\mathbf{x}(t) = k_1 \begin{bmatrix} 1 \\ 0 \\ 0 \end{bmatrix} e^{t} + k_2 \begin{bmatrix} 1 \\ -1 \\ 1 \end{bmatrix} e^{-t} + k_3 \begin{bmatrix} 1 \\ -1 \\ 2 \end{bmatrix} e^{-2t}$

11. $\mathbf{x}(t) = \frac{1}{4} \begin{bmatrix} 3 \\ 1 \end{bmatrix} e^{2t} + \frac{1}{4} \begin{bmatrix} 1 \\ -1 \end{bmatrix} e^{-2t}$

13. $\mathbf{x}(t) = \frac{5}{4} \begin{bmatrix} 1 \\ 1 \end{bmatrix} e^{2(t+1)} - \frac{1}{4} \begin{bmatrix} 1 \\ -3 \end{bmatrix} e^{-2(t+1)}$

15. $\mathbf{x}(t) = \frac{1}{2}\begin{bmatrix}3\\1\end{bmatrix}e^{2t} - \frac{1}{2}\begin{bmatrix}1\\-1\end{bmatrix}e^{-2t}$

17. $\mathbf{x}(t) = -\begin{bmatrix}1\\1\\0\end{bmatrix}e^{t-1} + \begin{bmatrix}1\\1\\1\end{bmatrix}e^{-t+1}$

19. $\mathbf{x}(t) = \begin{bmatrix}1\\1\\1\end{bmatrix}e^{3(t+1)}$ 21. $\mathbf{x}(t) = \begin{bmatrix}1\\0\\0\end{bmatrix}e^{t} - 2\begin{bmatrix}1\\-1\\1\end{bmatrix}e^{-t} + \begin{bmatrix}1\\-1\\2\end{bmatrix}e^{-2t}$

23. $\mathbf{x}(t) = -\begin{bmatrix}1\\-1\\1\end{bmatrix}e^{-t} + \begin{bmatrix}1\\-1\\2\end{bmatrix}e^{-2t}$

Section 2.4, Part 2

1. (a) $\mathrm{A}_1 = \begin{bmatrix}-2 & 1 & 1\\ 2 & -1 & -1\\ -2 & 1 & 1\end{bmatrix}$, $\mathrm{A}_1\mathbf{u}_1 = \begin{bmatrix}-2\\2\\-2\end{bmatrix} \neq \mathbf{0}$, $\mathrm{A}_1\mathbf{u}_2 = \mathbf{0}, \mathrm{A}_1\mathbf{u}_3 = \mathbf{0}$

(b) $\mathrm{A}_2 = \begin{bmatrix}2 & 3 & 1\\ 2 & 3 & 1\\ 2 & 3 & 1\end{bmatrix}$, $\mathrm{A}_2\mathbf{u}_1 = \mathrm{A}_2\mathbf{u}_3 = \mathbf{0}$, $\mathrm{A}_2\mathbf{u}_2 = \begin{bmatrix}6\\6\\6\end{bmatrix}$

3. $\mathrm{A}^{2k} = \mathrm{I}$ and $\mathrm{A}^{2k+1} = \mathrm{A} = \begin{bmatrix}0 & 1\\ 1 & 0\end{bmatrix}$ for $k = 0, 1, 2 \ldots$. Hence

$$e^{\mathrm{A}t} = \mathrm{I}\sum_{n=0}^{\infty}\frac{t^{2n}}{(2n)!} + \mathrm{A}\sum_{n=0}^{\infty}\frac{t^{2n+1}}{(2n+1)!}$$

$$= \mathrm{I}\cosh t + \mathrm{A}\sinh t$$

Thus $e^{\mathrm{A}t}\mathbf{k} = \mathbf{k}\cosh t + \mathrm{A}\mathbf{k}\sinh t$
Yes.

Section 2.5

1. (a) $\begin{bmatrix}1\\0\\0\end{bmatrix}, \begin{bmatrix}0\\1\\0\end{bmatrix}, \begin{bmatrix}0\\0\\1\end{bmatrix}$; (b) $\begin{bmatrix}1\\0\\0\end{bmatrix}, \begin{bmatrix}0\\1\\-\sqrt{2}\end{bmatrix}, \begin{bmatrix}0\\\sqrt{2}\\1\end{bmatrix}$; (c) $\begin{bmatrix}1\\0\\-1\end{bmatrix}, \begin{bmatrix}1\\-1\\0\end{bmatrix}, \begin{bmatrix}1\\1\\1\end{bmatrix}$

3. (a) $\mathbf{x}(t) = \begin{bmatrix}1\\1\\1\end{bmatrix}e^{3t}$; (b) $\mathbf{x}(t) = \frac{1}{3}\begin{bmatrix}2\\-1\\-1\end{bmatrix} + \frac{1}{3}\begin{bmatrix}1\\1\\1\end{bmatrix}e^{3t}$

(c) $\mathbf{x}(t) = \frac{1}{3}\begin{bmatrix}-1\\2\\-1\end{bmatrix} + \frac{1}{3}\begin{bmatrix}1\\1\\1\end{bmatrix}e^{3t}$; (d) $\mathbf{x}(t) = \frac{1}{3}\begin{bmatrix}-1\\-1\\2\end{bmatrix} + \frac{1}{3}\begin{bmatrix}1\\1\\1\end{bmatrix}e^{3t}$

5. Repeat the proof of Lemma 2.5.1, except that $(\mathbf{u}^{\mathrm{T}}\mathrm{A}\bar{\mathbf{v}})^{\mathrm{T}} = -\bar{\mathbf{v}}^{\mathrm{T}}\mathrm{A}\mathbf{u}$.
7. If λ is an eigenvalue of A, then so is $\bar{\lambda}$ according to Lemma 2.5.2. By Problem 5, the eigenvalues of A have zero imaginary part. If $\lambda = 0$ is not an eigenvalue of A then there is an even number of eigenvalues of A. This contradicts the hypothesis that A is a $(2n+1) \times (2n+1)$ matrix.

Section 2.6

1. Yes. The proof is analogous to the proof of Theorem 2.6.2.

Section 2.7, Part 1

1. $\mathbf{x}_3(t) = \text{Re}\left\{\begin{bmatrix} i \\ 1 \end{bmatrix} e^{-it}\right\}, \qquad \mathbf{x}_4(t) = \text{Im}\left\{\begin{bmatrix} i \\ 1 \end{bmatrix} e^{-it}\right\}$

They are identical: $\mathbf{x}_3 \equiv \mathbf{x}_2(t), \mathbf{x}_4(t) \equiv \mathbf{x}_1(t)$.

3. $\mathbf{x}(t) = c_1 \begin{bmatrix} 1 \\ 0 \\ 0 \end{bmatrix} e^t + c_2 \begin{bmatrix} 0 \\ \cos 2t \\ -\sin 2t \end{bmatrix} + c_3 \begin{bmatrix} 0 \\ \sin 2t \\ \cos 2t \end{bmatrix}$

5. $$\mathbf{x}(t) = c_1 \begin{bmatrix} 1 \\ -1 \\ 1 \end{bmatrix} + c_2 \left(\begin{bmatrix} 1 \\ 2 \\ 1 \end{bmatrix} \cos \sqrt{3}t - \sqrt{3} \sin \sqrt{3}t \begin{bmatrix} -1 \\ 0 \\ 1 \end{bmatrix}\right)$$
$$+ c_3 \left(\begin{bmatrix} 1 \\ 2 \\ 1 \end{bmatrix} \sin \sqrt{3}t + \sqrt{3} \cos \sqrt{3}t \begin{bmatrix} -1 \\ 0 \\ 1 \end{bmatrix}\right)$$

Section 2.7, Part 2

1. Since $\mathrm{A}\mathbf{u} = \lambda\mathbf{u}$, we have $\overline{\mathrm{A}\mathbf{u}} = \bar{\mathrm{A}}\bar{\mathbf{u}} = \bar{\lambda}\bar{\mathbf{u}}$, but A has real entries. Therefore $\bar{\mathrm{A}} = \mathrm{A}$ and the conclusion follows.

Section 2.8

1. (a) No solution of the form $\mathbf{u}e^{\lambda t}$. (b) $\mathbf{x}(t) = \begin{bmatrix} 1 \\ 0 \end{bmatrix} e^t$

3. (a) $\mathbf{x}(t) = 2 \begin{bmatrix} 1 \\ 0 \\ 0 \end{bmatrix} e^t - \begin{bmatrix} 0 \\ 0 \\ 1 \end{bmatrix} e^{2t}$

(b) No solution of the form

$$\mathbf{u}e^{\lambda_1 t} + \mathbf{v}e^{\lambda_2 t}$$

Chapter 3

Section 3.1

1. $\mathbf{x}(t) = \begin{bmatrix} 1 \\ \\ 2 \end{bmatrix} e^{2t}$ 3. $\mathbf{x}(t) = \begin{bmatrix} e^t \\ 0 \\ e^{-2t} \end{bmatrix}$ 5. $\mathbf{x}(t) = \begin{bmatrix} 1 \\ \\ \alpha \end{bmatrix} e^{\alpha t}$

7. $\mathbf{x}_0 = \begin{bmatrix} 2 \\ 3 \\ 5 \end{bmatrix}$ 9. $\begin{bmatrix} -3+t-t^2 \\ 1 - 2t \\ -2 \end{bmatrix} e^{-t} = \mathbf{x}(t)$ 11. $\mathbf{x}(t) = \begin{bmatrix} t^2+t \\ 2t+1 \\ 2 \end{bmatrix} e^{-t}$

Section 3.2, Part 1

5. $\mathbf{u}_2 = \begin{bmatrix} -2 \\ 3 \\ 1 \end{bmatrix}, \quad \mathbf{u}_1 = \begin{bmatrix} 3 \\ 0 \\ 1 \end{bmatrix}$ 7. $\mathbf{u}_2 = \begin{bmatrix} 0 \\ \\ -1 \end{bmatrix}, \quad \mathbf{u}_1 = \begin{bmatrix} 1 \\ -\alpha \\ 2 \end{bmatrix}$

9. $\mathbf{u}_3 = \begin{bmatrix} 0 \\ 0 \\ 1 \end{bmatrix}, \quad \mathbf{u}_2 = \begin{bmatrix} 0 \\ 1 \\ 0 \end{bmatrix}, \quad \mathbf{u}_1 = \begin{bmatrix} 1 \\ 0 \\ 0 \end{bmatrix}$

11. $\mathbf{u}_2 = \begin{bmatrix} 0 \\ 0 \\ 1 \\ 0 \\ 0 \end{bmatrix}, \quad \mathbf{u}_2 = \begin{bmatrix} 0 \\ 1 \\ 0 \\ 0 \\ 0 \end{bmatrix}, \quad \mathbf{u}_1 = \begin{bmatrix} 1 \\ 0 \\ 0 \\ 0 \\ 0 \end{bmatrix}$

$\mathbf{v}_2 = \begin{bmatrix} 0 \\ 0 \\ 0 \\ 0 \\ 1 \end{bmatrix}, \quad \mathbf{v}_1 = \begin{bmatrix} 0 \\ 0 \\ 0 \\ 1 \\ 0 \end{bmatrix}$

13. $\mathbf{u}_3 = \begin{bmatrix} 0 \\ 0 \\ 0 \\ 0 \\ 1 \end{bmatrix}, \quad \mathbf{u}_2 = \begin{bmatrix} 1 \\ 2 \\ 3 \\ 4 \\ 0 \end{bmatrix}, \quad \mathbf{u}_1 = \begin{bmatrix} 2 \\ 0 \\ 4 \\ 0 \\ 0 \end{bmatrix}$

Section 3.2, Part 2

3. Yes and of order k. Yes, if either a or b is not zero.
7. $(\mathrm{A}-\lambda \mathrm{I})^k(c\mathbf{u}) = c(\mathrm{A}-\lambda \mathrm{I})^k\,\mathbf{u} = \mathbf{0}$ if $\mathbf{u}$ is a root vector of order k. Also, similarly $(\mathrm{A}-\lambda \mathrm{I})^{k-1}(c\mathbf{u}) \neq \mathbf{0}$ if $c \neq 0$.

Section 3.3, Part 1

7. $\mathbf{x}(t) = \begin{bmatrix} c_1 \\ c_2 + c_3 t \\ c_3 \\ c_4 \end{bmatrix} e^t$

9. (a) $\mathbf{x}(t) = \begin{bmatrix} 1 \\ 0 \\ 0 \\ 0 \end{bmatrix} e^t$; (b) $\mathbf{x}(t) = \begin{bmatrix} 0 \\ 1 \\ 0 \\ 0 \end{bmatrix} e^t$; (c) $\mathbf{x}(t) = \begin{bmatrix} 0 \\ t \\ 1 \\ 0 \end{bmatrix} e^t$; (d) $\mathbf{x}(t) = \begin{bmatrix} 0 \\ t^2/2 \\ t \\ 1 \end{bmatrix} e^t$

11. $\mathbf{x}(t) = \begin{bmatrix} c + c_2 t \\ c_2 \\ c_3 \end{bmatrix} e^{2t}$ 13. $\mathbf{x}(t) = \begin{bmatrix} c_1 + c_2 t + \dfrac{c_3 t^2}{2} \\ c_2 + c_3 t \\ c_3 \end{bmatrix} e^{2t}$

15. $\mathbf{x}(t) = e^{-t} \begin{bmatrix} c_1 + c_2 t \\ c_2 \\ 0 \\ 0 \end{bmatrix} + e^t \begin{bmatrix} 0 \\ 0 \\ c_3 \\ c_4 \end{bmatrix}$

17. $\mathbf{x}(t) = e^{2t} \begin{bmatrix} c + (-2c_1 + c_2)t \\ c_2 + (-4c_1 + 2c_2)t \\ c_3 \\ c_4 \end{bmatrix}$

Section 3.3, Part 2

1. $$\mathbf{x}(t) = e^t \begin{bmatrix} 1+t+t^2 \\ 2t \\ 2+3t+2t^2 \\ 4t \\ 1 \end{bmatrix}$$

3. (a) $\begin{bmatrix} 2 & 0 & 0 & 0 & 0 \\ 0 & 2 & 1 & 0 & 0 \\ 0 & 0 & 2 & 0 & 0 \\ 0 & 0 & 0 & 2 & 1 \\ 0 & 0 & 0 & 0 & 2 \end{bmatrix}$ (b) $\begin{bmatrix} 2 & 0 & 0 & 0 & 0 \\ 0 & 2 & 0 & 0 & 0 \\ 0 & 0 & 2 & 1 & 0 \\ 0 & 0 & 0 & 2 & 1 \\ 0 & 0 & 0 & 0 & 2 \end{bmatrix}$ (c) $\begin{bmatrix} 2 & 1 & 0 & 0 & 0 \\ 0 & 2 & 0 & 0 & 0 \\ 0 & 0 & 2 & 1 & 0 \\ 0 & 0 & 0 & 2 & 1 \\ 0 & 0 & 0 & 0 & 2 \end{bmatrix}$

5. If $\mathbf{u}_k$ is a root vector of order k then $\mathbf{u}_k, \mathbf{u}_{k-1}, \ldots, \mathbf{u}_1$ is a chain and each vector is a root vector of order equal to its subscript. Apply Theorem 3.3.1 to each vector in the chain.

Section 3.4

1. If $\mathbf{x}$ and $\mathbf{y}$ are any two eigenvectors of A corresponding to the eigenvalue $\hat{\lambda}$ then
$$\mathrm{A}(c_1\mathbf{x}+c_2\mathbf{y}) = c_1\mathrm{A}\mathbf{x}+c_2\mathrm{A}\mathbf{y} = c_1\hat{\lambda}\mathbf{x}+c_2\hat{\lambda}\mathbf{y} = \hat{\lambda}(c_1\mathbf{x}+c_2\mathbf{y})$$

3. If $\mathbf{x}$ and $\mathbf{y}$ are root vectors of order k and j, respectively, $(k \geqslant j)$ of A corresponding to the eigenvalue $\hat{\lambda}$ then $(\mathrm{A}-\hat{\lambda}\mathrm{I})^k(c_1\mathbf{x}+c_2\mathbf{y}) = \mathbf{0}$ and thus $c_1\mathbf{x}+c_2\mathbf{y}$ is a root vector of A corresponding to $\hat{\lambda}$ or $c_1\mathbf{x}+c_2\mathbf{y} = \mathbf{0}$. In either case $c_1\mathbf{x}+c_2\mathbf{y}$ is a member of the given set. This proves the theorem.

Section 3.6

1. By hypothesis $(\mathrm{A}-\mu_1\mathrm{I})\ldots(\mathrm{A}-\mu_k\mathrm{I})\mathbf{x}_0 = \mathbf{0}$. Hence $(\mathrm{A}-\mu_2\mathrm{I})\ldots(\mathrm{A}-\mu_k\mathrm{I})\mathbf{x}_0$ is either $\mathbf{0}$ or an eigenvector. If it were $\mathbf{0}$ then $(z-\mu_2)(z-\mu_3)\ldots(z-\mu_k)$ would be an annihilating polynomial of degree $k-1$, contradicting the hypothesis. Thus $(\mathrm{A}-\mu_2\mathrm{I})\ldots(\mathrm{A}-\mu_k\mathrm{I})\mathbf{x}_0$ is an eigenvector, which means μ_1 is an eigenvalue of A. Similarly for $\mu_2, \mu_3, \ldots, \mu_k$.
3. By hypothesis λ_1 is the only eigenvalue of A. Hence, if $\mathbf{x}_0 \in \mathscr{R}^n$ is annihilated by $g(z)$ then $g(z) = (z-\lambda_1)^k$, $k \leqslant n$. But then $(\mathrm{A}-\lambda_1\mathrm{I})^k\mathbf{x}_0 = \mathbf{0}$, which proves that $\mathbf{x}_0$ is a root vector. ($g(z)$ is assumed to be the minimum annihilating monic polynomial of $\mathbf{x}_0$. Hence, $\mathbf{x}_0$ is a root vector of order k.)
5. Let $g(z)$ be the minimum monic annihilating polynomial of $\mathbf{x}_0 \neq \mathbf{0}$ and $h(z)$ any annihilating polynomial of $\mathbf{x}_0$. Then the degree of $h(z)$ is greater or equal to the degree of $g(z)$. By division
$$h(z) = \theta(z)g(z)+r(z)$$
where $\theta(z)$ and $r(z)$ are polynomials of degree less than the degree of $g(z)$. But $h(z)-\theta(z)g(z) = r(z)$ is another annihilating polynomial of $\mathbf{x}_0$. Hence $r(z) \equiv 0$ and $g(z)$ divides $h(z)$.
7. Since the roots of every annihilating polynomial are eigenvalues of A and cannot be repeated more often than once (since, by hypothesis the eigenvalues of A are distinct), $\mathrm{C}(z)$ annihilates every vector in $\mathscr{R}^n$. In particular then $\mathrm{C}(\mathrm{A})\mathbf{e}_i = \mathbf{0}$ which implies that the ith column of $\mathrm{C}(\mathrm{A})$ is a column of zeros, $i = 1, 2, \ldots, n$. Thus $\mathrm{C}(\mathrm{A}) = \mathrm{O}$.

Chapter 4

Section 4.1, Part 1

1. $\psi(t)=\frac{1}{4}\begin{bmatrix}3e^{2t}+e^{-2t} & 3e^{2t}-3e^{-2t}\\ e^{2t}-e^{-2t} & e^{2t}+3e^{-2t}\end{bmatrix}$

3. $\psi(t)=\frac{1}{3}\begin{bmatrix}2e^{t}+e^{4t} & -e^{t}+e^{4t}\\ -2e^{t}+2e^{4t} & e^{t}+2e^{4t}\end{bmatrix}$

5. $\psi(t)=\begin{bmatrix}e^{2t}+e^{t}-e^{-t} & -e^{2t}+e^{-t} & -e^{t}+e^{-t}\\ e^{t}-e^{-t} & e^{-t} & -e^{t}+e^{-t}\\ e^{2t}-e^{-t} & -e^{2t}+e^{-t} & e^{-t}\end{bmatrix}$

7. $\psi(t)=\begin{bmatrix}e^{t} & e^{2t}-e^{t} & e^{3t}-e^{2t}\\ 0 & e^{2t} & e^{3t}-e^{2t}\\ 0 & 0 & e^{3t}\end{bmatrix}$

9. $\psi(t)=\begin{bmatrix}e^{t} & e^{t}-2e^{-t}+e^{2t} & -e^{-t}+e^{-2t}\\ 0 & 2e^{-t}-e^{-2t} & e^{-t}-e^{-2t}\\ 0 & -2e^{-t}+2e^{-2t} & -e^{-t}+2e^{-2t}\end{bmatrix}$

11. $\psi(t)=\frac{1}{3}\begin{bmatrix}3e^{t} & 0 & 0\\ 0 & e^{t}+2e^{4t} & -\sqrt{2}e^{t}+\sqrt{2}e^{4t}\\ 0 & -\sqrt{2}e^{t}+\sqrt{2}e^{4t} & 2e^{t}+e^{4t}\end{bmatrix}$

13. $\psi(t)=e^{t}\begin{bmatrix}1 & t & 0\\ 0 & 1 & 0\\ 0 & t & 1\end{bmatrix}$ 15. $\psi(t)=e^{t}\begin{bmatrix}1 & t & 0 & 0\\ 0 & 1 & 0 & 0\\ 0 & 0 & 1 & t\\ 0 & 0 & 0 & 1\end{bmatrix}$

17. $\psi(t)=e^{2t}\begin{bmatrix}1 & t & 0\\ 0 & 1 & 0\\ 0 & 0 & 1\end{bmatrix}$ 19. $\psi(t)=e^{2t}\begin{bmatrix}1 & t & \frac{t^2}{2}\\ 0 & 1 & t\\ 0 & 0 & 1\end{bmatrix}$

21. $\psi(t)=\begin{bmatrix}e^{-t} & te^{-t} & 0 & 0\\ 0 & e^{-t} & 0 & 0\\ 0 & 0 & e^{t} & 0\\ 0 & 0 & 0 & e^{t}\end{bmatrix}$

Section 4.1, Part 2

1. $B\Phi'=BA\Phi$ which may not equal $AB\Phi$.
3. If $B\Phi$ is a fundamental matrix of $\mathbf{x}'=A\mathbf{x}$ then $AB=BA$. Yes. The hypothesis asserts $BA\Phi=AB\Phi$. Set $t=t_0$. Then $\Phi(t_0)$ is invertible and this implies $BA=AB$.

Section 4.2, Part 1

1. $\psi^{-1}(t)=\frac{1}{4}\begin{bmatrix}3e^{-2t}+e^{2t} & 3e^{-2t}-3e^{2t}\\ e^{-2t}-e^{2t} & e^{-2t}+3e^{2t}\end{bmatrix}$

3. $\psi^{-1}(t)=\frac{1}{3}\begin{bmatrix}2e^{-t}+e^{-4t} & -e^{-t}+e^{-4t}\\ -2e^{-t}+2e^{-4t} & e^{-t}+2e^{-4t}\end{bmatrix}$

5. $\psi^{-1}(t) = \begin{bmatrix} e^{-2t} & 0 & 0 \\ 0 & e^{-t} & 0 \\ 0 & 0 & e^{-t} \end{bmatrix}$

7. $\psi^{-1}(t) = \frac{1}{3}\begin{bmatrix} 2 & -1 & -1 \\ -1 & 2 & -1 \\ -1 & -1 & 2 \end{bmatrix} + \frac{1}{3}e^{-3t}\begin{bmatrix} 1 & 1 & 1 \\ 1 & 1 & 1 \\ 1 & 1 & 1 \end{bmatrix}$

9. $\psi^{-1}(t) = e^{-2t}\begin{bmatrix} 1 & -t & \dfrac{t^2}{2} \\ 0 & 1 & -t \\ 0 & 0 & 1 \end{bmatrix}$

Section 4.2, Part 2

5. We note that $(d/dt)\,\mathrm{Re}\,\mathbf{x}(t) = \mathrm{Re}\,\mathbf{x}'(t)$ and $\mathrm{Re}\,\{A\mathbf{x}(t)\} = A\,\mathrm{Re}\,\mathbf{x}(t)$. Then, $(d/dt)\ \mathrm{Re}\,\mathbf{x}(t) = \mathrm{Re}\,\mathbf{x}'(t) = \mathrm{Re}\,\{A\mathbf{x}(t)\} = A\ \mathrm{Re}\,\mathbf{x}(t)$. Also, $\mathrm{Re}\,\mathbf{x}(t_0) = \mathbf{x}(t_0) = \mathbf{x}_0$. The range of $\mathbf{x}(t)$ is real for all t.
7. No. $\mathrm{Re}\,\{A\mathbf{x}(t)\} = A\,\mathrm{Re}\,\mathbf{x}(t)$ would be false if A had even one entry which was not real.
9. Let Φ and Λ be fundamental matrices of $\mathbf{x}' = A\mathbf{x}$. Then $\Phi\Phi^{-1}(t_0)$ and $\Lambda\Lambda^{-1}(t_0)$ are normalized at t_0. Therefore, by Corollary 4.2.10, $\Phi(t)\Phi^{-1}(t_0) \equiv \Lambda(t)\Lambda^{-1}(t_0)$ which implies $\Phi(t)[\Phi^{-1}(t_0)\Lambda(t_0)] \equiv \Lambda(t)$. Thus $B = \Phi^{-1}(t_0)\Lambda(t_0)$ exists such that $\Phi(t)B = \Lambda(t)$. Clearly B is nonsingular. If $\hat{B}$ is such that $\Phi(t)\hat{B} \equiv \Lambda(t)$ then $\Phi(t_0)\hat{B} = \Lambda(t_0)$ and $B = \hat{B}$ is proved.

Section 4.3

1. $\Phi(t)\Phi(s) = \Phi(t+s) = \Phi(s+t) = \Phi(s)\Phi(t)$
3. By Theorem 4.3.6, $A\Phi(t_1) = \Phi(t_1)A$. But then

$$[\Phi(t_1)\Phi(t)]' = \Phi(t_1)\Phi'(t) = \Phi(t_1)A\Phi(t) = A[\Phi(t_1)\Phi(t)]$$

5. If $\mathbf{y}(t) = \mathbf{x}(-t)$, then $\mathbf{y}'(t) = -\mathbf{x}'(-t) = A\mathbf{x}(-t) = A\mathbf{y}(t)$.

Section 4.4, Part 1

1. (a) $x(t) = 2e^{t-1} - t - 1$, (b) $x(t) = -\frac{1}{2}(\cos t + \sin t) + \frac{1}{2}e^{t-\pi}$

3. $\mathbf{x}(t) = \begin{bmatrix} -1 + e^t \\ -\dfrac{1}{3} + \dfrac{e^t}{2} - \dfrac{e^{3t}}{6} \end{bmatrix}$ 7. (a) $\mathbf{x}(t) = e^t\begin{bmatrix} -1 + 2t \\ 1 + 2t \end{bmatrix} + \begin{bmatrix} 1 \\ 0 \end{bmatrix}$

(b) $\mathbf{x}(t) = e^t\begin{bmatrix} 1 + t \\ 2 + t \end{bmatrix} - \begin{bmatrix} 1 \\ 1 \end{bmatrix}$, (c) $\mathbf{x}(t) = e^t\begin{bmatrix} 1 - t \\ -t \end{bmatrix} + \begin{bmatrix} -1 \\ 0 \end{bmatrix}e^{-t}$

Section 4.4, Part 2

1. Set $\Phi(t) = \begin{bmatrix} \cosh 4t & -\sinh 4t \\ -\sinh 4t & \cosh 4t \end{bmatrix}$

Then

$$\mathbf{x}(t) = \Phi(t)\left\{\mathbf{k} + \begin{bmatrix} 1 \\ 1 \end{bmatrix}\int_0^t f(s)e^{4s}\,ds\right\}$$

2. If $\mathbf{x}(0) = [x_0, y_0]^T$, then

$$\mathbf{x}(t) = \Phi(t)\left\{\begin{bmatrix} x_0 \\ y_0 \end{bmatrix} + \begin{bmatrix} 1 \\ 1 \end{bmatrix} \int_0^t f(s)e^{4s}\,ds\right\}$$

If, also, $\mathbf{x}(t_0) = \mathbf{0}$, then since $\Phi(t)$ is nonsingular

$$\mathbf{0} = \begin{bmatrix} x_0 \\ y_0 \end{bmatrix} + \begin{bmatrix} 1 \\ 1 \end{bmatrix} \int_0^{t_0} f(s)e^{4s}\,ds$$

implies $x_0 = y_0 = -\int_0^{t_0} f(s)e^{4s}\,ds$.

Section 4.5

1. $\mathbf{x}(t) = \begin{bmatrix} -1 \\ -\frac{2}{3} \end{bmatrix}$ 3. $\mathbf{x}(t) = -\frac{1}{2}\begin{bmatrix} \sin t + \cos t \\ \cos t \end{bmatrix}$

5. $\mathbf{x}(t) = \begin{bmatrix} 2\cos t - \sin t \\ \cos t + \sin t \end{bmatrix}$

7. $x(t) = -\dfrac{K_1}{\alpha} + \dfrac{K_2}{-\alpha + a} e^{at} - \dfrac{\omega K_3}{\alpha^2 + \omega^2}\cos \omega t - \dfrac{\alpha K_3}{\alpha^2 + \omega^2}\sin \omega t$

9. $x(t) = -e^t - \frac{1}{4}\cos 2t + \frac{1}{4}\sin 2t$ 11. $\mathbf{x}(t) = e^t \sin t \begin{bmatrix} 1 \\ 1 \end{bmatrix}$

Chapter 5

Section 5.1

1. $\lambda = 0, \begin{bmatrix} 1 \\ 0 \\ 0 \end{bmatrix}$; $\lambda = 1, \begin{bmatrix} 1 \\ 1 \\ 1 \end{bmatrix}$; $\lambda = 5, \begin{bmatrix} 1 \\ 5 \\ 25 \end{bmatrix}$

3. $\mathbf{x}(t) = \begin{bmatrix} 1 & e^t & e^{5t} \\ 0 & e^t & 5e^{5t} \\ 0 & e^t & 25e^{5t} \end{bmatrix}\mathbf{k} + \dfrac{1}{26}\begin{bmatrix} 3\sin t - 2\cos t \\ 2\sin t + 3\cos t \\ -3\sin t + 2\cos t \end{bmatrix}$

5. $\mathbf{k} = \dfrac{-1}{20\cdot 26}\begin{bmatrix} -104 \\ 65 \\ -1 \end{bmatrix}$; yes; yes.

Section 5.2, Part 1

1. $\mathbf{x}' = \begin{bmatrix} 0 & 1 \\ 0 & 1 \end{bmatrix}\mathbf{x}$ 3. $\mathbf{x}' = \begin{bmatrix} 0 & 1 \\ 1 & 0 \end{bmatrix}\mathbf{x}$ 5. $\mathbf{x}' = \begin{bmatrix} 0 & 1 \\ 0 & 0 \end{bmatrix}\mathbf{x}$

7. $\mathbf{x}' = \begin{bmatrix} 0 & 1 & 0 & 0 \\ 0 & 0 & 1 & 0 \\ 0 & 0 & 0 & 1 \\ 0 & 0 & 0 & 1 \end{bmatrix}\mathbf{x}$ 9. $\mathbf{x}' = \begin{bmatrix} 0 & 1 & 0 & 0 \\ 0 & 0 & 1 & 0 \\ 0 & 0 & 0 & 1 \\ 1 & 0 & 0 & 0 \end{bmatrix}\mathbf{x}$

11. $\mathbf{x}' = \begin{bmatrix} 0 & 1 & 0 \\ 0 & 0 & 1 \\ \sin t & 0 & 0 \end{bmatrix}\mathbf{x} + \begin{bmatrix} 0 \\ 0 \\ f(t) \end{bmatrix}$ 13. $y^{(2)} = 0$ 15. $y^{(2)} - y^{(1)} = 0$

17. $y^{(2)} + y = 0$ 19. $y^{(3)} - y^{(2)} + y = 0$ 21. $y^{(3)} - y^{(1)} + y = 0$

Section 5.2, Part 2

1. $\mathbf{x}(t) = \begin{bmatrix} c_1t+c_2 \\ c_1 \\ 0 \\ 0 \end{bmatrix}$

3. Since $y_1 - y_2$ solves $y^{(2)} + a_1y^{(1)} + a_2y = 0$, $y(t_0) = y^{(1)}(t_0) = 0$, $y_1 - y_2 \equiv 0$ by the uniqueness theorem.

5. $\Phi = \begin{bmatrix} 1 & t & e^t & e^{-t} \\ 0 & 1 & e^t & -e^{-t} \\ 0 & 0 & e^t & e^{-t} \\ 0 & 0 & e^t & -e^{-t} \end{bmatrix}$ and $y(t) = c_1 + c_2t + c_3e^t + c_4e^{-t}$

7. $y(t) = c_1t + c_2t^2$

Section 5.3, Part 1

1. $y(t) = c_1 + c_2e^t - t$ 3. $y(t) = c_1e^{3t} + c_2e^{4t}$ 5. $y(t) = c_1 + c_2e^{-t} - t - te^{-t}$
7. $y(t) = c_1 \sin 2t + c_2 \cos 2t$ 9. $y(t) = c_1 + c_2e^t + c_3e^{-t} + c_4e^{2t}$
11. $y(t) = (c_1 \cos t + c_2 \sin t)e^t - \dfrac{t}{2}$ 13. $y(t) = c_1e^{-t} + c_2e^{-2t} - \frac{1}{20}\cos 2t + \frac{3}{20}\sin 2t$
15. $y(t) = c_1e^{-at} + c_2e^{-bt}$ 17. $y(t) = \cos 3t + \frac{1}{18}\sin 3t - \dfrac{t}{6}\cos 3t$
19. $y(t) = -\frac{1}{4}e^{-t} + \left(\dfrac{t}{2} - \dfrac{3}{4}\right)e^t$ 21. $y(t) = -\frac{1}{2}e^{-t} - \frac{1}{2}\cos t + \frac{1}{2}\sin t$
23. $y(t) = \frac{1}{2}(e^t \sin t - t)$ 25. $y(t) = t - 1 + e^{-t}$ 27. $y(t) = \frac{1}{2}(1+t) - \frac{1}{2}\cos te^t$
29. $y^{(2)} + 2y^{(1)} + 10y = 0$ 31. $\mathbf{x}(t) = \begin{bmatrix} 1 & e^t & e^{-t} \\ 0 & e^t & -e^{-t} \\ 0 & e^t & e^t \end{bmatrix}\mathbf{k}$

Section 5.3, Part 2

1. $y(t) = c_1e^{-\alpha t} + c_2e^{-\beta t} + \dfrac{te^{-\alpha t}}{\beta - \alpha}$

5. (a) $\Psi(t) = \begin{bmatrix} 2e^t - e^{2t} & -e^t + e^{2t} \\ 2e^t - 2e^{2t} & -e^t + 2e^{2t} \end{bmatrix}$, (b) $\mathbf{x}_p(t) = \displaystyle\int_0^t f(s)\begin{bmatrix} -e^{t-s} + e^{2(t-s)} \\ -e^{t-s} + 2e^{2(t-s)} \end{bmatrix} ds$

Section 5.4, Part 1

1. $y(t) = c_0 + c_1t + c_2e^t$ 3. $y(t) = (c_1 + c_2t)\cos t + (c_3 + c_4t)\sin t$
5. $y(t) = c_1 + (c_2 + c_3t)e^{2t} + \frac{1}{4}t^2e^{2t}$
7. $y(t) = (c_1 + c_2t)\cos t + (c_3 + c_4t)\sin t + \frac{1}{4}\cosh t$
9. $y(t) = c_0 + c_1t + c_2t^2$ 11. $y(t) = 1$
13. $y(t) = e^{-t}(\cos t + \sin t) = \sqrt{2}\,e^{-t}\cos\left(t - \dfrac{\pi}{4}\right)$
15. $(\alpha + (\beta + \alpha)t)e^{-t} = y(t)$ 17. $y(t) = 1$
19. $y(t) = \dfrac{-1}{(a+b)^2}(1 + (b-a)t)e^{-at} + \dfrac{e^{bt}}{(a+b)^2}$ 21. $y(t) = \dfrac{t^3}{6}e^{-t}$
23. (a) $y^{(3)} = 0$, (b) $y^{(5)} - y^{(4)} - 2y^{(3)} + 2y^{(2)} + y^{(1)} - y = 0$,
(c) $y^{(6)} + \frac{17}{2}y^{(4)} + 20y^{(2)} + 8y = 0$, (d) $y^{(2)} - 4y^{(1)} + 4y = 0$

Section 5.4, Part 2

1. $\begin{bmatrix} 0 & 1 \\ -\alpha^2 & 2\alpha \end{bmatrix}$; $\begin{bmatrix} 0 & 1 \\ -\alpha^2 & 2\alpha \end{bmatrix}\begin{bmatrix} 0 \\ 1 \end{bmatrix} = \begin{bmatrix} 1 \\ 2\alpha \end{bmatrix} = \alpha\begin{bmatrix} 0 \\ 1 \end{bmatrix} + \begin{bmatrix} 1 \\ \alpha \end{bmatrix}$

   where $\lambda = \alpha$ is an eigenvalue

3. $\Phi(t) = e^{\alpha t}\begin{bmatrix} 1 & t \\ \alpha & 1+\alpha t \end{bmatrix}$

   Therefore,

$$y_p(t) = \int_0^t (t-s)\, e^{\alpha(t-s)} f(s) \quad ds$$

   is a solution and

$$y(t) = e^{\alpha t}(k_1 + k_2 t) + y_p(t)$$

   is a general solution

Section 5.5

3. $\begin{bmatrix} 1 \\ \lambda_0 \\ \lambda_0^2 \end{bmatrix} e^{\lambda_0 t}$; $\left\{\begin{bmatrix} 0 \\ 1 \\ 2\lambda_0 \end{bmatrix} + \begin{bmatrix} 1 \\ \lambda_0 \\ \lambda_0^2 \end{bmatrix} t\right\} e^{\lambda_0 t}$; $\left\{\begin{bmatrix} 0 \\ 0 \\ 1 \end{bmatrix} + \begin{bmatrix} 0 \\ 1 \\ 2\lambda_0 \end{bmatrix} t + \begin{bmatrix} 1 \\ \lambda_0 \\ \lambda_0^2 \end{bmatrix} \frac{t^2}{2}\right\} e^{\lambda_0 t}$

   The first components are $e^{\lambda_0 t}$, $te^{\lambda_0 t}$, $(t^2/2)e^{\lambda_0 t}$. The second components are $\lambda_0 e^{\lambda_0 t}$, $(1+\lambda_0 t)e^{\lambda_0 t}$, $(t+\lambda_0 t^2/2)e^{\lambda_0 t}$.

Section 5.6

3. $\begin{bmatrix} m_1 & 0 \\ 0 & m_2 \end{bmatrix}\mathbf{x}'' = \begin{bmatrix} -k_1 - k_2 & k_2 \\ k_2 & -k_2 \end{bmatrix}\mathbf{x}$

Section 5.7, Part 1

1. $\mathbf{z}' = \begin{bmatrix} 0 & 0 & 1 & 0 \\ 0 & 0 & 0 & 1 \\ -2 & 1 & 0 & 0 \\ \frac{1}{2} & -\frac{1}{2} & 0 & 0 \end{bmatrix}\mathbf{z}$

3. $\mathbf{z}_1(t) = \begin{bmatrix} 0 \\ \cos 2t \\ 0 \\ -\sin 2t \end{bmatrix}$, $\quad \mathbf{z}_2(t) = \begin{bmatrix} 0 \\ \sin 2t \\ 0 \\ 2\cos 2t \end{bmatrix}$,

   $\mathbf{z}_3(t) = \begin{bmatrix} \cos t \\ 0 \\ -\sin t \\ 0 \end{bmatrix}$, $\quad \mathbf{z}_4(t) = \begin{bmatrix} \sin t \\ 0 \\ \cos t \\ 0 \end{bmatrix}$

7. $\mathbf{y}(t) = -\frac{1}{3}(\cos t + \sin t)\begin{bmatrix} -1 \\ 1 \end{bmatrix} + \frac{1}{6}(4\cos 2t + \sin 2t)\begin{bmatrix} 1 \\ 2 \end{bmatrix}$

Section 5.7, Part 2

5. Let $B = \begin{bmatrix} 1 & 2 \\ 0 & 1 \end{bmatrix}$ $\quad A = \begin{bmatrix} 2 & 2 \\ 0 & 2 \end{bmatrix}$

Then $\mathbf{u} = \begin{bmatrix} 0 \\ -1 \end{bmatrix}$ is not an eigenvector of A or B. However,

$$\mathbf{w} = \begin{bmatrix} 0 \\ -1 \\ 0 \\ 1 \end{bmatrix} = \begin{bmatrix} \mathbf{u} \\ \lambda\mathbf{u} \end{bmatrix}, \text{ where } \lambda = -1, \text{ is an eigenvector of}$$

$$C = \begin{bmatrix} 0 & I \\ -B & -A \end{bmatrix}. \text{ Here } \lambda = -1 \text{ is an eigenvalue of C.}$$

Section 5.8, Part 1

1. $\mathbf{y}(t) = \frac{1}{3}\begin{bmatrix} 1 \\ 3 \end{bmatrix}(a \cos t + b \sin t) + \frac{1}{6}\begin{bmatrix} -4 \\ 3 \end{bmatrix}(c \cos \sqrt{6}t + c \sin \sqrt{6}t) + \frac{1}{3}\begin{bmatrix} 1 \\ 3 \end{bmatrix}$

3. (a) If B is symmetric, then $B^T = B$ and $(ABA^T)^T = (A^T)^T B^T A^T = ABA^T$ proving ABA^T is also symmetric.
   (b) If A and B are symmetric then $B^T = B$ and $A^T = A$ and $(ABA)^T = A^T B^T A^T = ABA$. Thus ABA is also symmetric.

Section 5.8, Part 2

1. A symmetric matrix has n real linearly independent eigenvectors. (See Section 2.5.) In Exercise 2.4, Part 2, Problem 2, we see that $\mathbf{x}_i \cdot \mathbf{x}_j = 0$ if $i \neq j$. Hence

$$\begin{aligned} S\mathbf{x}\cdot\mathbf{x} &= S(\alpha_1\mathbf{x}_1 + \cdots + \alpha_n\mathbf{x}_n)\cdot(\alpha_1\mathbf{x}_1 + \cdots + \alpha_n\mathbf{x}_n) \\ &= (\alpha_1\lambda_1\mathbf{x}_1 + \cdots + \alpha_n\lambda_n\mathbf{x}_n)\cdot(\alpha_1\mathbf{x}_1 + \cdots + \alpha_n\mathbf{x}_n) \\ &= \lambda_1\alpha_1^2\|\mathbf{x}_1\|^2 + \cdots + \lambda_n\alpha_n^2\|\mathbf{x}_n\|^2 \end{aligned}$$

3. In general $A\mathbf{y}\cdot\mathbf{x} = \mathbf{y}\cdot A^T\mathbf{x}$; but since $A^T = A$, $A\mathbf{y}\cdot\mathbf{x} = \mathbf{y}\cdot A\mathbf{x}$. Hence $(ASA)\mathbf{x}\cdot\mathbf{x} = A(SA)\mathbf{x}\cdot\mathbf{x} = SA\mathbf{x}\cdot A\mathbf{x}$. Now set $\mathbf{z} = A\mathbf{x}$. Then $SA\mathbf{x} = A\mathbf{x}$ may be written $S\mathbf{z}\cdot\mathbf{z}$. From hypothesis and Problem 2, $S\mathbf{z}\cdot\mathbf{z} < 0$ which means that $ASA\mathbf{x}\cdot\mathbf{x} < 0$. By Problem 2, this implies that the eigenvalues of the symmetric matrix ASA are all negative.

Chapter 6

Section 6.1

3. Suppose $\mathbf{y}$ and $\mathbf{z}$ are solutions. Then from Problem 1, $\mathbf{y}-\mathbf{z}$ solves $\mathbf{x}' = A(t)\mathbf{x}$ and $\mathbf{y}(t_0) - \mathbf{z}(t_0) = \mathbf{0}$. Property (ii) combined with the observation that $\mathbf{w}(t) \equiv \mathbf{0}$ is a solution of $\mathbf{x}' = A(t)\mathbf{x}$, $\mathbf{x}(t_0) = \mathbf{0}$ leads to the conclusion that $\mathbf{y}(t) - \mathbf{z}(t) \equiv \mathbf{0}$.

11. $x(t) = ke^{\sin t}$ $\qquad$ 13. $x(t) = 4e^{\sin t} + e^{\sin t}\int_{\pi}^{t} e^{-\sin u} \sin u \, du$

15. $x(t) = k \exp\left[-\frac{(t-1)^2}{2}\right] + \exp\left[-\frac{(t-1)^2}{2}\right]\int_1^t u^2 \exp\left[\frac{(u-1)^2}{2}\right] du$

17. $x(t) = kt,\ t < 0$

Section 6.2, Part 1

1. $e^t = e^1 \cdot e^{t-1} = e\left(1 + (t-1) + \frac{(t-1)^2}{2!} + \cdots + \frac{(t-1)^n}{n!} + \cdots\right)$

$$= e \sum_{n=0}^{\infty} \frac{(t-1)^n}{n!}$$

3. $\frac{1}{1-t^2} = \sum_{n=0}^{\infty} t^{2n}$ 5. $ln\,(1-t) = -t - \frac{t^2}{2} - \frac{t^3}{3} - \cdots = \sum_{n=1}^{\infty} \frac{-t^n}{n}$

7. $\ln \frac{1+t}{1-t} = \ln(1+t) - \ln(1-t) = \sum_{n=1}^{\infty} \frac{t^n}{n} - \sum_{n=1}^{\infty} \frac{(-1)^n t^n}{n}$

$$= \sum_{n=1}^{\infty} [1-(-1)^n] \frac{t^n}{n} = 2 \sum_{n=1}^{\infty} \frac{t^{2n-1}}{2n-1}$$

9. $\cos^2 t = \frac{1}{2} + \frac{1}{2}\cos 2t = \frac{1}{2} + \frac{1}{2} \sum_{n=0}^{\infty} \frac{(-1)^n (2t)^{2n}}{(2n)!}$

17. (a) $\begin{bmatrix}1\\1\\1\end{bmatrix} + \sum_{n=1}^{\infty} \begin{bmatrix}0\\0\\1\end{bmatrix} \frac{t^n}{n!}$ (b) $\begin{bmatrix}1\\1\\1\end{bmatrix} \sum_{n=0}^{\infty} \frac{t^n}{n!}$

(c) $\begin{bmatrix}0\\1\\0\end{bmatrix} + \begin{bmatrix}1\\1\\0\end{bmatrix} t + \begin{bmatrix}0\\0\\1\end{bmatrix} t^2 + \begin{bmatrix}1\\0\\0\end{bmatrix} \sum_{n=1}^{\infty} \frac{(-1)^n\, t^{2n+1}}{(2n+1)!}$

Section 6.2, Part 2

1. $\tan t = t + \left(\frac{1}{2!} - \frac{1}{3!}\right) t^3 + \left(\frac{1}{5!} - \frac{1}{2!3!} - \frac{1}{4!} + \frac{1}{2!2!}\right) t^5 + \cdots$

3. The identity

$$1 + t + t^2 + \cdots + t^k = \frac{1}{1-t} - \frac{t^{k+1}}{1-t}$$

may be established by induction. For $|t| < 1$

$$\lim_{k\to\infty} (1 + t + \cdots + t^k) = 1 + t + t^2 + \cdots = \frac{1}{1-t}$$

since $\lim_{k\to\infty} \frac{t^{k+1}}{1-t} = 0$ when $|t| < 1$.

5. Yes. The polynomial $a_0 + a_1 t + \cdots + a_n t^n$ trivially satisfies Definition 6.2.1, where $a_m = 0$ if $m > n$ and hence for *every* $r > 0$, (6.2.1) holds.

Section 6.3, Part 1

1. $\mathbf{x}_0 = \begin{bmatrix}1\\0\end{bmatrix}$, $\mathbf{x}_1 = \begin{bmatrix}1\\0\end{bmatrix}$, $\mathbf{x}_2 = \frac{1}{2}\begin{bmatrix}1\\1\end{bmatrix}$, $\mathbf{x}_3 = \frac{1}{6}\begin{bmatrix}1\\3\end{bmatrix}$

3. $\mathbf{x}_0 = \begin{bmatrix}a\\b\end{bmatrix}$, $\mathbf{x}_1 = \begin{bmatrix}a\\b\end{bmatrix}$, $\mathbf{x}_2 = \frac{a+b}{2}\begin{bmatrix}1\\1\end{bmatrix}$, $\mathbf{x}_3 = \frac{1}{6}\begin{bmatrix}a+3b\\3a+b\end{bmatrix}$

5. $\mathbf{x}_0 = \begin{bmatrix}1\\0\\0\end{bmatrix}, \quad \mathbf{x}_1 = \begin{bmatrix}1\\0\\1\end{bmatrix}, \quad \mathbf{x}_2 = \frac{1}{2}\begin{bmatrix}1\\0\\1\end{bmatrix}, \quad \mathbf{x}_3 = \frac{1}{6}\begin{bmatrix}1\\0\\3\end{bmatrix}$

7. $\mathbf{x}_0 = \begin{bmatrix}1\\1\\1\end{bmatrix}, \quad \mathbf{x}_1 = \begin{bmatrix}1\\1\\0\end{bmatrix}, \quad \mathbf{x}_2 = \frac{1}{2}\begin{bmatrix}0\\-1\\1\end{bmatrix}, \quad \mathbf{x}_3 = \frac{1}{6}\begin{bmatrix}-3\\1\\-4\end{bmatrix}$

9. $x_k = \dfrac{1}{k!}, \quad k = 0, 1, 2, \ldots$

Section 6.3, Part 2

9. (a) $$y_1(t) = 1 + \sum_{n=1}^{\infty} \frac{t^{4n}}{(3.4)(7.8)\cdots(4n-1)(4n)}$$

$$y_2(t) = t + \sum_{n=1}^{\infty} \frac{t^{4n+1}}{(4.5)(8.9)\cdots(4n)(4n+1)}$$

(b) $$y_1(t) = 1 + \sum_{n=1}^{\infty} \frac{t^{3n}}{(2.3)(5.6)\cdots(3n-1)(3n)}$$

$$y_2(t) = t + \sum_{n=1}^{\infty} \frac{t^{3n+1}}{(3.4)(6.7)\cdots(3n)(3n+1)}$$

(c) $y_1(t) = 1, \quad y_2(t) = t$

(d) Let

$$a_{n+2} = \frac{-1}{(n+1)(n+2)} \sum_{k=0}^{n} \frac{a_k}{(n-k)!}$$

Let $a_0 = 1$, $a_1 = 0$, and solve the above to obtain $y_1(t) = 1 + \Sigma_{n=1}^{\infty} a_n t^n$. Let $a_0 = 0$, $a_1 = 1$ and solve to obtain, $y_2(t) = t + \Sigma_{n=1}^{\infty} a_n t^n$.

Section 6.4, Part 1

1. The substitution $\alpha = 2r+1$ in Equation (6.4.4) yields

$$y_{k+2} = \frac{(k-2r-1)(k-2r)}{(k+2)(k+1)} y_k \quad (*)$$

(a) Since $y_0 = y(0) = 0$ by hypothesis, we deduce $y_2 = 0$. This leads in turn to $y_4 = y_6 = \cdots = 0$.
(b) Set $k = 2r+1$. Then (*) yields $y_{2r+3} = 0$. From this $y_{2r+5} = y_{2r+7} \cdots = 0$ follows.
(c) Since $y_1 = y'(0) = 1$, $y_3, y_5, \ldots, y_{2r+1}$ are all unequal to zero as (*) shows.

3. $$y(t) = 1 + \sum_{n=1}^{\infty} \frac{t^{2n}}{2^n n!} = \sum_{n=0}^{\infty} \frac{(t^2/2)^n}{n!} = e^{t^2/2}$$

5. $$y_{2k} = \frac{(-\alpha)(2-\alpha)\cdots(2k-2-\alpha)(\alpha+1)(\alpha+3)\cdots(\alpha+2k-1)}{(2k)!}$$

7. Suppose $\alpha = 2r$, r a positive integer. Then

$$y(t) = a_0 \left\{ 1 + \sum_{k=1}^{r} \frac{2^{2k}(k-1-r)\cdots(-r)}{(2k)!} t^{2k} \right\}$$

If $\alpha = 2r+1$, r a nonnegative integer, then

$$y(t) = a_1 \left\{ t + \sum_{k=1}^{r} \frac{2^{2k}(k-1-r)\cdots(-r)}{(2k+1)!} t^{2k} \right\}$$

Section 6.4, Part 2

1. If $u(t) = (t^2-1)^k$ then $u'(t) = k2t(t^2-1)^{k-1}$. Hence $(t^2-1)u' = 2tk(t^2-1)^k = 2ktu$ as required.
5. When $t = 1$ the term in the sum

$$k! \sum_{j=0}^{k} \binom{k}{j}^2 (t-1)^{k-j}(t+1)^j$$

which is not zero occurs when $k = j$. Thus

$$z(1) = k! \binom{k}{k}^2 (2)^k = k!\,2^k$$

Section 6.5, Part 1

1. Since $A = I$, $A^k = I$ and thus

$$\mathbf{x}(t) = \mathbf{x}_0 \sum_{k=0}^{\infty} \frac{t^k}{k!} = \begin{bmatrix} 1 \\ 0 \end{bmatrix} e^t$$

3. For the companion system $A^{2k} = (-1)^k I$ and $A^{2k+1} = (-1)^k A$. Since

$$\mathbf{x}(0) = \begin{bmatrix} 0 \\ 1 \end{bmatrix}, (-1)^k A \begin{bmatrix} 0 \\ 1 \end{bmatrix} = (-1)^k \begin{bmatrix} 1 \\ 0 \end{bmatrix} = A^{2k+1}$$

while $A^{2k} = (-1)^k \begin{bmatrix} 0 \\ 1 \end{bmatrix}$ Thus,

$$\mathbf{x}(t) = \sum_{k=0}^{\infty} \frac{(-1)^k t^{2k}}{(2k)!} \begin{bmatrix} 0 \\ 1 \end{bmatrix} + \sum_{k=0}^{\infty} \frac{(-1)^k t^{2k+1}}{(2k+1)!} \begin{bmatrix} 1 \\ 0 \end{bmatrix}$$

Hence

$$y(t) = \sum_{k=0}^{\infty} \frac{(-1)^k t^{2k+1}}{(2k+1)!} = \sin t$$

5. $y(t) = -e^{2t}$

Section 6.6, Part 1

5. (a) $y(t) = kte^t$; (b) $y(t) = \dfrac{kt}{1-t}$ for $t \neq 1$;

(c) $y(t) = \dfrac{k}{\cos t}$, for $t \neq \pm\dfrac{\pi}{2}, \pm\dfrac{3\pi}{2}, \ldots$

Section 6.6, Part 2

3. We find

$$\sum_{k=0}^{\infty} \mathbf{x}_k (k+c) t^{k+c+1} = \sum_{k=0}^{\infty} A\mathbf{x}_k t^{k+c}$$

where $A = \begin{bmatrix} 1 & 1 \\ 0 & 1 \end{bmatrix}$. This leads to $A\mathbf{x}_0 = \mathbf{0}$, which in turn implies that $\lambda = 0$ is an eigenvalue of A or $\mathbf{x}_0 = \mathbf{0}$. Neither is possible.

Section 6.7

3. If $y(t)$ and $y^*(t)$ were linearly dependent on $(0, \infty)$ they would be proportional; that is $y(t)/y^*(t) = k \neq 0$, for all t in $(0, \infty)$. However, $\lim_{t\to 0} y(t)/y^*(t)$ is either zero or infinity as can be seen from Equations (6.7.4) and (6.7.6).
5. Suppose $\alpha = v$, v a positive integer. Then

$$y(t) = a_0\left\{1 + \sum_{n=1}^{v} \frac{(-v)(1-v)\cdots(n-1-v)}{(n!)^2} t^n\right\}$$

Chapter 7

Section 7.1

1. (b), (e)
3. Let $\mathbf{x} = \begin{bmatrix} x_1 \\ x_2 \end{bmatrix}$, $\quad \mathbf{f}(t) = \begin{bmatrix} f_1(t) \\ f_2(t) \end{bmatrix}$,

$$\mathbf{z} = \begin{bmatrix} z_1 \\ z_2 \\ z_3 \end{bmatrix} = \begin{bmatrix} t \\ x_1 \\ x_2 \end{bmatrix}, \qquad \mathbf{F} = \begin{bmatrix} 1 \\ a_{11}z_2 + a_{12}z_3 + f_1(z_1) \\ a_{21}z_2 + a_{22}z_3 + f_2(z_1) \end{bmatrix}$$

Then $\mathbf{z}' = \mathbf{F}(\mathbf{z})$.

Section 7.2

1. (a) $y' = f'(v)x'$ where $v = x(t)$. From the given system $3xy = y' = f'(v)x' = f'(v)xy$. Thus, if $y \neq 0, f'(v) = 3$.
   (b) From (a) $f(v) = y = 3v + c$. From the initial condition $y(0) = 3x(0) + c$ implies $c = -2$.
   (c) Since $x' = xy$, from (b), $x' = x(3x-2)$.
   (d) This is an integration of (c).
   (e) $\mathbf{x}(t) = 2/(3-e^{2t})$ follows from (d) and $y(t) = [6/(3-e^{2t})] - 2$ follows from (b).

Section 7.3, Part 2

1. If $z = y^{1-n}$, $z' = (1-n)y^{-n}y' = (1-n)y^{-n}\{-P(t)y + Q(t)y^n\} = -(1-n)P(t)z + (1-n)Q(t)$. Thus z satisfies $z' + (1-n)P(t)z = (1-n)Q(t)$.

Chapter 8

Section 8.1

1. As t varies from 0 to π, x_1 decreases continuously from 1 to -1 while, simultaneously, x_2 first increases from 0 to 1 as t varies from 0 to $\pi/2$ and then

decreases to 0 as t varies from $\pi/2$ to π. The upper semicircle is traversed counterclockwise. A similar analysis for t between π and 2π completes the argument.

3. A circle centered at (0, 0) with radius $(\alpha^2+\beta^2)^{1/2}$. The orientation is clockwise.

Section 8.2, Part 1

1. The points on the line $2x_1 - x_2 = 0$ 3. (0, 0) and (1, 0)

Section 8.2, Part 2

1. Theorem 4.3.1 and Corollary 4.1.5.
3. If A has a zero eigenvalue then $\mathrm{A}\mathbf{x} = \mathbf{0}$ is satisfied by the eigenvector $\mathbf{u} = \begin{bmatrix} u_1 \\ u_2 \end{bmatrix}$. The critical points are the points on the line $u_2x_1 - u_1x_2 = 0$.

Section 8.3, Part 1

1. (0, 0) is the only critical point. There are no noncritical periodic orbits.
3. (0, 0) is the only critical point. Every circle centered at (0, 0) is a periodic orbit and these are the only ones.
5. Zero velocity at t^* means $\mathbf{x}'(t^*) = \mathbf{0} = \mathbf{f}(\mathbf{x}(t^*))$ and thus $\mathbf{x}(t^*)$ is a critical point. By Theorem 8.2.4 $\mathbf{x}(t) = \mathbf{x}(t^*)$ for all t and hence for all t, $\mathbf{x}' = \mathbf{0}$; the particle is always at rest.
7. No. Since the particle must pass through $(0, \frac{1}{2})$ (at least) once in both directions, x' at this point is not unique. This is forbidden by hypothesis.
9. Since

$$\det\begin{bmatrix} 0 & 1 \\ -4 & 0 \end{bmatrix} \neq 0, \qquad \begin{bmatrix} 0 & 1 \\ -4 & 0 \end{bmatrix} = \mathrm{A}$$

is nonsingular and hence $\mathrm{A}\mathbf{x} = \mathbf{0}$ has the unique solution, $\mathbf{x} = \mathbf{0}$.

Section 8.3, Part 2

3. By the "mean-value" theorem,

$$x_1(t_2) - x_1(t_1) = x_1'(\xi)(t_2 - t_1)$$

for some ξ such that $t_1 < \xi < t_2$. Thus,

$$|x_1(t_2) - x_1(t_1)| = |x_1'(\xi)|\,|t_2 - t_1| > \mathrm{K}|t_2 - t_1|$$

for all t_1, t_2, such that $\mathrm{T} < t_1 < t_2$.

5. By the continuity of $\|\mathbf{f}\|$, we have

$$\lim_{t\to+\infty} \|\mathbf{x}'(t)\| = \lim_{t\to+\infty} \|\mathbf{f}(\mathbf{x})\| = \|\mathbf{f}(\lim_{t\to+\infty} \mathbf{x}(t))\| = \|\mathbf{f}(\mathbf{a})\| > 0$$

Therefore, at least one component of $\mathbf{x}'$, say $x_1'(t)$ satisfies $\lim_{t\to+\infty} |x_1'(t)| = \mathrm{K} > 0$.

Section 8.4, Part 1

1. (a) X (b) VIII (c) I (d) XV (e) XIX (f) XII (See Figure 8.4.15.)

INDEX

72 73 74 75 76 9 8 7 6 5 4 3 2 1